Oscillations of Disks

Astrophysics and Space Science Library

More information about this series at http://www.springer.com/series/5664

Shoji Kato

Oscillations of Disks

Springer

Shoji Kato
Emeritus professor
Department of Astronomy
Kyoto University
Kyoto, Japan

ISSN 0067-0057 ISSN 2214-7985 (electronic)
Astrophysics and Space Science Library
ISBN 978-4-431-56206-1 ISBN 978-4-431-56208-5 (eBook)
DOI 10.1007/978-4-431-56208-5

Library of Congress Control Number: 2016948204

Cover image: Drawing by Jun Fukue

Printed on acid-free paper

This Springer imprint is published by Springer Nature
The registered company is Springer Japan KK

Preface

Accretion disks are one of the most important ingredients in the Universe. Recognition of their importance is, however, rather recent in the long history of astrophysics. Discovery of quasars in the early 1960s was a trigger for the start of studying accretion disks. This is because the enormous power of energy generation of quasars is found to be related to accretion disks surrounding black holes. Accretion disks are now known in various active objects, including active galactic nuclei, stellar-mass black holes, ultraluminous X-ray sources, γ-ray bursts, and stellar and galactic jets.

Almost all astrophysical objects have periodic and quasiperiodic time variabilities in various timescales. This gives us the important tools to study physical and dynamical states of the objects. Typical examples are helioseismology and asteroseismology, where internal structures of the Sun and stars are examined by analyzing their time variabilities. Astrophysical objects with accretion disks also have, in many cases, time variabilities, and they are an important tool for studying the structures of those objects. Obvious observational evidence which shows time-periodic or quasiperiodic phenomena in accretion disks was, however, limited until near the end of the 1990s. In that era, however, studies on long-term time variations in Be stars and superhumps in dwarf novae had been made with much progress.

The launch of the Rossi X-ray Timing Explorer (RXTE) in 1996 changed the situation. RXTE discovered quasiperiodic variations in X-ray binaries. They are kHz quasiperiodic oscillations (QPOs) (kHz QPOs) in neutron-star low-mass X-ray binaries and high-frequency QPOs (HF QPOs) in black-hole low-mass X-ray binaries. Since that time, quasiperiodic time variations have been observed in various objects with accretion disks, including micro-quasars, ultraluminous sources, active galactic nuclei, and the Galactic Center. This observational evidence stimulated theoretical studies on disk oscillations in order to explore disk structures and environments around the disks. This opened a new field of "diskoseismology" (or "discoseismology").

Much attention has been paid especially to high-frequency quasiperiodic oscillations in neutron-star and black-hole binaries, because their frequencies are close to the Keplerian frequency in the innermost part of relativistic disks and also they appear sometimes with a pair whose frequency ratio is close to 3:2. Studies of these quasiperiodic oscillations are important because they may directly present us with dynamical phenomena in strong gravitational fields and may lead to a new way to evaluate spins of central compact objects. The spin of the central black holes is usually evaluated by comparing observed spectra of accretion disks with theoretically derived ones. The purpose of this book is to review the present state of studies of disk oscillations.

This book consists of two parts. In Part I, we first briefly summarize observational evidence that shows or suggests disk oscillations. Then, after presenting basic properties of disk oscillations, we derive, in an approximate way, wave equations describing disk oscillations and classify the oscillations into types. Our attention is particularly given to the trapping of disk oscillations in the radial direction. Finally, an attempt to improve wave equations is presented. In Part II, excitation processes of disk oscillations are presented. Three important processes are reviewed with additional comments on other possible processes. A process of wave-wave resonant instability and its application are presented somewhat in detail.

I thank Professor Wasaburo Unno, under whom I started my research career on astrophysics in the1960s at the University of Tokyo, and Professor Donald Lynden-Bell for his hospitality at Cambridge University in 1976–1977, where I started my studies on disk oscillations. I also thank Professor Tomokazu Kogure at Kyoto University for having encouraged me to write a book. I appreciate many colleagues with whom I collaborated and held invaluable discussions on various stages on studies of disk oscillations. Among them, I especially thank Marek A. Abramowicz, Omer M. Blaes, Jun Fukue, Fumio Honma, Jiri Horák, Wlodek Kluźniak, Dong Lai, Jufu Lu, Stephen H. Lubow, Ryoji Matsumoto, Shin Mineshige, Ramesh Narayan, Atsuo T. Okazaki, Zdenek Stuchlik, Gabriel Török, and Robert V. Wagoner (honorific omitted). Chapter 8 especially is based on discussions on one-armed oscillations of Be-star disks with Atsuo T. Okazaki. Finally, many thanks are due to Doctors Hisako Niko and Akiyuki Tokuno and Ms. Risa Takizawa of Springer Japan for their helpful editorial support.

Nara, Japan Shoji Kato
24 March 2016

Contents

List of Symbols

(x, y, z)	Cartesian coordinates
(r, φ, z)	cylindrical coordinates
$\boldsymbol{B} = (B_r, B_\varphi, B_z)$	magnetic flux density
D	distance between primary and secondary
E	wave energy
H	disk half-thickness
$\mathscr{H}_0, \mathscr{H}_1, \mathscr{H}_2, \ldots$	Hermite polynomials
$\Im$	imaginary part
M	mass (of primary star)
$M_\odot$	solar mass
$M_{\rm s}$	mass of secondary star
$\boldsymbol{N}$	viscous stress force per unit mass
R	radius in spherical coordinates, $R^2 = r^2 + z^2$
R_*	stellar radius
$\Re$	real part
a	mean binary separation
a_*	normalized black-hole spin parameter
$\boldsymbol{b} = (b_r, b_\varphi, b_z)$	perturbed magnetic field
c	speed of light
$c_{\rm A}$	Alfvén speed
$c_{\rm s}$	acoustic speed
e	eccentricity
f	defined by $h_1(r, \eta) = f(r)g(\eta, r)$ [Chaps. 5, 6, 7, and Sect. 8.1]
	defined by $u_r(r, \eta) = f(r)g(\eta, r)$ [Sect. 8.2]
f_h	defined by $h_1(r, \eta) = f_h(r)g_h(\eta, r)$ [Sect. 8.2]
f_u	defined by $u_r(r, \eta) = f_u(r)g(\eta, r)$ [Chap. 5]
f_z	defined by $u_z(r, \eta) = f_z(r)g_z(\eta, r)$
h_1	$h_1 = p_1/\rho_0$
m	wavenumber in the azimuthal direction
n	number of node(s) in the vertical direction
n_r	number of node(s) in the horizontal direction

p	pressure
p_0, p_1	unperturbed pressure and Eulerian variation of pressure
q	$q = M_s/M$
r	radius (cylindrical coordinates)
r_c	corotation radius
r_c	capture radius of disk oscillation, which is the same as r_{out}
r_D	disk radius due to tidal truncation
r_{IL}	radius of inner Lindblad resonance
r_L	radius of Lindblad resonance
r_{OL}	radius of outer Lindblad resonance
r_{out}	outer radius of trapped region of disk oscillations
r_{in}	inner edge of disk
r_g	Schwarzschild radius= $2GM/c^2$
r_{out}	outer edge of disk
r_t	disk radius due to tidal truncation (= r_D)
t_{ij}	ij-component of viscous stress tensor
$\boldsymbol{u} = (u_r, u_\varphi, u_z)$	Eulerian variation of velocity by perturbations
$\hat{\boldsymbol{u}} = (\hat{u}_r, \hat{u}_\varphi, \hat{u}_z)$	velocity perturbations defined by $\boldsymbol{u} = \Re[\hat{\boldsymbol{u}}\exp(i\omega t)]$
$\breve{\boldsymbol{u}} = (\breve{u}_r, \breve{u}_\varphi, \breve{u}_z)$	velocity perturbations defined by $\boldsymbol{u} = \Re[\breve{\boldsymbol{u}}\exp(i\omega t - im\varphi)]$
$\boldsymbol{v} = (v_r, v_\varphi, v_z)$	Eulerian variation of velocity by perturbations
z_s	vertical thickness of truncated disk
Ω	angular velocity of disk rotation
$\Omega_\perp$	vertical epicyclic frequency
Ω_K	angular velocity of Keplerian rotation
Ω_{orb}	orbital angular velocity observed from inertial frame
Ω^*_{orb}	orbital angular velocity observed from primary star
δ	Lagrangian variation
δ_{ij}	Kronecker's delta
η	damping rate by turbulent viscosity
η	$\eta = z/H$
η_s	$\eta_s = z_s/H$
κ	horizontal epicyclic frequency
κ_{max}	maximum value of κ
λ	characteristic wavelength of oscillations in radial direction
λ_D	characteristic radial scale of disk structure
$\boldsymbol{\xi} = (\xi_r, \xi_\varphi, \xi_z)$	Lagrangian displacement vector
$\hat{\boldsymbol{\xi}}$	displacement vector defined by $\boldsymbol{\xi} = \Re[\hat{\boldsymbol{\xi}}\exp(i\omega t)]$
$\breve{\boldsymbol{\xi}}$	displacement vector defined by $\boldsymbol{\xi} = \Re[\breve{\boldsymbol{\xi}}\exp(i\omega t - im\varphi)]$
$\boldsymbol{\xi}_D$	displacement vector due to tidal force
ρ	density
ρ_0	unperturbed density
ρ_{00}	unperturbed density on equator
ρ_1	perturbed density
φ	azimuthal angle

ψ	gravitational potential
ψ_{D}	tidal potential
ψ_{eff}	effective potential
ω	angular frequency of waves
ω_i	imaginary part of ω, ($-\omega_i$ is growth rate)
$\tilde{\omega} \equiv \omega - m\Omega$	angular frequency of waves in corotating frame

Part I
Basic Properties and Disk Oscillations

Chapter 1
Introduction

Abstract We start this introduction by presenting a brief history of appearance of accretion disks in astrophysical studies and by pointing out importance of diskoseismology. Then, we review astrophysical objects which have accretion disks, and present a brief survey of observational evidences of time variations in accretion disks, focusing on those which will be related to disk oscillations. The main oscillatory phenomena which we focus our attention are V/R variations in Be stars, positive and negative superhumps in dwarf novae, high-frequency quasi-periodic oscillations in neutron-star and black-hole X-ray binaries.

Keywords Disk oscillations • Diskoseismology • High-frequency QPOs • Superhumps • V/R variations

1.1 Brief History of Emergence of Accretion Disks in Astrophysics

Accretion disks are relatively new research subjects in a long history of astronomy and astrophysics. Until the discovery of quasars in early 1960s and subsequent theoretical studies on their energy sources, accretion disks are little known in astronomers and astrophysicists, although they are now well-known to be one of important ingredients in the Universe. Because the discovery was one of epoch making events in astronomy, it is introduced in various articles. Here, following a book "Black-Hole Accretion Disks" by Kato et al. (2008), we briefly summarize the outline of story of the discovery of accretion disks.

After the World War II, several sky survey projects in radio wavelength band were started, and the so-called 3C catalogue (the third Cambridge radio-source catalogue) was published in 1959 by Cambridge University. Based on the catalogue, the identification of their optical counterparts was started. Matthews and Sandage found in 1960 a 16-magnitude "star" at the position of 3C 48, the 48th object in the 3C catalogue (Mathews and Sandage 1963). The "star" is extremely blue compared with a normal star and changes its luminosity within 1 year or on a much shorter timescale. Subsequently, Greenstein obtained the spectrum of 3C 48 and revealed that it is extraordinary in the sense that it exhibits broad emission features (Greenstein and Matthews 1963).

S. Kato, *Oscillations of Disks*, Astrophysics and Space Science Library 437,
DOI 10.1007/978-4-431-56208-5_1

Using lunar occultation, Hazard and his colleagues identified another strong radio source, 3C 273, as a 13-mag "star" (Hazard et al. 1963). They determined accurate position of two components, A and B. The latter coincided with a point-like source and the former showed a jet-like structure. Then, Schmidt observed its spectrum, and noticed in February of 1963 that these emission features are just the hydrogen Balmer lines, although they are shifted toward long wavelengths (i.e., redshifted) (Schmidt 1963). This was the moment of the discovery of *quasars*. The redshifts of 3C 48 and 3C 273 are now found to be 0.368 and 0.158, respectively.

There was a long discussion as to whether quasars are extragalactic or Galactic objects. After a long discussion, we now believe that quasars are cosmological objects and that their redshifts represent the cosmological expansion. The redshift $z = 0.158$ of 3C 273, for example, means that its distance is about 1.9×10^9 light years (if $H_0 = 71\ \mathrm{km\,s^{-1}\,Mpc^{-1}}$ and the Universe is flat).

If so, quasars should release an enormous amount of energy, which was a puzzle in these days. In the case of 3C 273, for example, the radiant energy, evaluated based on the apparent luminosity and distance, is up to $10^{47}\ \mathrm{erg\,s^{-1}}$, which is a thousand-times more luminous than a normal galaxy. In addition, this tremendous energy is radiated from the very center of the quasar.

Since the discovery in 1963, the energy source of quasars has been a great enigma in astronomy. This was finally solved (at least energetically) by the concept of supermassive black hole and surrounding accretion disks. This is the first time when *accretion disks* move into limelight in astrophysics.

One of the reasons why the accretion disks took limelight is that it can release energy for a long time in a rate stronger efficiency than the nuclear energy burning. Let us first consider the efficiency of nuclear energy release. In nuclear burning the energy release rate is the highest when hydrogen is burned. When hydrogen of mass M is perfectly converted into helium, release energy E_N is

$$E_\mathrm{N} = 0.007Mc^2, \tag{1.1}$$

where c is the speed of light.

Let us next consider gravitational energy released by accretion disks. The gravitational energy E_G of an object of mass M and size R is approximately

$$E_\mathrm{G} \sim \frac{GM^2}{R}, \tag{1.2}$$

where G is the gravitational constant. In normal stars in which gravitational force ($\sim GM/R^2$) is balanced by pressure force ($\sim c_\mathrm{s}^2/R$), i.e., $R \sim GM/c_\mathrm{s}^2$, the gravitational energy E_G is on the order of

$$E_\mathrm{G} \sim Mc_\mathrm{s}^2, \tag{1.3}$$

where c_s is the sound speed inside stars. Even if temperature is taken as high as 10^9K, c_s^2 is less than 10^{17} cm^{2}s^{-2}, and we have

$$E_G \sim 0.0001Mc^2, \tag{1.4}$$

which is much smaller than E_N given above.

In the case where a compact object is surrounded by an accretion disk, however, the accretion disk can release much energy. For example, if the compact source is a black hole or a neutron star and the accretion disk penetrates toward the central compact object till the radius of $3r_g$, where r_g is the Schwarzshild radius defined by $r_g = 2GM/c^2$,[1] the gravitational energy given by equation (1.2) is

$$E_G \sim \frac{1}{6}Mc^2, \tag{1.5}$$

which is much larger than the release energy by hydrogen burning given by equation (1.1). This means that if gas falls till the radius of $3r_g$ from infinity, a part of the energy given by equation (1.5) will be radiated away from the accretion disks,[2] which is higher than the nuclear energy release.

1.2 Importance of Studying Disk Oscillations and Diskoseismology

Arguments in Sect. 1.1 suggest that in many energetic astrophysical phenomena accretion disks are involved as the places of energy release. Nowadays, furthermore, accretion disks are known to exist in a wide range of objects from proto-planetary systems to galaxies. Accretion disks are one of most important and elementary ingredients in the Universe as well as stars.

Many accretion disks have time variations, and thus studies of oscillatory phenomena of accretion disks are of importance to clarify the disk structures and environments around the disks. To understand this situation, let us remember the history of relations between studies of stellar structures and those of stellar oscillations. Studies of stellar structure began in the beginning of twentieth century, and Eddington wrote the famous book "The Internal Constitution of Stars" in 1926.

---

[1]In the case of the Schwarzschild metric, the circular particle orbit around a central object is dynamically stable till the radius $3r_g$ (see, for example, Kato et al. 2008). This means that the gas can fall, by gradually losing angular momentum by viscosity, till the radius of $3r_g$ with roughly keeping circular orbit.

[2]Under certain situations, gravitational energy released is swallowed into central black holes as advection energy or outflows without being radiated away as thermal energy from disks. They are advetion-dominated accretion flows (ADAF) or radiatively inefficient accretion flows (RIAF) (see Sect. 1.6).

At that time theoretical studies on stellar pulsation already began, since various types of variable stars have been observed from long years ago. Eddington had been interested in the mechanism of Cepheid variables until just before his death.

A modern development of the relation between stellar structure and stellar oscillation began by the discovery of 5 min oscillations in the solar photosphere (Leighton et al. 1962). They found that the solar surface is covered by nearly vertical oscillations almost everywhere whose frequencies are close to 5 min. They are called "five minute oscillations". In their early studies, the 5 min oscillations are thought to be locally excited by granules (convective elements) in the solar photosphere and chromosphere as trapped oscillations. As observational data are accumulated, however, it became clear that the 5 min oscillations are independent of each granules, and are global eigenmode oscillations on the solar surface. The early observations and various theoretical models of the 5-min oscillations have been reviewed by Stein and Leibacher (1974).

Subsequently, many non-radial oscillations in addition to the 5 min ones are observed on the solar surface, and they are found to be well described as global eigenmode oscillations (nonradial oscillations) in the Sun. The excitation of the non-radial oscillations is now considered to be due to stochastic processes of turbulent convections in the solar convection zone (Goldreich and Keeley 1977a,b).[3]

Since eigenfrequencies of solar non-radial oscillations reflect the inner structure of the Sun, detailed comparisons between observed oscillations and calculated behaviors of oscillation modes can clarify the internal structure of the Sun. By this comparison the structure of the Sun is now rather clarified, although we cannot observe the inner structure of the Sun directly. For example, the internal rotation of the Sun is found to be almost uniform in the inner radiative zone, but in the outer convection zone the rotation is differential, i.e., the equatorial region rotates faster than in the polar region (see a review by Thompson et al. 2003). The transition region between the uniformly rotating inner region and the outer convective region is called the tachocline. This research field examining the internal structure of the Sun by use of solar oscillations is called *helioseismology*. Recently, a field called *local helioseismology* is also developing, where propagating waves in parts of the Sun (not global eigenmodes) are used to examine the structure of the Sun. This latter method is closer (than helioseimology) to the technique used to study the inertial structure of the Earth by use of seismic (earthquake) waves. The

[3]Ando, H. & Osaki, Y. (1975) proposed that the solar oscillations are excited by the κ-mechanism, which is widely known as the major excitation mechanism of radial pulsation of stars and of some non-radial oscillations in stars. In solar non-radial oscillations, however, many oscillation modes are observed simultaneously. Hence, it is difficult to consider that all of them are excited by the κ-mechanism alone, because the mechanism requires a proper phase relation between the change of κ (opacity) and oscillation motion. This is one of the reasons why the excitation processes of solar non-radial oscillations are considered to be due to stochastic processes of turbulent motions. The stochastic processes will be also one of prominent mechanisms of excitation of disk oscillations, which will be discussed in Sect. 13.2.

local helioseimology contributes on understanding of large-scale flow and magnetic structure in the Sun.

In studies of stellar variabilities, necessity of nonradial oscillations to describe observational phenomena in β Cephei stars, δ Scute stars, and so on had been recognized, from long years ago. Nowadays, however, almost all stars are known to have nonradial oscillations. Basics of nonradial oscillation of stars have been reviewed by Unno et al. (1989). The technique of helioseismology is now applied to stars, and the field is called *astroseismology*.

The helioseismology, local helioseismology, and astroseismology are making great progress in our understanding on the internal structure of the Sun and stars. Similar contributions are expected on studies of disk oscillations in understanding disk structures and environments around disks. The term *diskoseismology* or *discoseismology* was thus introduced.[4] One of the main progresses expected by diskoseismology is to explore the spin of central black-hole sources of relativistic accretion disks.

The launch of Rossi X-ray Timing Explorer (RXTE) in 1996 found high-frequency quasi-periodic oscillations (HFQPOs) in X-ray binaries, which stimulated studies of disk oscillations. The frequencies of these quasi-periodic oscillations are as high as the Keplerian frequency of the innermost region of relativistic disks and thus they are supposed to be related to disk oscillations in the innermost part of relativistic disks, where gases are in strong gravitational fields. Hence, HFQPOs will be an important tool to clarify directly gas motions in strong gravitational fields. Furthermore, studies of HFQPOs have a potential importance in exploring the spin of the central sources.

Spin of black-hole X-ray binaries is usually evaluated by comparing observed Fe K_α line or continuous X-ray spectra with those of disks models (McClintock et al. 2011). The results show that in some black hole sources their spins are close to the possible maximum value. The estimate of spin of black hole objects from the spectrum fitting is, however, not so robust because of various ambiguities of models and interpretation of observations. Estimate of the spin of black holes from disk oscillations is thus expected as one of independent way to evaluate the spin of black hole objects.

---

[4]The term of diskoseismology was introduced first perhaps by R.V. Wagoner and his groups in the middle of 1980s.

1.3 Astrophysical Objects with Disks

Accretion disks exist around various types of gravitating objects. They are formed by gasses falling from outside to the central objects by loosing angular momentum.[5] In this section we briefly summarize the central objects surrounded by accretion (excretion) disks. There are many excellent reviews on these objects, e.g., Wheeler (1993), Blandford et al. (1995), Lewin et al. (1995), Warner (1995), and Frank et al. (2002); and so on.

Typical objects with accretion (or excretion) disks and characteristics of these disks are summarized in Tables 1.1 and 1.2, which are duplication (with slight addition) of Tables 1.1 and 1.2 in "Black-Hole Accretion Disks" by Kato et al. (2008).

Gaseous disks are formed in various astrophysical circumstances. Well-known disks are those in binary systems, since gases falling to a primary star from a secondary cannot fall straightly to the primary because of angular momentum of

Table 1.1 Central objects with disks (Modified Table 1.1 of Kato et al. 2008).

Object	Central "star"	Mass	Size
YSO	PS/TTS	$\sim M_\odot$	$\sim R_\odot$
CV/SSXS	WD	$\sim M_\odot$	$\sim 10^{-2}\,R_\odot$
XB(NS)	NS	$\sim M_\odot$	$\sim 10\,\mathrm{km}$
XB(BHB)	BH	$\geq 3\,M_\odot$	$(r_g \gtrsim 10\,\mathrm{km})$
Be	Be stars	~ 3–$17\,M_\odot$	~ 4–$10\,R_\odot$
ULX	IMBH	$\sim 10^{2-4}\,M_\odot$	$(r_g \sim 300$–$30{,}000\,\mathrm{km})$
AGN	SMBH	$\sim 10^{5-9}\,M_\odot$	$(r_g \sim 0.002$–$20\,\mathrm{AU})$

Note: *YSO* young stellar object, *PS* protostar, *TTS* T Tauri star, *CV* cataclysmic variable, *SSXS* supersoft X-ray source, *WD* white dwarf, *XB* X-ray binary, *NS* neutron star, *BHB* black hole binary, *BH* black hole, *Be* Be star, *ULX* ultra-luminous X-ray source, *IMBH* intermediate-mass black hole, *AGN* active galactic nucleus, *SMBH* supermassive black hole

Table 1.2 Disks in various objects (Modified Table 1.2 of Kato et al. 2008).

Object	Mass	Size	Temperature
YSO	$\sim M_\odot$	~ 100 AU	$\sim 10^{1-4}$ K
CV/SSXS	$\ll M_\odot$	$\sim R_\odot$	$\sim 10^{4-6}$ K
XB(BHB)	$\ll M_\odot$	$\sim R_\odot$	$\sim 10^{4-9}$ K
Be (single)	$\ll M_\odot$	$\sim 10R_*(\mathrm{H}_\alpha)$, $\gg R_*$(radio)	$\sim 10^4 - 2\times 10^4$ K
(Binary)		$\sim 0.5r_\mathrm{p}(4 \sim 30R_*)$	
AGN	$\leq 10^6\,M_\odot$	~ 1 pc	$\sim 10^{3-5}$ K

Note: R_*: radius of Be star, r_p: distance of periastron

[5]There is another type of disks, called excretion disks. In these disks, gases are ejected from central objects by getting angular momentum. The disks surrounding Be stars belong this types of disks (Lee et al. 1991).

orbital motions. They form ring around the primary, and the inner part of the ring falls gradually onto the central object by losing angular momentum to the outer part, while the outer part extends outward by getting angular momentum. This is an accretion disk of binary systems. They can be classified into several types by differences of primary and secondary stars.

Even in single stars, gases can surround around the equatorial plane of the stars in the case where the stars rotate so rapidly that gases are ejected from the equator. Such disks are called *excretion disks*.

In galactic nuclei the central engines will be surrounded by accretion disks as mentioned before in relation to quasars. The gases will come from tidally disrupted stars and gaseous clouds, or collisions of galaxies.

The remainings of this section are devoted to rough survey of objects with accretion disks, following a review by Kato et al. (2008).

1.3.1 Young Stellar Objects

In star forming regions where stars are born from interstellar molecular clouds, the central part shrinks to form a new star, called a *protostar*, while the envelope settles down as a gaseous disk around the newly born star. Such a disk is often called a *protoplanetary disk*, since a planet is often formed there. Observational and theoretical studies on protoplanetary disks are one of the most developing fields in accretion disks in the present decade. In the present book, however, possible oscillatory phenomena in protoplanetary disks are outside of our scope.

1.3.2 Cataclysmic Variables

There are many types of close binary systems. The systems consisting of a white dwarf (primary star) and a red companion (from G type to M type star) show cataclysmic light variations. They are classified as *cataclysmic variables* (CVs), which include novae, dwarf novae, recurrent novae, nova-like variables, and polars as subclasses. When the mass accretion rate is higher than that of CVs, they are supposed to be *supersoft X-ray sources* (SSXSs).

In these systems the gas of the companion overflows through the Lagrange point toward the white dwarf (Roche overflow). Since the gas has angular momentum due to orbital motion, it forms an accretion disk around the primary compact star (unless the star is strongly magnetized).

Rougly speaking, the mass accretion rate from the secondary determines the type of time variabilities. In cases where the accretion rate on the surface of the primary is low, hydrogen burning on the surface gives rise to thermal instability (due to shell burning) and burst-like luminosity changes occur (novae). If mass accretion rate is sufficiently high, however, the hydrogen burning on the surface of the primary occurs steadily, and no burst-like luminosity variation occurs (supersoft X-ray sources). In polars, white dwarfs have strong magnetic fields and the accretion gas falls to the polar caps guided by dipole fields.

If the mass accretion rate is moderate, small outburst frequently occur (dwarf novae). There are many excellent reviews concerning dwarf novae (e.g,. Kahabka and van der Heuvel 1997; Warner 1995; Wheeler 1993). In the outburst phase of dwarf novae the brightness increases by 2–5 magnitudes compared with the quiescent phase. The duration is a few days to a couple of weak, while the interval is several months. Distinct from novae, the origin of this dwarf-novae outburst is thought to be a thermal limit-cycle instability in the accretion disks (Hōshi 1979; Meyer and Meyer-Hofmeister 1981). In addition to this (normal) outburst, in dwarf novae, superoutbursts are present, which occur less frequently compared with normal outburst. During the superoutburst, periodic humps, called *superhumps*, always appear with a period slightly longer than the orbital periods by a few percent. The superoutburst- superhump phenomena are understood by tidal instability (Hirose and Osaki 1990; Lubow 1991; Whitehurst 1988a,b). The whole set of the cycle of outburst-superoutburst is understood by the thermal-tidal instability model by Osaki (1989) (see a review by Osaki 1996).

In addition to superhumps, sometimes small amplitude oscillatory phenomena whose period is slightly shorter than the orbital period are observed in cataclysmic variables. This is called *negative superhumps*, and normal superhumps are then called *positive superhumps*.

The superhump phenomena are considered to be due to disk oscillations induced by the tidal instability. Hence, they will be reviewed more in detail in Sect. 1.5.1.

1.3.3 X-Ray Binaries and Ultra-luminous Sources

In close binary systems in which the primary star is a neutron star or a black hole, the gravitational potential is so deep that the activity of the systems can be observed not only in the optical range, but also in the X-ray range. They are called *X-ray binaries*. They are classified into *neutron-star X-ray binaries* (NSXBs) and *black-hole X-ray binaries* (BHXBs) by the difference of gas-accreting primary stars. X-ray binaries are also roughly classified into *low-mass X-ray binaries* (LMXBs) and *high-mass X-ray binaries* (HMXBs), according to the mass of a Roch-lobe filling secondary star. X-ray binaries are also classified into X-ray bursters, X-ray pulsars, and so on from the observational point of view.

1.3.3.1 Neutron-Star X-Ray Binaries

LMXBs, where the secondary star is a red dwarf, will be old systems, and the magnetic fields of the neutron star are weak. The accretion disk thus penetrates into near or on the surface of the neutron star. Some of LMXBs show X-ray bursts and called *X-ray Bursters*. The burst phenomena will be due to shell burning, and is similar to novae in CVs. Some group of LMXBs show repetitive long-term light variations. They are called *X-ray transients* (or *X-ray novae*). The origin of the outbursts is thought to be an accretion-disk instability similar to that in dwarf novae (e.g., Cannizzo et al. 1995; Mineshige and Wheeler 1989).

Roughly speaking, LMXBs with neutron-star are also classified into Atoll and Z-sources by behaviors on color-color diagram (CD) or hardness-intensity diagram (HID) (see, e.g., van der Klis 2004). The Atoll sources are though to have lower luminosity compared with the Z-sources, the latter having luminosities close to the Eddington critical one. Depending on positions on CD/HID, LMXBs present characteristic time variations. Excellent reviews on time variabilities of LMXBs are presented by van der Klis (2000, 2004).

Since some of time variations in LMXBs will be related to disk oscillations, they will be reviewed more in detail in Sect. 1.4.

High-mass X-ray binaries (HMXBs) are X-ray binaries in which the secondary star is a high-mass star. *Be/X-ray binaries* belong to this class. Be/X-ray binaries consist of a Be star (primary) and a compact object (secondary, a neutron star in general) (Reig 2011). Be stars are rapidly rotation B type stars with emission lines. The spin of Be stars is so high that gases are ejected in the equatorial region and excretion disks are formed in the equatorial region (Lee et al. 1991). In Be/X-ray binaries, the orbit of the secondary star is usually eccentric, and dynamical interaction between the disk and the secondary brings about time variations.

1.3.3.2 Black-Hole Binaries and Microquasars

Some X-ray binaries consisting of a normal star and a black hole are often called *black-hole binaries* (BHBs). They exhibit relativistic twin jets and called *microquasars* because of an analogy to quasars. An excellent review on black-hole binaries is written by Remillard and McClintock (2006). Some black-hole binaries exhibit quasi-periodic oscillations as neutron-star LMXBs do. Roughly speaking the quasi-periodic oscillations are classified into *high-frequency quasi-periodic oscillations* (HFQPOs) and *low-frequency quasi-periodic oscillations* (LFQPOs). The former will be mentioned more in Sects. 1.4.1 and 1.4.2.

1.3.3.3 Ultra-luminous X-Ray Sources

Ultraluminous X-ray sources (ULXs) are bright, point-like X-ray sources in nearby galaxies with apparent luminosities in the range of a few times 10^{39-41} $\mathrm{ergs\,s^{-1}}$ (Fabbiano 1989; Swartz et al. 2011). Their energy output exceeds the Eddington limit of stellar-mass black holes ($\sim 20\,M\odot$), if radiation is isotropic. The nature of ULXs is a puzzle, and three possibilities are conceivable. (1) The objects are intermediate-mass BHs (Colbert and Mushotzky 1999), (2) they are stellar-mass BHs with beaming (King et al. 2001), or (3) they are stellar mass BHs emitting above their Eddington limit (Begelman 2006; Ohsuga et al. 2005). We think that the true nature is still controversial. They might be an inhomogeneous sample with various types of sources present. Quasi-periodic oscillations are observed in some of ULXs, as will be mentioned in Sect. 1.4.3.

1.3.4 Galactic Nuclei

As discussed in history of quasars in Sect. 1.1, the presence of accretion disks around supermassive black holes is one of most energetic sources in the Universe. Almost all galactic centers, including our Galactic center, are now believed to consist of supermassive black holes and disks surrounding them. *Active galactic nuclei* (AGNs) will be systems with active phase of such systems.

Several possibilities are conceivable concerning the origin of disks around massive black holes. In processes of formation of a massive black hole by coalescence of less massive objects some part of gases remains around a coalesced object to form accretion disks. Collisions or tidal disruptions of clouds and stars orbiting around central black hole will be also sources of gas supply to accretion disks .

Recently, much attention has been taken to the time variations of Sgr A*, the center of our Galaxy, since a quasi-periodic oscillations of $P = 16.8\,\mathrm{min}$ were detected at IR (Genzel et al. 2003), which will be discussed in Sect. 1.4.4.

1.4 Quasi-periodic Oscillations in Various Objects

As mentioned in Sect. 1.3, many astrophysical systems have accretion disks. In these objects various types of small amplitude quasi-periodic oscillations are often observed. Their frequencies depend on the type of objects. One of typical examples is quasi-periodic oscillations in LMXBs.

The launch of RXTE (Rossi X-ray Timing Explorer) in 1996 led to the discovery of quasi-periodic oscillations ($\gtrsim$100 Hz) in low-mass X-ray binaries (LMXBs). Some of them are found to appear in twin. Especially, the discovery of twin QPOs in black-hole binaries whose frequency ratio is close to the ratio of 3:2 took much attention of many researchers in both theoretical and observational fields. A resonance model by Abramowicz and Kluźniak group (Abramowicz and Kluźniak 2001; Kluźniak and Abramowicz 2001) stimulated development of this field. This is because if the mechanism of the QPOs is clarified, it provides indispensable information on the properties of black holes, accretion disks, and strong gravity in general.

Variabilities of LMXBs are classified by van der Klis (2004) into four groups; (1) high-frequency phenomena ($\geq$100 Hz), (2) a low-frequency complex (a group of correlated 10^{-2}–10^2 Hz phenomena), (3) power-law components, and (4) other phenomena. The high-frequency phenomena are further classified into three subclasses

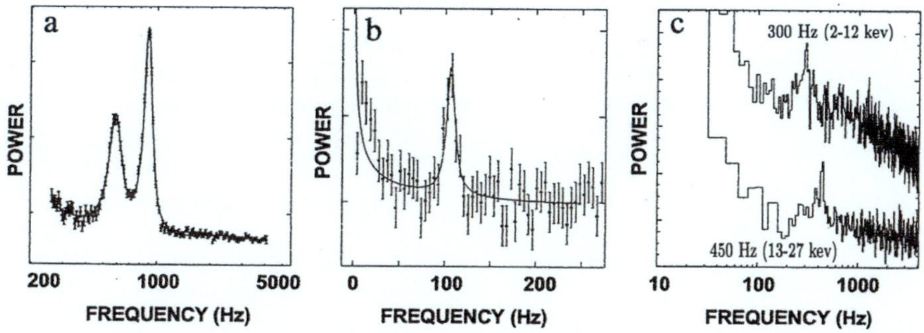

Fig. 1.1 Various high-frequency QPOs in X-ray binaries. (**a**) Twin kHz QPOs in Sco X−1 (van der Klis et al. 1997), (**b**) hectohertz QPO in 4U 0614+09, (**c**) HFQPOs in GRO J1655−40 (Strohmayer 2001a). In panel (**c**), the *lower curve* is the power in the energy range of 13–27 keV and a QPO can be seen at 450 Hz, while the *upper curve* is the power in the range of 2–12 keV and a QPO is seen at 300 Hz (After M. van der Klis (2004), in *Compact Stellar X-ray Sources*, eds. W.H.G. Lewin and M. van der Klis (Cambridge University Press; Cambridge), 39, CUP ©; panels (**a**) and (**c**) Astrophys. J. ©).

(van der Klis 2004); (i) neutron-star kilohertz QPOs (kHz QPOs); (ii) black-hole high-frequency QPOs (HFQPOs), and (iii) neutron-star hectohertz QPOs (hHz QPOs). Examples of these three subclasses of high frequency QPOs are shown in Fig. 1.1. As shown in Fig. 1.1, kHz QPOs in neutron-star binaries and HFQPOs in black-hole binaries often appear in twin. There are excellent reviews on variabilities in X-ray binaries by van der Klis (2004) and on black-hole X-ray binaries by Remillard and McClintock (2006).

In Sects. 1.4.1 and 1.4.2, we focus our attention only on the high-frequency QPOs, because they have frequencies comparable with the Keplerian frequency in the innermost region of disks. Examination of high-frequency QPOs will thus present us a key information on the structure of the innermost region of relativistic disks and spins of the central sources, as mentioned before.

1.4.1 HFQPOs in Black-Hole Binaries

Although the existing database of RXTE observations is very large, HFQPOs in black hole sources are detected only in a few ones. The sources and observed frequencies are summarized in Table 1.3.

It is noted that HFQPOs are observed at particular X-ray state of disks. The X-ray states can be classified into three ones; low (or thermal), hard, and very high (or steep power-low) states. HFQPOs are known to be associated with the steep power-low state (Remillard 2005, and see figure 1.2). Furthermore, the lower HFQPO of a pair QPOs is seen when the source has high X-ray luminosity, while the higher HFQPO of the pair is seen at lower luminosity (Remillard 2005, and see figure 1.2).

Table 1.3 Frequencies of HFQPOs in black-hole binaries.

Source name	Frequencies (Hz)	References
GRS 1915+105	41, 67, 113, 168	Morgan et al. (1997), Strohmayer (2001b), Belloni et al. (2001), Remillard et al. (2002), and Belloni et al. (2006)
GRO J1665−40	300, 450	Remillard et al. (1999) and Strohmayer (2001a)
XTE J1550−564	92, 184, 276	Homan et al. (2001), Miller et al. (2001), and Remillard et al. (2002)
H1743−322	166, 242	Homan et al. (2005) and Remillard et al. (2006)
XTE J1650−500	250	Homan et al. (2003)
4U 1630−47	184	Klein-Wolt et al. (2004)
XTE J1859+226	190	Cui et al. (2000)
IGR J17091−3624	66, 164	Altamirano and Belloni (2012)

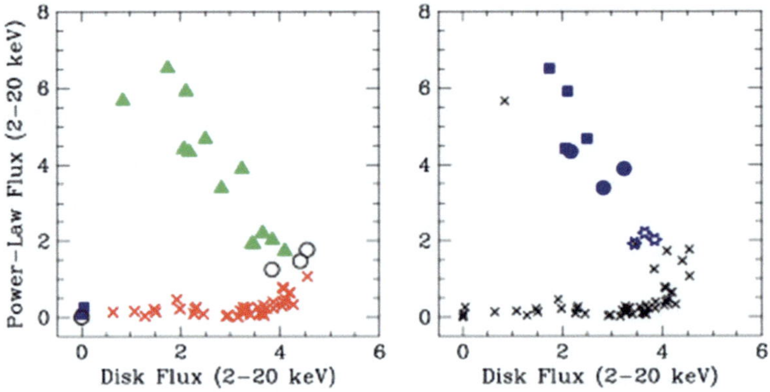

Fig. 1.2 X-ray states and HFQPOs during 1996–1997 outburst of GRO J1655-40. The *left panel* shows the energy diagram, where flux from the accretion disk is plotted versus flux from the power-law component. Here, the symbol type denotes the X-ray state: thermal (X), hard (*square*), steep power-law state (*triangle*), any type of intermediate state (*circle*). The *right panel* shows the same data points, while the symbol choice denotes HFQPOs detections: 300 Hz (*square*), 450 Hz (*star*), both HFQPOs (*circle*), and no HFQPOs (X). The HFQPOs detections are clearly linked to the steep power-low state, and the HFQPO frequency is clearly correlated with power-law luminosity (After Remillard 2005, Astron. Nachr. 326, 804, reprinted with permission of Astronomische Nachrichten ©).

This suggests that both of hot and cool components are necessary for appearance of HFQPOs (Fig. 1.2).

Attempts to describe HFQPOs in black-hole binaries by disk oscillations will be presented in Chaps. 7 and 12.

1.4.2 kHz QPOs in Neutron-Star Binaries

Compared with HFQPOs in black-hole binaries (BHBs), the kHz QPOs are observed in many neutron-star binaries (NSBs), but their frequencies are not robust compared with the cases of HFQPOs in black-hole binaries. The oscillations occur often in pairs and their frequencies are correlated (Belloni et al. 2002; Psaltis et al. 1999). The correlation between frequencies of the upper and lower kHz QPOs are shown in Fig. 1.3 for some typical neutron-star binaries. As shown in this figure, the frequency regions are different for Z-sources and Atoll sources. Roughly speaking, however, they are along a common line.

Attempts describing kHz QPOs by disk oscillations will be presented in Chaps. 7 and 8.

In a rough sense, NS LMXBs are classified into Z-sources and Atoll ones by their behaviors on color-color diagram. In Z-sources, for example, they have three branches of horizontal, normal, and flaring ones on color-color diagram. In each branch, they have proper low-frequency QPOs. For example, as sources move from left to right across the horizontal branch, the frequency of horizontal branch oscillations increases from $\sim$20 to $\sim$50 Hz. The frequency of the normal branch oscillations is around 6 Hz. The frequencies of these low-frequency QPOs are roughly correlated with those of high-frequency ones (see figure 2.9 of van der Klis 2004). Frequency correlations between high-frequency QPOs and low-frequency ones are also known in BH LMXBs.

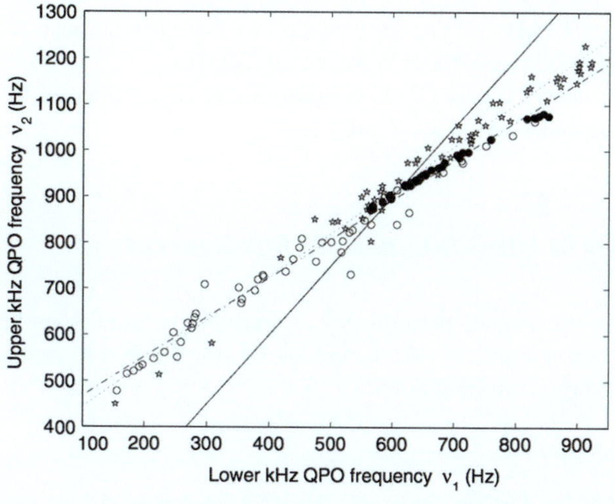

Fig. 1.3 Frequency correlation between *lower* and *upper* kHz QPO frequencies for the Sco X-1 (*filled circles*), for Z sources (*open circles*) and for Atoll sources (*stars*). The *dotted line* is the best linear fit to the Atoll points, the *dot-dashed line* is the best linear fit to the Z sources (excluding Sco X-1), and the *thick line* represents a fixed 3:2 ratio (After Belloni et al. 2005, Astron. Astrophys. 437, 209, reproduced with permission ©ESO).

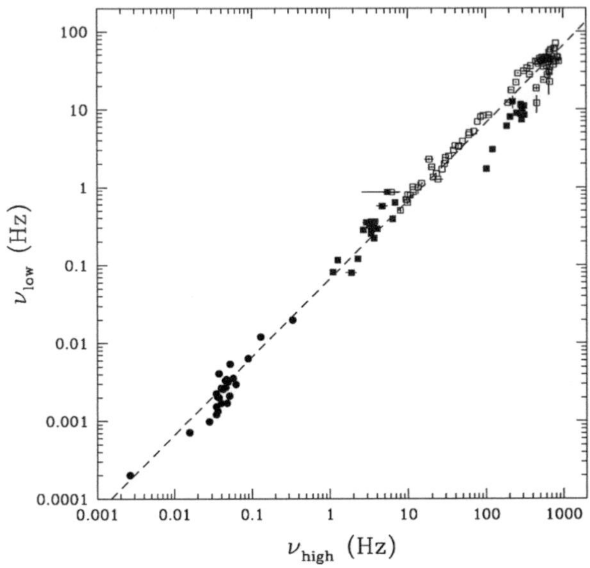

Fig. 1.4 Frequency correlation between the high and low QPOs from CVs to X-ray binaries. The *filled circles* are for CVs, the *open squares* are for neutron-star binaries, and the *filled squares* are for black-hole binaries (After Warner et al. 2003, Mon. Not. R. Astron. Soc. 344, 1193, reproduced with permission of Oxford University Press ©).

Finally, one of interesting observational evidences is that the correlation between lower- and higher-frequency QPOs in twin kHz QPOs are phenomenologically extended to oscillations in cataclysmic variables beyond oder five. In cataclysmic variables we have so-called dwarf novae oscillations (DNOs) of $P_{\rm DNO} \sim 20$ s as well as quasi-periodic oscillations (QPOs) of $P_{\rm QPO} \sim 32$–165 s, i.e., $P_{\rm QPO} \sim 15 P_{\rm DNO}$. The correlations in kHz QPOs are extended to this correlation in CVs, which is shown in Fig. 1.4 (Mauche 2002; Warner et al. 2003).

Attempts to describe kHz QPOs in neutron-star X-ray binaries by disk oscillations will be presented in Chaps. 7 and 12.

1.4.3 QPOs in Ultra-luminous X-Ray Sources

Some of ULXs show quasi-periodic X-ray oscillations. Studies on the oscillations will give important clues on the nature of ULXs. Table 1.4 summarizes quasi-periodic oscillations observed in ULXs.

*1.4.4 QPOs in Active Galactic Nuclei and Sgr A**

Ozernoy and Usov (1977) have pointed out that long-term quasi-periodic variations from several hundred days to a few years are observed in optical wavelength in some AGNs. In X-ray wavelengths, active galactic nuclei also have time variations,

Table 1.4 Frequencies of QPOs in ULXs.

Source name	Frequencies (Hz)	References
IC 342 X−1	642 mHz	Agrawal and Nandi (2014)
M82 X−1	50–170 mHz, 46 ± 2 mHz	Strohmayer and Mushotzsky (2003), Dewangan et al. (2006), and Caballero-García et al. (2013)
NGC 5408 X−1	10–40 mHz	Strohmayer et al. (2007), Strohmayer and Mushotzsky (2009), and Dheeraj and Strohmayer (2012)
NGC 6946 X−1	8.5 mHz	Rao et al. (2010)
M82 X42.3+59	3–4 mHz	Feng et al. (2010)

Table 1.5 Frequencies of QPOs in active galactic nuclei.

Source name	Frequencies (Hz)	References
RE J1034+396	$\sim 2.6 \times 10^{-4}$	Gierliński et al. (2008) and Alston et al. (2014)
Swift J164449.3+573,451	~200 s	Reis et al. (2012)
2XMM J123103.2+110,648	~3.8 h	Lin et al. (2013)
MS 2254.9–3712	~2 h	Alston et al. (2015)

but they had been considered usually to have no periodic nor quasi-periodic variations. Recently, however, X-ray observations show the presence of quasi-periodic variations in a frequency range scaled down from those of black-hole binaries.[6] The first robust AGN HFQPOs detection came from Seyfert galaxy RE J1034+396, with ~1 h periodicity (Gierliński et al. 2008). A recent work by Alston et al. (2014) confirms this: The QPO frequency remains persistent at $\sim 2.6 \times 10^{-4}$ Hz over years. The QPO is only detected when the source has a low flux and is spectrally harder, observed in a hard band. The QPOs observed so far in AGNs are summarized in Table 1.5.

The time variations of our Galactic center, Sgr A*, was first reported by Genzel et al. (2003). They observed quasi-periodic oscillations of $P = 16.8$ min in a flaring phase at IR. Variabilities of similar timescale have been found in the phases of X-ray flares (Achenbach et al. 2004). Furthermore, Miyoshi et al. (2012) have detected spatially resolved QPOs close to 16.8 min from the data of VLBA observations at 43 GHz, including other QPOs. They were observed 1.5 days after a millimeter-wave shorter-time flare. The detection of oscillations by VLBA is important, because the places where oscillations exist are resolved. Observations by VLBA may open a way to a new field of *local discoseismology* (cf., local helioseismology). Observations of QPOs from Sgr A* are summarized in Table 1.6.

[6]For the meaning of "scaled down", see the next footnote.

Table 1.6 Frequencies of QPOs detected in Sgr. A*.

Obs. epoch (UT)	Obs. band	Period (min)	References
2003/06/15–16	K-band	16.8±2, 28.0	Genzel et al. (2003)
2004/09	1.60, 1.87, 1.90 (μm)	33 ± 2	Yusef-Zadeh et al. (2006)
2002/10, 2004.08	2–10 (keV)	22.2	Bélanger et al. (2006)
2007/04/04	L-band	22.6	Hamaus et al. (2009)
2007/07/22	L-band	45.4	Hamaus et al. (2009)
2004/03/08 09:30–16:30	43 (GHz)	16.8 ± 1.4, 22.2 ± 1.4, 31.2 ± 1.5, 56.4 ± 6	Miyoshi et al. (2012)

1.5 Long-Term Variations in Disks

Up to the present, we have considered quasi-periodic oscillations whose frequencies are on the order of the Keplerian frequency in the innermost region of disks. In this sense they are short-term oscillations.[7] Distinct from these oscillations, there are long-term oscillations in the sense whose frequencies are comparable with the Keplerian frequency at the outer edge of disks. Two of these examples are described below.

1.5.1 Positive and Negative Superhumps in Dwarf Novae

In Sect. 1.3.2, time variations of dwarf novae were described. Among their various time variations, superhumps are of interest here, because they are related to disk oscillations. Figure 1.5 shows light variation and two-dimensional power spectrum of superhump of V1504 Cygni, obtained by Osaki and Kato (2013) by analyzing data of the Kepler satellite. The upper panel of this figure shows that superoutbursts occasionally occur in addition to outbursts.

The middle and lower panels of Fig. 1.5 show that in the superoutburst stage disk luminosity has small amplitude periodic variations whose frequencies are slightly lower than the orbital frequency of the revolution of the binary system. This is called *(positive) superhump*. This means that the disk has a slowly precessing prograde

[7]The frequencies observed at Sgr A* are much lower than those in LMXBs, but they are short-term oscillations in the following sense. If the frequencies of QPOs are roughly equal to the Keplerian frequency of the innermost region of relativistic disks, they are on the order of $(GM/r_g^3)^{1/2}$, where M is the mass of the central source and r_g is the Schwarzschild radius defined by $r_g = 2GM/c^2$. Hence, the frequency is proportional to $1/M$. If this scaling is applied, the frequencies of QPOs of Sgr A* are roughly on the line of the $1/M$ scaling from those of QPOs of LMXBs. In this sense, we called the QPOs in Sgr A* short-term variations.

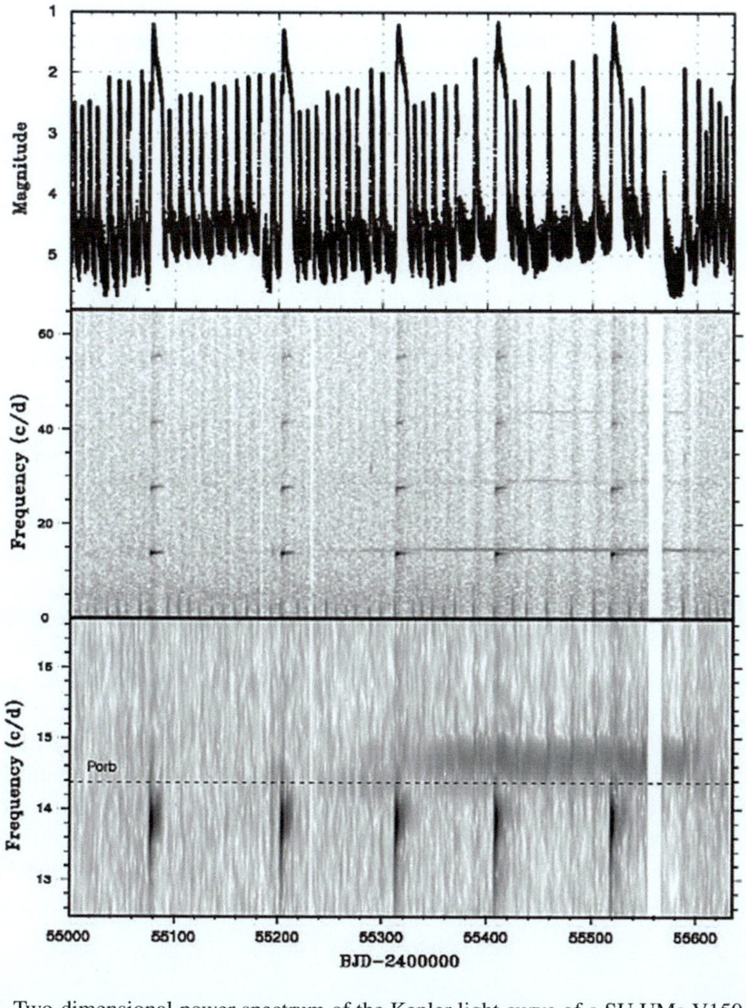

Fig. 1.5 Two-dimensional power spectrum of the Kepler light curve of a SU UMa V1504 Cygni. The *upper* is the light curve, the *middle* is the power spectrum, and the *lower* is an enlargement for the frequency region around the orbital one. It is noticed that in the superoutburst state there are small amplitude oscillations whose frequency is slightly low compared with the orbital frequency of the binary system (After Osaki and Kato 2013, Publ. Astron. Soc. Jpn., 65, 59, reproduced with permission of Publ. Astron. Soc. Jpn. ©).

deformation. Recently, the superhumps whose frequencies are higher than the revolution frequency are also taken attention, which are called *negative superhump*. In the lower panel of Fig. 1.5, negative superhumps are shown like widely spreaded clouds. This means that retrograde precession waves are also present sometimes on disks of DNs. It is noted that the negative superhumps appear not only at the superoutburst stage but also at the outburst stage.

Positive and negative superhumps can be interpreted as typical low frequency oscillations which are excited in nearly Keplerian disks. This issue will be discussed later in detail in Chaps. 11 and 12.

1.5.2 V/R Variations in Be Stars

Be stars are rapidly rotating B-type stars with emission lines and equatorial disks. The equatorial disks are geometrically thin, high-density gases in nearly Keplerian rotation (e.g., Porter and Rivinius 2003). Concerning the origin of the disks, several competing scenarios are presented, but the most promising one is the scenario where the disks are formed by viscous excretion of gas ejected from the central rapidly rotating B-type stars (Lee et al. 1991; see also Porter and Rivinius 2003)

Many Be stars are known to show long-term variations in their spectra over years to decades. Typical ones are variations of the ratio of relative intensity of violet (V) and red (R) peaks of a double-peaked emission line profile, which is called long-term V/R variations. Figure 1.6 shows the V/R variations of Hα emission lines in a Be star γ Cas (Horaguchi et al. 1994).

The long-term variations of Be stars is considered to be due to a slow precession of one-armed, global oscillations in disks (Kato 1983). Many observations suggest that the observed one-armed oscillations are prograde, i.e., the one-armed deformation of disks slowly precesses in the same direction as the disk rotation. For example, Telting et al. (1994) derived the conclusion that it is prograde in the case of a nearly edge-on Be star β^1 Mon. Vakili et al. (1998) also derived the same conclusion for a Be Shell star ζ Tauri from interpretation of observations made by interferometry. Mennickent et al. (1997) analyzed photometric data of a sample of 6 V/R variable Be stars, and leads to the conclusion that observational data are compatible with prograde global disk oscillations. Mon et al. (2013), however, suggest that in a Be-shell star EW Lac the V/R variation is of retrograde structure.

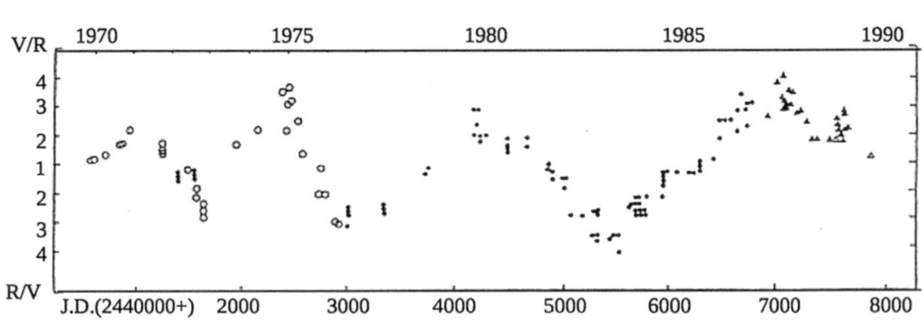

Fig. 1.6 Long-term variations of γ Cas in 1969–1989 (Adapted from Horaguchi et al. (1994), Publ. Astron. Soc. Jpn., 46, 9, reproduced with permission of Publ. Astron. Soc. Jpn. ©).

The V/R variations will be due to one-armed oscillations of disks, but there are discussions whether the oscillations can describe the prograde nature of the observed V/R variations, This issue will be argued in Sect. 8.2.

1.5.3 Long-Term Variations in Be/X-Ray Binaries

In Be/X-ray binaries the secondary star usually has an eccentric orbit whose rotational axis is misaligned with the spin axis of the primary. Two types of long-term variations have been observed in these Be/X-ray binaries. i.e., normal (type 1) outbursts and giant (type 2) outbursts (Moritani et al. 2013; Negueruela et al. 2001; Reig et al. 2007). The normal outbursts occur around the periastron passage of the compact object, and are considered to come from enhancement of mass transfer (Negueruela and Okazaki 2001; Okazaki and Negueruela 2001), but the latter seems to be less understood and many models are proposed. The giant outburst might be partially related to oscillatory phenomena in disks.

1.6 Brief History and Summary on Accretion Disk Models

Before closing this chapter, we will briefly summarize what types of accretion disks are known up to the present, and what types of disks we concerned with in this book.

As mentioned in Sect. 1.1, importance of accretion disks took footlights in 1960s by discovery of quasars and by efficiency of gravitational energy release by accretion disks. Lynden-Bell (1969) pointed out explicitly the importance of gradual mass accretion processes as energy sources of quasars and suggested the presence of died quasars at the centers of galaxies.

The importance of mass accretion processes in energy liberation was, however, already pointed out by Zel'dovich and Novikov (1964) and Salpeter (1964), independently. They considered an isolated system. As an energy source of X-ray stars, Hayakawa and Matsuoka (1964) proposed accretion processes in a binary system. Shklovsky (1967) explained ScoX-1 (the first X-ray source observed in the Universe except for the Sun) as accretion onto a neutron star. After these studies, in 1970s the importance of accretion processes and the concept of accretion disks began to be widely recognized in fields of studies of binary systems.

After several attempts to construct accretion disk models by many investigators, Shakura and Sunyaev (1973) proposed an accretion disk model, which is now called *the standard accretion disk model*, or *α-disk model*. The naming of α-model comes from the parameter α introduced to describe the magnitude of turbulent viscosity. The relativistic version is given by Novikov and Thorne (1973). In this standard model the gravitational energy released by accretion through viscosity is assumed to be radiated away at the place of the release as thermal energy. Hence, the disk temperatures are relatively low. As observational evidences concerning

X-ray binaries and galactic nuclei are accumulated, however, it was recognized that there are many observational evidences which cannot be simply described by the standard model. In order to describe high temperature disks, an optically thin radiation-dominated disk model was proposed by Shapiro et al. (1976), soon after the standard model. The model was, however, found to be thermally unstable and will not be realized.

In the 1970s–1980s, as a limiting case of no accretion, geometrically thick tori are considered by Abramowicz and his group (Abramowicz et al. 1978; Kozlówski et al. 1978, and others). Furthermore, ion tori, which are optically thin and have such a low density that they cannot radiate efficiently, are examined by Rees et al. (1982).

In the late 1980s the presence of new type of hot, optically thick disks was recognized by Abramowicz et al. (1988). In this model the gravitational energy released through viscosity is transported inwards as advection energy instead of being radiated away outside the disks as thermal energy. Since the released energy is stored as thermal energy, the disk temperature is high. This disk model was called *slim disks*. The presence of such disk models was soon conformed by an independent numerical code (Abramowicz et al. 1989). Subsequently, slim disk models of supercritical accretion flows with photon trapping (accretion rate being more than the Eddington critical rate) have been extensively examined by Mineshige's group (e.g., Ohsuga et al. 2005). In the 1990s, the importance of another radiatively inefficient disk modes (optically thin models) was recognized by Narayan and Yi (1994) and Abramowicz et al. (1995) after a pioneering work by Ichimaru (1977). These models are called *advection-dominated accretion flows* (ADAFs) and now are included in a group of disks with a wider notion of *radiatively inefficient accretion flows* (RIAFs).

In the 2000s another important disk model has been proposed by Matsumoto's group, e.g., (Machida et al. 2006; Oda et al. 2007, 2009, 2010, 2012). This is *magnetic-fields-supported accretion disks*, and can bridge between optically thin disks (RIAFs) and the standard disks so that these two disk states can be continuously changed.

As described above there are various types of disk models, corresponding to describing various kinds of objects and their various states. In this monograph, however, we examine disk oscillations in the most simplified situations, since our present interest is still at the level of knowing basic properties of disk oscillations. That is, the disks we treat in this monograph are geometrically thin standard disks.

References

Abramowicz, M. A., Jaroszyński, M., & Sikora, M. 1978, Astron. Astrophys., 63, 221
Abramowicz, M. A., Czerny, B., Lasota, J.-P., & Szuskiewicz, E. 1988, Astrophys. J., 332, 646
Abramowicz. M. A., Kato, S., & Matsumoto, R. 1989, Publ. Astron. Soc. Jpn., 41, 1215
Abramowicz, M. A., Chen, X., Kato, S., Lasota, J.-P., & Regev, O. 1995, Astrophys. J., 438, L37
Abramowicz, M. A., & Kluźniak, W. 2001, Astron. Astrophys., 374, L19
Achenbach, B., Gross, N., Proquet, D., & Predehl, P. 2004, Astron. Astrophys., 417, 71
Agrawal, V. K.,& Nandi, A. 2014, Mon. Not. R. Astron. Soc., 446
Alston, W. N., Markevičiūtė, J., Kara, E., Fabian, A. C., & Middleton, M. 2014, Mon. Not. R. Astron. Soc., 445, 16
Alston, W., Fabian, A., Markevičiūtė, J., Parker, M., Middleton, M., & Kara, E., 2015, Astro-ph 1510.01111
Altamirano, D., & Belloni, T. 2012, Astrophys. J., 747, 4
Ando, H. & Osaki, Y. 1975, Publ. Astron. Soc. Jpn., 27, 581
Begelman, M. C. 2006, Astrophys. J., 643, 1065
Bélanger, G., Terrier, R., de Jager, O, C., Goldwurm, A., & Melia, F. 2006, J. Phys.: Conf. Ser., 54, 420
Belloni, T., Méndez, M., & Sanchez-Fernandez, C. 2001, Astron. Astrophys., 372, 551
Belloni T., Méndez M., & Homan, J. 2005, Astron. Astrophys., 437, 209
Belloni, T., & Soleri, P., et al. 2006, Mon. Not. R. Astron. Soc., 369, 305
Belloni, T., Psaltis, D., & van der Klis, M. 2002, Astrophys. J., 572, 392
Blandford, R., Netzer, H., & Woltjer, L. 1995, *Active Galactic Nuclei* (Springer-Verlag, Berlin)
Caballero-García, M. D., Belloni, T., & Zampieri, L. 2013, Mon. Not. R. Astron. Soc. 436, 3262
Cannizzo, J. K., Chen, W., & Livio, M. 1995, Astrophys. J., 454, 880
Colbert, E. J., & Mushotzky, R. F. 1999, Astrophys. J., 519, 89
Cui, W., Shrader, C. R., et al. 2000, Astrophys. J., 535, L123
Dewangan, G. C., Titarchuk, L., & Griffiths, R. E. 2006, Astrophys. J., 637, 21
Dheeraj, P. R., & Strohmayer, T. E., 2012, Astrophys. J., 753, 139
Fabbiano, G., 1989, Annual Rev. Astron. Astrophys., 27, 87
Frank, J., King, A., & Raine, D. 2002, *Accretion Power in Astrophysics*, 3rd ed. (Cambridge University Press, Cambridge)
Feng, H., Rao, F., & Kaaret, P. 2010, Astrophys. J., 710, L137
Genzel, R., Schödel, R., Ott, T., Eckart, A., Alexander, T., Lacombe, F., Rouan, D., & Aschenbach, B. 2003, Nature, 425, 934
Gierliński, M., Middleton, M., Ward, M., & Done, C. 2008, Nature, 455, 369
Greenstein, J. L., & Matthews, T. A. 1963, Astron. J., 68, 279
Goldreich, P., & Keeley, D. A. 1977a, Astrophys. J., 211, 934
Goldreich, P., & Keeley, D. A. 1977b, Astrophys. J., 212, 243
Hamaus, N., Paumard, T., Müller, T., Gillessen, S., Eisenhauer, F., Tripple, S., & Genzel, R. 2009, Astrophys. J., 692, 1075
Hayakawa, S., & Matsuoka, M. 1964, Prog. Theor. Phys. Suppl., 30, 204
Hazard, C., Mackey, M. B., & Shimmins, A. J. 1963, Nature, 197, 1037
Hirose, M., & Osaki, Y. 1990, Publ. Astron. Soc. Jpn., 42, 135
Homan, J., & Wijnands, R., et al. 2001, Astrophys. J. Suppl, 132, 377
Homan J., Klein-Wolt M., Rossi S., Miller J. M., Wijnands R., Belloni, T., van der Klis M., & Lewin W. H. G. 2003, Astrophys. J. 586, 1262
Homan, J., Miller, J. M., et al. 2005, Astrophys. J., 623, 383
Horaguchi, T., Kogure, T., Hirata, R., Kawai, N., Matsuoka, M., Murakami, T., Doazan, V., Slettebak, A., et al. 1994, Publ. Astron. Soc. Jpn., 46, 9
Hōshi, R., 1979, Prog. Theor. Phys., 61, 1307
Ichimaru, S. 1977, Astrophys. J. , 214, 840
Kahabka, P., & van der Heuvel, E. P. J. 1997, Annu. Rev. Astron. Astrophys., 35, 69

Kato, S. 1983, Publ. Astron. Soc. Jpn., 35, 249
Kato, S., Fukue, J., & Mineshige, S. 2008, *Black-Hole Accretion Disks — Towards a New Paradigm*, (Kyoto University Press, Kyoto)
King, A. R., Davis, B., Ward, M. J., Fabbiano, G., & Elvis, M. 2001, Astrophys. J., 552, L109
Klein-Wolt, M., Homan, J., & van der Klis, M. 2004, Nucl. Phys. B-Proc. Suppl., 132, 381
Kluźniak W., & Abramowicz M. A. 2001, Acta Phys. Pol. B32, 3605
Kozlówski , M., Jaroszyński, M., & Abramowicz, M. A., 1978, Astron. Astrophys., 63, 209
Lee. U., Saio, H., & Osaki, Y. 1991, Mon. Not. R. Astron. Soc. , 250, 432,
Leighton, R.B., Noyes, R.W., & Simon, G. W., 1962, Astrophys. J., 135, 474
Lewin, W. H. G., van Paradijs, J., & van der Heuvel, E. P. J. 1995, *X-ray Binaries* (Cambridge University Press, Cambridge)
Lin, D., Irwin, J. A., Godet, O., Webb, N. A., & Barret, D. 2013, Astrophys. J., 776, L10
Lubow, S. H., 1991, Astrophys. J., 381, 259
Lynden-Bell, D. 1969, Nature, 223, 690
Machida, M., Nakamura, K. E., & Matsumoto, R. 2006, Publ. Astron. Soc. Jpn., 58, 193
Matthews, T. A., & Sandage, A. R. 1963, Astrophys. J., 138, 30
Mauche C. W. 2002, Astrophys. J., 580, 423
McClintock, J. E., Narayan, R., Davis, S. W., Gou, L., Kulkarni, A., Orosz, J. A., Penna, R. F., Remillard, R. A., & Steiner, J. F. 2011, Class. Quantum Gravity, 28, 4009
Mennickent, R. E., Sterken, C., & Vogt, N. 1997, Astron. Astrophys., 326, 1167
Meyer, F., & Meyer-Hofmeister, E. 1981, Astron. Astrophys., 104, L10
Miller, J. M., Wijnands, R., Homan, J., Belloni, T., Pooley, D., Corbel, S., Kouveliotou, C., van der Klis, M., & Lewin, W. H., G. 2001, Astrophys. J., 563, 928
Mineshige, S., & Wheeler, J. C., 1989, Astrophys. J. , 343, 241
Miyoshi, M., Shen, Z. Q., Oyama, T., Takahashi, R., & Kato, Y. 2012, Publ. Astron. Soc. Jpn., 63, 1093
Mon, M., Suzuki, M., Moritani, Y., & Kogure, T. 2013, Publ. Astron. Soc. Jpn., 65, 77
Morgan, E. H., Remillard, R. A., & Greiner, J. 1997, Astrophys. J. , 482, 993
Moritani, Y., et al. 2013, Publ. Astron. Soc. Jpn., 65, 83
Narayan, R., & Yi, I. 1994, Astrophys. J., 428, L13
Negueruela , I., & Okazaki, A. T. 2001, Astron. Astrophys., 369, 108
Negueruela, I., Okazaki, A. T., Fabregat, J., Coe, M. J., Murari, U., & Tomov, T. 2001, Astron. Astrophys., 369, 117
Novikov, I. D., & Thorne, K. S., 1973, in *Black Holes*, ed., C.DeWitt & B. DeWitt (Gordon and Breach, New York)
Oda, H., Machida, M., Nakamura, K. E., & Matsumoto, R. 2007, Publ. Astron. Soc. Jpn., 59, 457
Oda, H., Machida, M., Nakamura, K. E., & Matsumoto, R. 2009, Astrophys. J., 697, 16
Oda, H., Machida, M., Nakamura, K. E., & Matsumoto, R. 2010, Astrophys. J., 712, 639
Oda, H., Machida, M., Nakamura, K. E., Matsumoto, R., & Narayan, R. Publ. Astron. Soc. Jpn., 2012, 65, 15
Ohsuga, K., Mori, M., Nakamoto, T., & Mineshige, S. 2005, Astrophys. J., 628, 368
Okazaki, A.T., & Negueruela, I. 2001, Astron. Astrophys., 2001, 377, 161
Osaki, Y. 1989, Publ. Astron. Soc. Jpn., 41, 1005
Osaki, Y. 1996, Publ. Astron. Soc. Pac., 108, 39
Osaki, Y., & Kato, T. 2013, Publ. Astron. Soc. Jpn., 65, 50
Ozernoy L. M., & Usov V. V. 1977, Astron. Astrophys. 56, 163
Porter, J. M., & Rivinius, T. 2003, Publ. Astron. Soc. Pac., 115, 1153
Psaltis D., Belloni T., & van der Klis M. 1999, Astrophys. J., 520, 262
Rao, F., Feng, H., & Kaaret, P. 2010, Astrophys. J., 722, 620
Rees, M.J., Begelman, M. C., Blandford, R. D., & Phinney, E. S., 1982, Nature, 295, 17
Reig, P., Larionov, V., Negueruela. I., Arkharov, A. A., & Kudryavtseva, N. A., 2007, Astron. Astrophys., 462, 1081
Reig, P. 2011, Astrophys. Space Sci., 332, 1

Reis, R. C., Miller, J. M., Reynolds, M. T., Gültekin, K., Maitra, D., King, A. L., & Strohmayer, T. E., 2012, Science, 337, 949
Remillard, R. A. 2005, Astro. Nachr., 326, 804
Remillard, R. A., Muno, M., McClintock, J. E., & Orosz, J. A. 2002, Astrophys. J., 580, 1030
Remillard, R. A., McClintock, J. E., Orosz, J., & Levin, A. M. 2006, Astrophys. J., 637, 1002
Remillard, R. A., & McClintock, J. E., et al. 1999, Astrophys. J., 517, L127
Remillard, R. A., & McClintock, J. E., 2006, Annu. Rev. Astron. Astrophys., 44, 49
Salpeter, E. E. 1964, Astrophys. J., 140, 796
Schmidt, M. 1963, Nature, 197, 1040
Shakura, N. I., & Sunyaev, R. A. 1973, Astron. Astrophys., 24, 337
Shapiro, S. L., Lightman, A. P., & Eardley, D. H. 1976, Astrophys. J., 204, 187
Shklovsky, I. S. 1967, Astrophys. J., 148L, 1
Stein, R. F.,& Leibacher, J. 1974, Annu. Rev. Astron. Astrophys., 12, 407
Strohmayer, T. E. 2001a, Astrophys. J., 552, L49
Strohmayer, T. E. 2001b, Astrophys. J., 554, L169
Strohmayer, T. E., & Mushotzsky, R. F. 2003, Astrophys. J., 586, L61
Strohmayer, T. E., Mushotzsky, R. F., & Winter, L et al. 2007, Astrophys. J., 660, 580
Strohmayer, T. E., & Mushotzsky, R. F. 2009, Astrophys. J., 703, 1386
Swartz, D. A., Soria, R., Tennant, A., F., & Yukita, M. 2011, Astrophys. J., 741, 49
Telting, J. H., Heemskerk, M. H. M., Henrichs, H. F., & Savonije, G. J. 1994, Astron. Astrophys., 288,558
Thompson, M. J., Christensen-Dalsgaad, J., Miesch, M. S., & Toomre, J. 2003, Annual Rev. Astron. Astrophys., 41, 599
Unno, W., Osaki, Y., Ando, H., Saio, H., & Shibahashi, H. 1989, *Nonradial Oscillations of Stars* (Univ. Tokyo Press, Tokyo)
Vakili, F., Mourard, D., Stee, Ph., Bonneau, D., Berio, P., Chesneau, O., Thureau, N., Morand, F., Labeyrie. A., & Tallon-Bosc, I. 1998, Astron. Astrophys., 335, 261
van der Klis, M. 2000, Ann. Rev. Astron. Astrophys. 38, 717
van der Klis, M. 2004, in *Compact Stellar X-ray Sources*, eds. W. H. G. Lewin and M. van der Klis (Cambridge University Press, Cambridge), 39
van der Klis, M., Wijnands, R. A. D., Herne, K., & Chen, W. 1997, Astrophys. J., 481, L97
Warner, B. 1995, *Cataclysmic Variable Stars* (Cambridge: Cambridge University Press)
Warner, B., Woudt, P. A., & Pretorius, M. L. 2003, Mon. Not. R. Astron. Soc., 344, 1193
Whitehurst, R. 1988a, Mon. Not. R. Astron. Soc., 232, 35
Whitehurst, R. 1988b, Mon. Not. R. Astron. Soc., 233, 529
Yusef-Zadeh, F., Bushouse, H., et al. 2006, Astrophys. J., 644, 198
Zel'dovich, Ya. B., & Novikov, L. D. 1964, Dokl. Akad. Nauk., 155, 1033
Wheeler, J.C., (cd) 1993, *Accretion Disks in Compact Stellar Systems* (World Scientific, Singapore)

Chapter 2
Basic Quantities Related to Disk Oscillations

Abstract After presenting basic assumptions and approximations which we adopt in this book, we describe important concepts and dynamical quantities which are basic in studying disk oscillations. They are horizontal and vertical epicyclic frequencies, Lindblad resonances, and coronation resonance.

Keywords Corotation resonance • Epicyclic frequencies • Lindblad resonances

2.1 General Remarks on Subjects of Our Studies

Before describing basic dynamical quantities which are important in studying disk oscillations, we briefly notice basic approximations which we adopt in this book.

After the discovery of quasars, studies on accretion disks became one of main subjects in astronomy, because they are found to appear in various objects and to have important contributions to their evolutions. In this book, however, we do not treat all kinds of disks. We treat here only nonself-gravitating and geometrically thin disks. Effects of global magnetic fields on disk oscillations are studied partially. Furthermore, effects of accretion flows on oscillations are neglected.

2.1.1 Nonself-Gravitating Disks

In proto-planetary disks and galactic disks, self-gravity of disks has important effects on dynamical behaviors of the disks. For example, one of main purposes of Lin & Shu's density wave theory in 1964 was, we think, to put in effects of self-gravity into their dispersion relation of density waves, although the formulation was presented in the appendix of their paper.[1]

The effects of self-gravity on disks are twofolds. One is on disk structure and the other is on perturbations. Let us first consider the former. The disk is assumed to be

[1]It is noted that the propagation region of self-gravitating galactic density waves is different from that of nonself-gravitating oscillations considered in this book.

S. Kato, *Oscillations of Disks*, Astrophysics and Space Science Library 437,
DOI 10.1007/978-4-431-56208-5_2

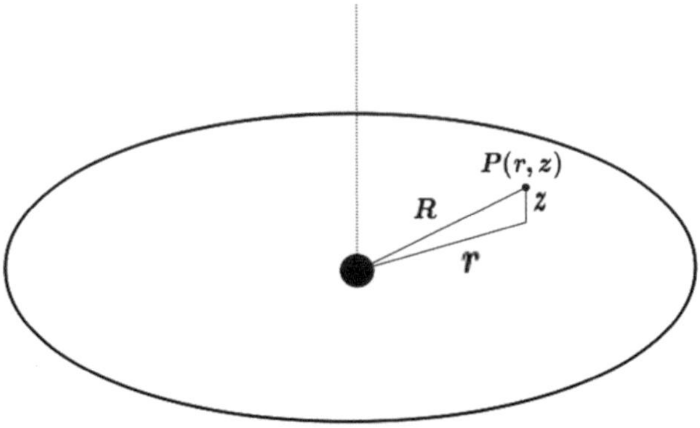

Fig. 2.1 Representing the position of a point P(r, z) in disk by coordinates (r, z), where (r, z) is a part of cylindrical coordinates whose center is at the disk center, the z-axis being perpendicular to the disk plane. The symbol R represents the distance between the disk center and the position P(r, z).

geometrical thin and surrounds a central object of mass, M, with nearly Keplerian rotation (see Fig. 2.1). Then, the vertical component of gravitational force acting at $P(r, z)$ is $(GM/R^2)(z/R) \sim (GM/r^3)z$ towards the equator, in the case where $z \ll r$. On the other hand, the force acting towards the equator by the self-gravity of disks is $\sim 4\pi G\rho z$, where ρ is the gas density in the disk and G the gravitational constant.[2] The effects of self-gravity on disk structure are negligible if $4\pi G\rho < (GM/r^3)$. Because surface density, Σ, of disks is related to ρ through disk thickness, H, as $\Sigma \sim 2\rho H$, the above inequality leads to

$$\Sigma < \frac{M}{2\pi r^2}\frac{H}{r}. \tag{2.1}$$

Roughly speaking, this inequality implies that the disk mass ($\sim \pi r^2 \Sigma$) needs to be smaller than the mass of the central source by H/r. This condition is satisfied in almost all disks except in planetary disks and galactic disks.

Next, even if the above inequality is satisfied, the self-gravity can have non-negligible effects on perturbations. In wavy perturbations in the radial direction, the pressure restoring force makes the perturbations oscillatory, when their radial wavelength is sufficiently short (acoustic oscillations). As the wavelength increases, however, the effects of self-gravity increase and act in the direction to make the disks unstable (gravitational instability). If the wavelength increases further, the restoring force due to disk rotation acts in the direction to make the perturbation again an oscillatory motion (inertial oscillations). That is, for perturbations with intermediate wavelengths the disks have tendency to become unstable. The disks can

[2] If the gravitational potential is written as $\psi(r, z)$, the Poisson equation is $\nabla^2\psi = 4\pi G\rho$. This gives $\partial\psi/\partial z \sim 4\pi G\rho z$.

be regarded as stable if they are stable against the above-mentioned intermediate-size perturbations. To express this condition of stability, Toomre (1964) introduced so-called Q-parameter defined by

$$Q = \alpha \frac{\kappa c_s}{G\Sigma}, \tag{2.2}$$

where α is dimensionless quantity of the order of unity, κ is the radial epicyclic frequency which will be defined later in Sect. 2.2, and c_s is the acoustic speed in disks. The disks are gravitationally stable if

$$Q > 1. \tag{2.3}$$

The disks which we treat in this book are those where the above two conditions, equations (2.1) and (2.3), are satisfied. In disks surrounding normal stars, white dwarf, neutron stars, and stellar-mass black holes the above conditions are satisfied.

2.1.2 Geometrically Thin Disks

In advection-dominated accretion flows (ADAFs) and slim disks, disk thickness, H, is not sufficiently small compared with r. In standard disks, however, the disk can be taken to be thin in the sense of $H/r \ll 1$. In this book we are mainly interested in such geometrically thin disks.

2.1.3 Neglect of Accretion Flows on Wave Motions

In constructing disk models, accretion flows are essentially important in angular momentum balance in the radial direction, and important in some cases in thermal energy balance (ADAFs and slim disks). Even in such disks as the standard disks, accretion flows have important roles around the inner edge of disks, since the flow becomes transonic there. In studies of disk oscillations in this book, however, effects of accretion flows on oscillations are neglected. This is mainly because of simplicity. Careful considerations will be necessary in the future.

2.1.4 Effects of Global Magnetic Fields

Effects of global magnetic fields on disk oscillations are of importance. There are, however, many ambiguities in configuration of global magnetic fields in disks. In this book, we examine only supplementary the effects of magnetic fields in some special cases (Sects. 4.2.4 and 8.1).

2.1.5 General Relativity

Effects of general relativity on behaviors of disk oscillations are of very importance, when high-frequency QPOs in NS LMXBs and BH LMXBs are examined. Full examination of effects of general relativity on oscillations is, however, complicated with many subsidiary terms. In this book, to extract the essential parts of general relativity on oscillations, we adopt general relativistic expressions for epicyclic frequencies. Except this, we adopt the Newtonian formulations.

2.2 Epicyclic Frequencies

When we consider motions in differentially rotating disks, there are some important quantities characterizing the motions. One of them is *epicyclic frequency*. There are two kinds of epicyclic frequency, which are called *radial epicyclic frequency* and *vertical epicyclic frequency*. Let us consider first the radial epicyclic frequency.

2.2.1 Radial Epicyclic Frequency in Pressureless Disks

First, we consider an infinitesimally thin, axisymmetric disk (with no temperature), subject to an external force, $-F(r)$ (per unit mass), towards the disk center, where r is the distance from the disk center. Then, the disk is at a dynamical equilibrium when $r\Omega^2(r) = F(r)$, where $\Omega(r)$ is the angular velocity of disk rotation.

Let us now consider a fluid element on this disk and perturb it from the circular orbit, keeping its motion on the disk plane without changing angular momentum. The element rotates with an elliptical orbit around the disk center. The radial coordinate changes periodically around the radius of the original circular orbit, when the deviation from the original circular orbit is small. The frequency of this radial oscillation around the original radius is called (radial) epicyclic frequency.

This frequency can be calculated by the following consideration. If a particle at radius, r, is moved to a radius, $r + \xi$, holding its specific angular momentum, the angular velocity of the displaced particle, Ω^*, becomes to $\Omega^* = [r/(r+\xi)]^2\Omega(r)$. Then, the centrifugal force acting on the particle is $(r+\xi)(\Omega^*)^2$, which is $\Omega^2(r) - 3\xi\Omega^2(r)$ when the displacement ξ is small. On the other hand, the force toward the central source at the radius $r + \xi$ is $F(r + \xi)$, which is $(r + \xi)\Omega^2(r + \xi)$. In the case where the displacement is small, this force toward the disk center is given by $r\Omega^2(r) + (\Omega^2 + rd\Omega^2/dr)\xi$. Hence, the net restoring force to return the particle to the original radius is the difference between the above two forces, which is $(4\Omega^2 + rd\Omega^2/dr)\xi$ and proportional to ξ. This means that the particle makes a harmonic oscillation in the radial direction, if $[4\Omega^2 + rd\Omega^2/dr]$ is positive. That is, the radial coordinate of the fluid element oscillates around the original radius r with frequency, $[4\Omega^2 + rd\Omega^2/dr]^{1/2}$. This frequency is (radial) epicyclic frequency, $\kappa(r)$, and expressed as

$$\kappa^2 = 2\Omega\left(2\Omega + r\frac{d\Omega}{dr}\right) = \frac{2\Omega}{r}\frac{d}{dr}(r^2\Omega). \tag{2.4}$$

In the case where the force, $-F(r)$, is the gravitational one coming from a point source with mass M, $F(r)$ is GM/r^2 and we have $\Omega = (GM/r^3)^{1/2}$. Such disks are called *Keplerian disks* and $\Omega = (GM/r^3)^{1/2}$ is called *Keplerian angular velocity* and is written as $\Omega_{\rm K}$, i.e., $\Omega_{\rm K} = (GM/r^3)^{1/2}$. In Keplerian disks, we have

$$\kappa = \Omega = \Omega_{\rm K}. \tag{2.5}$$

The above argument can be made in a mathematical way. Let us consider, on the equator, a test particle rotating around a central object under axisymmetric potential force, $-d\psi/dr$. The radial component of equation of motion is given by

$$\frac{d^2r}{dt^2} = -\frac{d\psi}{dr} + \frac{\ell^2}{r^3}, \tag{2.6}$$

where $\ell(=\text{const.})$ is the specific angular momentum of the particle. By introducing an effective potential, $\psi_{\rm eff}$, equation (2.6) is written as

$$\frac{d^2r}{dt^2} = -\frac{d\psi_{\rm eff}}{dr}, \tag{2.7}$$

where $\psi_{\rm eff} = \psi + \ell^2/2r^2$. A circular orbit exist at the radius, say r_0, where $d\psi_{\rm eff}/dr = 0$. If this radius is a local minimum of $\psi_{\rm eff}$, i.e., $d^2\psi_{\rm eff}/dr^2 > 0$ there, this circular orbit is stable against a small perturbation of the orbit. This is because for $r = r_0 + \xi$, the radial displacement ξ is governed by

$$\frac{d^2\xi}{dt^2} = -\left(\frac{d^2\psi_{\rm eff}}{dr^2}\right)_0\xi, \tag{2.8}$$

where the subscript 0 denote the value at $r_{\rm c}$. Equation (2.8) shows that the particle makes a harmonic oscillation around r_0 with frequency, $(d^2\psi_{\rm eff}/dr^2)_0^{1/2}$. This frequency is epicyclic frequency and we have

$$\kappa^2 = \left(\frac{d^2\psi_{\rm eff}}{dr^2}\right)_0 = \left[\frac{d}{dr}\left(\frac{d\psi}{dr} - \frac{\ell^2}{r^3}\right)\right]_0 = 2\Omega\left(2\Omega + r\frac{d\Omega}{dr}\right). \tag{2.9}$$

It is noted that the disk is dynamically unstable, if κ^2 is negative. This is the well-known *Rayleigh instability*. The quantity $r^2\Omega$ is specific angular momentum, and thus the Rayleigh criterion implies that the disk is unstable if specific angular momentum decreases outwards.

In the followings, some cases where the force towards the disk center is not the gravitational force of a point mass, GM/r^2, are discussed.[3] In such cases we have $\kappa \neq \Omega_{\rm K}$.

2.2.1.1 Case of Rotating Central Star

One of examples of $\kappa \neq \Omega$ is the case where the central star has an oblate shape by rotation. In this case the gravitational potential, ψ, is deviated from the Keplerian one, $-GM/(r^2+z^2)^{1/2}$, with an additional quadrupole gravitational potential (Ogilvie 2008):

$$\psi(r,z) = -\left(\frac{GM}{R}\right)\left[1 + \frac{Q}{3}\frac{(r^2-2z^2)R_*^2}{R^4}\right], \tag{2.10}$$

where $R^2 = r^2+z^2$ and R_* is the radius of the central star and dimensionless quantity Q represents the strength of rotational deformation of the star.[4] In the equatorial plane $\psi = -(GM/r)[1+(Q/3)(R_*/r)^2]$, and the radial force balance between the gravitational force and the centrifugal force gives

$$\Omega = \Omega_{\rm K}\left[1 + Q\left(\frac{R_*}{r}\right)^2\right]^{1/2}. \tag{2.11}$$

The associated epicyclic frequency $\kappa(r)$ is then

$$\kappa(r) = \Omega_{\rm K}\left[1 - Q\left(\frac{R_*}{r}\right)^2\right]^{1/2}. \tag{2.12}$$

Oscillations in disks with the above deviation of epicyclic frequency from the Keplerian one have been examined by Papaloizou et al. (1992), Savonije and Heemskerk (1993), and Ogilvie (2008) in relation to one-armed oscillations in Be-star disks.

[3]The disk is still assumed to be pressureless. The effects of gas pressure on epicyclic frequency is considered in Sect. 2.2.2.

[4]For a star with uniform angular velocity Ω^*, we have $Q = k_2\Omega^{*2}R_*^3/GM$, k_2 being the apsidal motion constant (Ogilvie 2008).

2.2.1.2 Case of Binary System

Another important case of $\kappa \neq \Omega$ is tidally deformed disks. Well-known objects with tidally deformed disks are dwarf novae, where the disk of primary star (white dwarf) is deformed by the tidal force of secondary star. A general expression for tidal potential, $\psi_{\rm tidal}$, is given in appendix D (see equation (D.6)). In the case where the orbital plane of secondary star coincides with the disk plane (coplanar) and the orbit is circular, the time-averaged part of the tidal potential, $\overline{\psi}_{\rm tidal}$, is given by[5]

$$\overline{\psi}_{\rm tidal}(r,z) = -\frac{GM_{\rm s}}{a}\left[1+\frac{1}{4}\left(\frac{R}{a}\right)^2\left(1-3\frac{z^2}{R^2}\right)+\dots\right], \tag{2.13}$$

where $R=(r^2+z^2)^{1/2}$ and a is binary separation. The ratio R/a is assumed to be smaller than unity, and $\overline{\psi}_{\rm tidal}(r,z)$ has been expanded by a series of R/a. In equation (2.13) $M_{\rm s}$ is the mass of the secondary star.

If the pressure force is neglected, the angular velocity of disk rotation on the equator, $\Omega(r)$, is given by

$$\Omega = \Omega_{\rm K}\left[1-\frac{1}{2}q\left(\frac{r}{a}\right)^3\right]^{1/2}, \tag{2.14}$$

where $q=M_{\rm s}/M$. In the case where q is smaller than unity (this is really the case of dwarf novae) and we are interested in the radial region of $r/a<1$ (in the case of dwarf novae, the disk size, r_t, is really smaller than a by tidal truncation), we have approximately

$$\Omega = \Omega_{\rm K}\left[1-\frac{1}{4}q\left(\frac{r}{a}\right)^3\right]. \tag{2.15}$$

The epicyclic frequency defined by equation (2.4) is then given by

$$\kappa = \Omega_{\rm K}\left[1-2q\left(\frac{r}{a}\right)^3\right]^{1/2} \sim \Omega_{\rm K}\left[1-q\left(\frac{r}{a}\right)^3\right]. \tag{2.16}$$

[5]Equation (D.5) leads to

$$\overline{\psi}_{\rm tidal} = -\frac{GM_{\rm s}}{a}\left[1+\left(\frac{R}{a}\right)^2\overline{P_2(\cos\vartheta)}+\dots\right],$$

where the overline means time average. In the simplified case considered in the text, equation (D.8) shows

$$\overline{P_2(\cos\vartheta)} = -\frac{1}{4}\left(3\frac{z^2}{R^2}-1\right).$$

2.2.2 Radial Epicyclic Frequency in Fluid Disks

In hydrodynamical disks with temperature, pressure force contributes to the force balance in the radial direction. In the unperturbed equilibrium state the radial force balance on a fluid element of unit mass is

$$-\frac{1}{\rho_0}\frac{dp_0}{dr} - r\Omega_{\rm K}^2 + r\Omega^2 = 0, \tag{2.17}$$

where $\rho_0(r)$ and $p_0(r)$ are the density and pressure in the unperturbed disks.

When a fluid element is perturbed from the equilibrium position, pressure force acting on the fluid element varies. This makes gaseous motion in fluid complicated. If the variation of the pressure force on fluid motions is neglected, however, the fluid element makes epicyclic oscillations under $\Omega(r)$ given by equation (2.17). In this sense, it is helpful to define the epicyclic frequency by equation (2.4) even in the case of fluid disks.

For simplicity, we assume that the disk is isothermal with acoustic speed $c_{\rm s}$. Then, the above force balance gives

$$\Omega^2 = \Omega_{\rm K}^2 + c_{\rm s}^2\frac{d\ln\rho_0}{rdr}. \tag{2.18}$$

If the density decreases outwards as $\rho_0 \propto r^{-\alpha}$, α being taken to be constant, and $c_{\rm s}$ is expressed as $c_{\rm s} = H\Omega_{\rm K}$ (see equation (4.6)), where H is the half-thickness of disks, equation (2.18) can be written as[6]

$$\Omega = \Omega_{\rm K}\left[1-\alpha\left(\frac{H}{r}\right)^2\right]^{1/2} \sim \Omega_{\rm K}\left[1-\frac{\alpha}{2}\left(\frac{H}{r}\right)^2\right]. \tag{2.19}$$

In deriving the final relation in equation (2.19), $H/r \ll 1$ has been used. By using equation (2.19), we can derive from equation (2.4) the epicyclic frequency in the form:

$$\kappa = \Omega_{\rm K}\left[1-2\alpha\left(\frac{H}{r}\right)^2\right]^{1/2} \sim \Omega_{\rm K}\left[1-\alpha\left(\frac{H}{r}\right)^2\right], \tag{2.20}$$

where the disk has been assumed to be isothermal in the radial direction, and thus $H = c_{\rm s}/\Omega \propto r^{3/2}$. Since the pressure force is usually directed outwards, the angular velocity of disk rotation, $\Omega(r)$, is smaller than $\Omega_{\rm K}(r)$. The epicyclic frequency is also smaller than $\Omega_{\rm K}(r)$, as shown in equation (2.20).

---

[6]In disks with non-zero temperature, the disk has vertical thickness due to the presence of the pressure force in the vertical direction. The disk thickness, H, is related to $c_{\rm s}$ by $c_{\rm s} = H\Omega_\perp$ (see Chap. 4).

2.2.3 *Vertical Epicyclic Frequency*

Next, we consider an another kind of displacement of a disk particle from equilibrium position. If a particle is displaced from the disk plane in the vertical direction, a restoring force acts toward the original disk plane. As shown below, this restoring force is proportional to displacement, ξ_z, in the vertical direction. Hence, the displaced particle makes a harmonic oscillation around the equatorial plane. This frequency is called *vertical epicyclic frequency.*

If the gravitational potential resulting from central source is written as ψ, the gravitational force acting on a fluid element in the vertical direction is $-\partial\psi/\partial z$, which is written as f_z, for simplicity. We are particularly interested in a fluid element on the equator ($z = 0$) in the unperturbed state. Then, $(f_z)_0 = 0$, where the subscript 0 shows the value at $z = 0$. The vertical force acting on a displaced element, say δf_z, is $(f_z)_1 + \xi_z(\partial f/\partial z)_0$, where ξ_z is the vertical displacement of the fluid element. Here, the subscript 1 denotes the Eulerian variation. In the present problem, the potential is externally given, and thus $(f_z)_1 = 0$. This means that a perturbed element displaced in the vertical direction makes a harmonic oscillation as

$$\frac{\partial^2 \xi_z}{\partial t^2} = \left(\frac{\partial f}{\partial z}\right)_0 \xi_z. \tag{2.21}$$

The frequency of the vertical oscillation, $\Omega_\perp$, is given by

$$\Omega_\perp^2 = -\left(\frac{\partial f}{\partial z}\right)_0 = \left(\frac{\partial^2 \psi}{\partial z^2}\right)_0. \tag{2.22}$$

This frequency, $\Omega_\perp$, is called vertical epicyclic frequency.

In the case of the Keplerain disks with sherically symmetric central star, we have $\psi = -GM/R$, where $R^2 = r^2 + z^2$. Hence, we have

$$\Omega_\perp = \Omega_{\rm K}. \tag{2.23}$$

The difference between $\Omega_\perp$ and $\Omega_{\rm K}$ occurs in the cases where (i) the central star is not spherically symmetric due to rotation, and (ii) the disk is subject to tidal force of secondary star.

2.2.3.1 Case of Rotating Central Star

If the central star is oblate by rotation, the gravitational potential of the central star is given by equation (2.10) as mentioned before. In this case, equation (2.22) gives

$$\Omega_\perp = \Omega_{\rm K}\left[1 + 3Q\left(\frac{R_*}{r}\right)^2\right]^{1/2}. \tag{2.24}$$

2.2.3.2 Case of Binary System

Since vertical epicyclic frequency is defined by equation (2.22), from equation (2.13) we have

$$\Omega_\perp = \Omega_{\rm K}\left[1+q\left(\frac{r}{a}\right)^3\right]^{1/2} \simeq \Omega_{\rm K}\left[1+\frac{1}{2}q\left(\frac{r}{a}\right)^3\right]. \tag{2.25}$$

It is noted that in the present approximations of $r/D \ll 1$ and of no eccentricity of the orbit, i.e., $e = 0$, we see that $\kappa - \Omega$ and $\Omega_\perp - \Omega$ are the same with the opposite signs. This character is preserved even if we proceed to the next order of approximations where the effects of $e \neq 0$ are taken into account, although it is not presented here.

2.2.4 *General Relativistic Versions of Epicyclic Frequencies*

The general relativistic versions of κ and $\Omega_\perp$ are very important, since in the innermost region of relativistic disks, trapping of various modes of oscillations is expected by the specific nature of radial distributions of epicyclic frequencies. These trapped oscillations may be one of possible causes of QPOs observed in neutron-star and black-hole X-ray binaries.

2.2.4.1 Radial Epicyclic Frequency

Here, we present a simple derivation of κ in the case where the central object has no rotation (the Schwarzschild metric), based on a Newtonian analogue (see Kato et al. 2008). A mathematically rigorous derivation of full relativistic versions of κ and $\Omega_\perp$ is given in appendix B.

Let us write the equation describing the radial oscillation of a test particle in the form:

$$\frac{d^2r}{dt^2} = -\frac{d\psi_{\rm eff}}{dr}, \tag{2.26}$$

where the effective potential, $\psi_{\rm eft}(r)$, is expressed as

$$\psi_{\rm eff}(r) = -\frac{GM}{r}\left(1+\frac{\ell^2}{c^2r^2}\right)+\frac{\ell^2}{2r^2}. \tag{2.27}$$

The difference of equation (2.27) from that in the Newtonian case is the presence of the term of $-(GM/r)\ell^2/c^2r^2$. This quantity $\ell^2/c^2r^2(= r^2\Omega^2/c^2)$ can be interpreted as showing the effect of a mass increase of a test particle by rotational motion. The

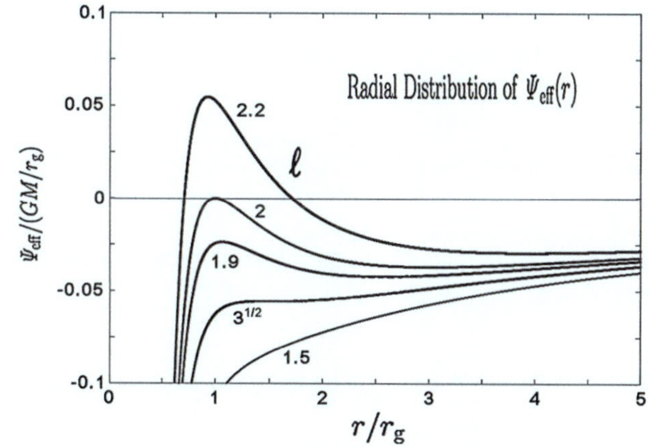

Fig. 2.2 Radial distribution of $\psi_{\rm eff}$ as functions of ℓ. $r_{\rm g}$ represents the Schwarzschild radius defined by $2GM/c^2$.

radial dependence of $\psi_{\rm eff}$ is shown in Fig. 2.2. Circular orbits of a test particle exist at the radii where $d\psi_{\rm eft}/dr = 0$.

It is noted that in the case of Newtonian dynamics, the centrifugal force increases inward with $1/r^2$ and this always dominates over the gravitational force with $1/r$ when r is small. Due to this, there is always a potential minimum at a certain radius, implying the present of stable circular orbit. In the present case of equation (2.27), however, the effective potential does not always have a potential minimum because of the presence of the term with $1/r^3$. This term predominates when specific angular momentum ℓ is small beyond a certain limit.

As is well known in stability theory, if the radius where $d\psi_{\rm eff}/dr = 0$ is a point of potential minimum, the circular orbit at the radius is stable, while if it is not so, this circular orbit is unstable. This consideration suggests that we can derive an expression for κ^2 by examining the expression for $d^2\psi_{\rm eff}/dr^2$ (see Sect. 2.2.1).

Equation (2.26) shows that if a particle is perturbed from the equilibrium radius, the deviation, say ξ, is governed by (see equation (2.8))

$$\frac{d^2\xi}{dt^2} = -\left(\frac{d^2\psi_{\rm eff}}{dr^2}\right)\xi \tag{2.28}$$

With the help of equation (2.27), we have

$$\frac{d^2\psi_{\rm eff}}{dr^2} = \left(1 - \frac{r_{\rm g}}{r}\right)\frac{GM/r^2}{r - 3r_{\rm g}/2}, \tag{2.29}$$

where $r_{\rm g}$ is the Schwarzschild radius defined by

$$r_{\rm g} = \frac{2GM}{c^2}. \tag{2.30}$$

Equation (2.28) shows that the quantity, $d^2\psi_{\rm eff}/dr^2$, given by equation (2.29) is the square of the epicyclic frequency observed in proper time. Since the redshift factor for the circular motion is

$$\left(1-\frac{3r_{\rm g}}{2r}\right)^{-1/2}, \tag{2.31}$$

we divide $d^2\psi_{\rm eff}/dr^2$ by the square of the redshift factor to obtain the epicyclic frequency observed at infinity (Kato and Fukue 1980):

$$\kappa^2=\frac{GM}{r^2}\left(1-\frac{3r_{\rm g}}{r}\right). \tag{2.32}$$

Of course, in the limit of $r_{\rm g}=0$, κ^2 given above tends to the Newtonian expression of $\kappa^2=\Omega_{\rm K}^2$.

As mentioned before, a relativistic derivation of κ is given in appendix B.2. Here, the results are presented. We consider the Kerr metric, which is specified by mass, M, and spin of the central black hole. In order to express the magnitude of the spin by dimensionless parameter, the spin parameter, a_*, is introduced. The case of $a_*=0$ corresponds to no spin, and $a_*=1$ is the case of the maximum spin. Then, the epicyclic frequency in disks in the Kerr metric, κ, is given by

$$\kappa^2=\frac{GM}{r^2}\frac{1-3r_{\rm g}/r\pm 8a_*(r_{\rm g}/2r)^{3/2}-3a_*^2(r_{\rm g}/2r)^2}{[1\pm a_*(r_{\rm g}/2r)^{3/2}]^2}, \tag{2.33}$$

which was derived in necessity of studying disk oscillations by Okazaki et al. (1987), but had been derived more generally by Aliev and Galtsov (1981) (see also Fukue 1980). In equation (2.33), $\pm$ sign corrspends to a direct/retrograde orbit.

If $a_*=0$, equation (2.33) tends to equation (2.32), and it tends further to $\kappa^2=GM/r^3$ in the Newtonian limit ($r_{\rm g}=0$). The radial dependence of epicyclic frequency, $\kappa(r)$, is shown in Fig. 2.3, as a function of $r/r_{\rm g}$ for some values of a_*.

2.2.4.2 Vertical Epicyclic Frequency

In the case of the Kerr metric, $\Omega_\perp$ differs from the angular velocity of the relativistic Kepler orbits, $\Omega_{\rm K}$, and also from the relativistic epicyclic frequency, κ. Kato (1990) derived an expression for $\Omega_\perp$ for studying disk oscillations, but its general expression had been more rigorously derived by Aliev and Galtsov (1981):

$$\Omega_\perp^2(r)=\Omega_{\rm K}^2\left[1\mp 4\left(\frac{r_{\rm g}}{2r}\right)^{2/3}a_*+3\left(\frac{r_{\rm g}}{2r}\right)^2a_*^2\right], \tag{2.34}$$

where a minus/plus sign corresponds to a direct/retrograde orbit.

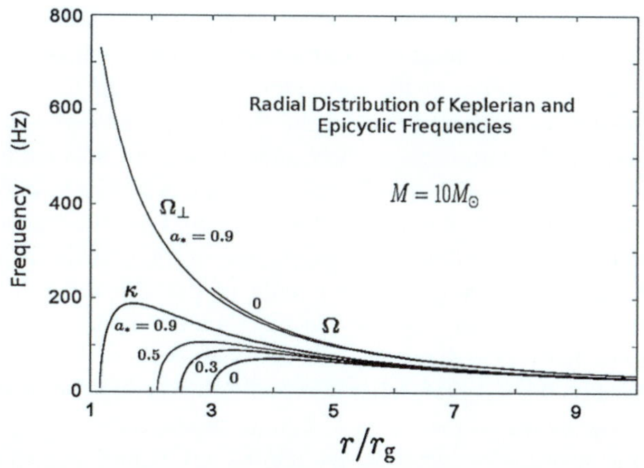

Fig. 2.3 Radial distribution of κ as a function of r/r_g for some a_*. For comparison, Ω_K and $\Omega_\perp$ for $a_* = 0.9$ are shown.

It is noted that there is a relation among Ω_K, $\Omega_\perp$, and κ, which is $\Omega_K > \Omega_\perp > \kappa$.

2.3 Corotation and Lindblad Resonances

There are another important quantities in the cases where disk oscillations are considered. They are corotation and Lindblad resonances.

2.3.1 Corotation Resonance

The rotation of geometrically thin disks is differential. On such differentially rotating disks we superpose normal mode oscillations (not shearing modes), i.e., waves which rotate uniformly in the azimuthal direction with a constant angular velocity. That is, we consider a wave whose frequency is ω and the azimuthal wavenumber m. The wave is proportional to $\exp[i((\omega t - m\varphi)]$. The phase velocity of the wave in the azimuthal direction is ω/m. Since the disk is differentially rotating, the phase velocity, ω/m, of the wave in the azimuthal direction becomes equal to the angular velocity of disk at a certain radius. Then, the disk rotation and the wave resonantly interact. This is called *corotation resonance*. The corotation radius is defined as the radius where

$$\frac{\omega}{m} = \Omega, \quad \text{or} \quad \omega = m\Omega. \tag{2.35}$$

As will be shown later, if we derive a dispersion relation describing wave motions (oscillations), the corotation point (radius) appears as a singular point.

Perhaps, systematic studies on the wave motions with $\exp[i((\omega t - m\varphi)]$, m being a constant in disks, were started first in the field of galactic dynamics (Lin and Shu 1964). One of the purposes of these studies was to solve the dilemma of winding problem of spiral pattern of spiral galaxies. Statistical studies show that geometrical differences of spiral pattern of galaxies and physical nature of galaxies are correlated. This suggests that the spiral pattern of the galaxies is not a material pattern which changes in the short time scale of galactic rotation (a few times 10^8 years) but a wave pattern (*density wave*) maintained for a longer time. This was the start of density wave theory in galactic dynamics.

Density waves on the galactic stellar disks are propagating waves and eventually disappear by propagating in the radial direction, unless there are some sustaining processes of the waves (for example, excitation, reflection, and so on) (Toomre 1969). Mark (1976 and a series of his papers) showed that the corotation resonance between density waves and disk rotation amplifies the density waves, which is called *corotation amplification* or *corotation instability*. His analyses are very complicated, because the analyses are made for collisionless systems.[7] Later, non-axisymmetric instability of thick gaseous tori was found by Papaloizou and Pringle (1984; 1985; 1987) (see also Drury 1985), which is called *Papaloizou-Pringle instability*. Subsequently, many researchers studied this instability not only for tori but also for disks. The results show that the instability is the corotation instability. The reason why the instability was found first for tori and not for disks is that in the case of disks the growth rate is much smaller than that in tori because of small penetration factor of waves to the point of corotation radius. Recently, the corotation instability again has been taken much attention in relation to one of possible excitation processes of quasi-periodic oscillations observed in X-ray binaries (Lai and Tsang 2009; Tsang and Lai 2009). Corotation instability of gaseous disks will be presented in detail in Chap. 10.

It is noticed that the corotation resonance had been known in geophysics (meteorology and oceanography) and fluid dynamics, before in astrophysics. In meteorology, for example, the corotation radius is known as critical layer, where the phase velocity of a wave becomes equal to flow velocity. The phenomena of "sudden stratospheric warming" are related to this resonance.

2.3.2 Lindblad Resonances

Let us consider again wave motions whose time and azimuthal dependences are $\exp[i(\omega t - m\varphi)]$. In addition to the radius where $\omega/m = \Omega$, the wave equation describing oscillations have apparently singularities at the radii of $\omega/m = \Omega \pm \kappa/m$

[7] Galactic disks consisting of stars are collisionless systems.

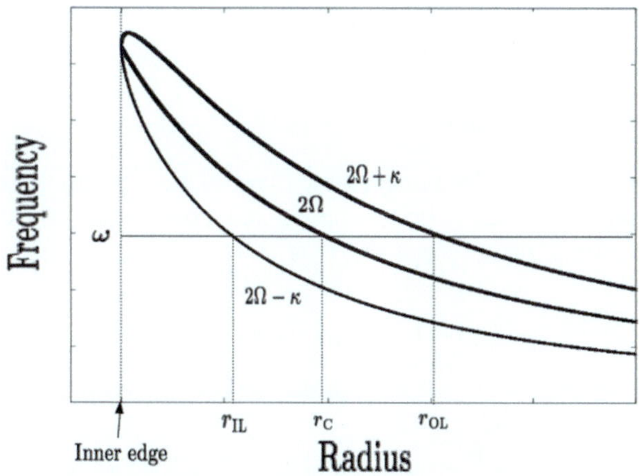

Fig. 2.4 Schematic picture showing $r_{IL} < r_c < r_{OL}$ in the case of a two-armed oscillation ($m = 2$) with frequency ω in relativistic disks.

(i.e., $\omega = m\Omega \pm \kappa$), (e.g., see equation (3.50)). These resonances are called *Lindblad resonances*, and the radii where $\omega/m = \Omega \pm \kappa/m$ are called the radii of Lindblad resonances. In disks where Ω decreases outwards, $\omega/m = \Omega - \kappa/m$ occurs at a radius smaller than the corotation radius, while $\omega/m = \Omega + \kappa/m$ does at a radius larger than the corotation radius. The former is called *inner Lindblad resonance* and the radius is expressed as r_{IL}, and the latter is called *outer Lindblad resonance* and the radius is represented by r_{OL}. The three radii, r_{IL}, r_c, and r_{OL} are in the order of $r_{IL} < r_c < r_{OL}$, and their positions on frequency – radius relation are schematically shown in Fig. 2.4.

Concerning the Lindblad resonances we should notice a difference between particle disks (e.g., stellar disks) and gaseous disks. In particle disks, a resonance between a wave and particles really occurs at the Lindblad resonances, and the resonances are important on disk dynamics. For example, the resonance leads to damping of the density wave (Lynden-Bell and Kalnajs 1972). The density waves in stellar disks are excited at the corotation resonance and absorbed at a Lindblad resonance. Even in gaseous disks, external forces can have resonant effects on disks at Lindblad resonances (e.g., Goldreich and Tremaine 1979; Lubow and Ogilvie 1998). In gaseous disks with no external forces, however, the Lindblad resonances are not real resonances. That is, the radii of Lindblad resonances, i.e., r_{IL} and r_{OL}, are points of apparent singularities, and not points of real singularities.

2.3.2.1 Comments on Singular Points

Here, we mention briefly a difference between wave motions near Lindblad resonances and those near corotation resonance, following the standard theory of differential equations.

We start from a differential equation of the form:

$$\frac{d^2f}{dx^2}+p(x)\frac{df}{dx}+q(x)f=0, \tag{2.36}$$

where f is a perturbed quantity associated with oscillations, and x is a dimensionless radial coordinate and $x=0$ is a point where p and q have poles. The coefficient p is taken to have a pole of first order at $x=0$ as

$$p(x)=\frac{1}{x}\biggl(\alpha_0+\alpha_1x+\alpha_2x^2+\ldots\biggr), \tag{2.37}$$

and $q(x)$ has a pole of second order as

$$q(x)=\frac{1}{x^2}\biggl(\beta_0+\beta_1x+\beta_2x^2+\ldots\biggr). \tag{2.38}$$

Our purpose is to obtain solutions of equation (2.36) in the region near $x=0$. To do so, we write solutions in vicinity of $x=0$ as

$$f=x^{\rho}\sum_{\nu=0}^{\infty}a_{\nu}x^{\nu}\quad (a_0\neq 0) \tag{2.39}$$

and determine ρ and a_ν's by equating the terms of the same power of x in equation (2.36), after equation (2.39) being substituted into equation (2.36). From the lowest order terms with respect to x, i.e., the terms of $x^{\rho-2}$ ($\nu=0$), we have

$$\rho(\rho-1)+\rho\alpha_0+\beta_0=0. \tag{2.40}$$

This equation has two solutions of ρ, say ρ_1 and ρ_2, and we have formally two set of solutions of equation (2.36):

$$f_1=x^{\rho_1}\sum_{\nu=0}^{\infty}a_{\nu}(\rho_1)x^{\nu}, \tag{2.41}$$

$$f_2=x^{\rho_2}\sum_{\nu=0}^{\infty}a_{\nu}(\rho_2)x^{\nu}. \tag{2.42}$$

In the cases of $\rho_1=\rho_2$ and $\rho_1=\rho_2+n$ (n is a positive integer), however, the above two solutions are not independent (see text books of differential equations). In these cases, we must derive another independent solution. Derivation of the solution is complicated, but is presented in many text books of differential equations.

In the case of $\rho_1 = \rho_2$, one of the solution, say f_1, is equation (2.41). The other solution is given by

$$f_2 = f_1 \ln x + x^{\rho_1} \sum_{\nu=0}^{\infty} \left[\frac{d}{d\rho} a_\nu(\rho) \right]_{\rho=\rho_1} x^\nu. \tag{2.43}$$

It is important to note that a logarithmic singularity appears in f_2.

In the case where $\rho_1 = \rho_2 + n$ ($n = 1, 2, \ldots$), the second solution, f_2, is given by

$$f_2 = A f_1 \ln x + x^{\rho_2} \sum_{\nu=0}^{\infty} \left[\frac{d}{d\rho} \Big((\rho - \rho_2) a_\nu(\rho) \Big) \right]_{\rho=\rho_2} x^\nu, \tag{2.44}$$

where

$$A = \frac{1}{a_0} \Big[(\rho - \rho_2) a_n(\rho) \Big]_{\rho=\rho_2}. \tag{2.45}$$

Different from the case of $\rho_1 = \rho_2$, the term of logarithmic singularity can vanish, because $A = 0$ in some cases.

Now, we apply the above results to Lindblad resonances and corotation resonance. In the case of Lindblad resonances, we have (see, for example, equation (5.15))

$$\alpha_0 = -1, \quad \text{and} \quad \beta_0 = 0, \tag{2.46}$$

and equation (2.40) gives

$$\rho_1 = 2, \quad \text{and} \quad \rho_2 = 0. \tag{2.47}$$

The difference of ρ_1 and ρ_2 is an integer, i.e., $n = 2$. Hence, a logarithmic singularity appears in one of the solutions, say f_2, unless $A = 0$. A detailed calculations, however, show that in the present case we have $A = 0$. Hence, no singularity appears at the radii of Lindblad resonances.

Distinct from the case of Lindblad resonances, in the case of corotation resonance, we have (see again, e.g., equation (5.15))

$$\alpha_0 = 1, \quad \text{and} \quad \beta_0 = 0, \tag{2.48}$$

for p-mode oscillations.[8] Equation (2.40) then shows that in this case we have

$$\rho_1 = \rho_2 = 0. \tag{2.49}$$

[8]It is noted that n in equation (5.15) denotes the number of node(s) of oscillations in the vertical direction, and is different from n in this subsection. If n in equation (5.15) is taken to be zero, the equation represents p-mode oscillations.

Because ρ_1 and ρ_2 are equal, the logarithmic singularity does not disappear (see equation (2.43)). This logarithmic singularity at the coronation resonance is the cause of wave amplification or damping at the corotation resonance (see also Chap. 10).

References

Aliev, A. N., & Galtsov, D. V. 1981, Gen. Relativ. Gravit., 13, 899
Drury, L. O'C. 1985, Mon. Not. R. Astron. Soc., 217, 821
Fukue, J. 1980, Master Theses of Kyoto University (in Japanese)
Goldreich, P., & Tremaine, S. 1979, Astrophys. J., 233, 857
Kato, S., & Fukue, J. 1980, Publ. Astron. Soc. Jpn., 32, 377
Kato, S. 1990, Publ. Astron. Soc. Jpn., 42, 99
Kato, S., Fukue, J., & Mineshige, S. 2008, *Black-Hole Accretion Disks — Towards a New Paradigm* (Kyoto University Press, Kyoto)
Lai, D., & Tsang, D. 2009, Mon. Not. R. Astron. Soc., 393, 979
Lin, C. C., & Shu, F. H. 1964, Astrophys. J., 140, 646
Lubow, S. H., & Ogilvie, G. 1998, Astrophys. J., 504, 983
Lynden-Bell, D., & Kalnajs, A. J. 1972, Mon. Not. R. Astron. Soc., 157, 1
Mark, J.W-K., 1976, Astrophys. J., 205, 363
Papaloizou, J. C. B., & Pringle, J. E. 1984, Mon. Not. R. Astron. Soc., 208, 721
Papaloizou, J. C. B., & Pringle, J. E. 1985, Mon. Not. R. Astron. Soc., 213, 799
Papaloizou, J. C. B. & Pringle, J. E., 1987, Mon. Not. R. Astron. Soc., 225, 267
Papaloizou, J. C. B., Savonije, G. J., & Henrichs, H. F. 1992, Astron. Astrophys., 265, L45
Ogilvie, G. I. 2008, Mon. Not. R. Astron. Soc., 388, 1372
Okazaki, A. T., Kato, S., & Fukue, J. 1987, Publ. Astron. Soc. Jpn., 39, 457
Savonije, G. J. & Heemskerk, M. H. M. 1993, Astron. Astrophys. 276, 409
Toomre, A. 1964, Astrophys. J., 139, 1217
Toomre, A. 1969, Astrophys. J., 158, 899
Tsang, D, & Lai, D. 2009, Mon. Not. R. Astron. Soc., 400, 470

Chapter 3
Derivation of Linear Wave Equations and Wave Energy

Abstract We derive wave equations describing periodic oscillations in geometrically thin disks. There are two ways describing wave equations: Lagrangian and Eulerian descriptions. The former is better to examine some basic properties of oscillations, but the latter is convenient to study practical problems. In addition to wave equation, in this chapter, we derive expressions for wave energy, since the wave energy is one of important quantities which are conserved in periodic oscillations.

We treat here rather simplified disks. That is, they are geometrically thin nonself-gravitating, and have no radial accretion flows. We are interested in disks in strong gravitational fields, since high-frequency quasi-periodic oscillations observed in neutron stars and black-hole sources may come from the innermost part of relativistic disks. In this book, however, we use the conventional way that the effects of the general relativity are taken into account within the framework of Newtonian formulation by adopting general relativistic expressions for epicyclic frequencies. Oscillations are taken to be adiabatic and inviscid. The effects of dissipative processes on oscillations are considered in part II (Chaps. 9 and 13).

Keywords Energy density • Energy flux • Orthogonality • Wave energy • Wave equations

3.1 Lagrangian Description of Oscillations and Wave Energy

In Lagrangian formulation in the Newtonian frame, the hydrodynamical equation describing perturbations over equilibrium states can be generally written as (Lynden-Bell and Ostriker 1967)

$$\frac{D_0^2\boldsymbol{\xi}}{Dt^2} = \delta\left(-\nabla\psi - \frac{1}{\rho}\nabla p\right), \tag{3.1}$$

where $\boldsymbol{\xi}$ is a displacement vector associated with perturbations, and D_0/Dt is the time derivative along an unperturbed flow, $\boldsymbol{u}_0$, and is related to the Eulerian time

S. Kato, *Oscillations of Disks*, Astrophysics and Space Science Library 437,
DOI 10.1007/978-4-431-56208-5_3

derivative, $\partial/\partial t$, by

$$\frac{D_0}{Dt} = \frac{\partial}{\partial t} + \boldsymbol{u}_0 \cdot \nabla. \tag{3.2}$$

In equation (3.1), δ represents the Lagrangian variation of the quantities in the subsequent parentheses, and ψ is the gravitational potential. Other notations in equation (3.1) have their usual meanings. The term $D_0^2\boldsymbol{\xi}/Dt^2$ in equation (3.1) comes from the fact that the operator δ and D/Dt (the time derivative along the perturbed flow $\boldsymbol{u}$) can be commuted with each other (Lynden-Bell and Ostriker 1967) as

$$\delta\left(\frac{D\boldsymbol{u}}{Dt}\right) = \frac{D_0}{Dt}\delta\boldsymbol{u} = \frac{D_0^2}{Dt^2}\boldsymbol{\xi}. \tag{3.3}$$

Equation (3.1) is valid for nonlinear self-gravitating perturbations, but hereafter we restrict our attention only on linear nonself-gravitating perturbations. For quasi-nonlinear description of equation (3.1), see Chap. 11.

We assume here that the perturbations are adiabatic. Then, the Lagrangian variation of pressure, δp, can be expressed in terms of $\delta\rho$ by use of adiabatic relation:

$$\delta p = c_s^2\delta\rho = \Gamma_1\frac{p_0}{\rho_0}\delta\rho, \tag{3.4}$$

where $\rho_0(\boldsymbol{r})$ and $p_0(\boldsymbol{r})$ are the density and pressure in the unperturbed state, and Γ_1 is the barotropic index specifying the linear part of the relation between Lagrangian variations δp and $\delta\rho$.

Furthermore, $\delta\rho$ can be expressed in terms of $\boldsymbol{\xi}$, which is the equation of continuity:

$$\delta\rho + \rho_0\mathrm{div}\boldsymbol{\xi} = 0. \tag{3.5}$$

As mentioned above, self-gravity of the disk gases is neglected. This means that the gravitational potential, $\psi(\boldsymbol{r})$, comes from external sources, and there is no Eulerian perturbation of $\psi(\boldsymbol{r})$, i.e., $\psi_1(\boldsymbol{r}) = 0$. Then, after lengthly manipulation, we can reduce equation (3.1) to (Lynden-Bell and Ostriker 1967)

$$\rho_0\frac{\partial^2\boldsymbol{\xi}}{\partial t^2} + 2\rho_0(\boldsymbol{u}_0\cdot\nabla)\frac{\partial\boldsymbol{\xi}}{\partial t} + \mathscr{L}(\boldsymbol{\xi}) = 0, \tag{3.6}$$

where $\mathscr{L}(\boldsymbol{\xi})$ is a linear operator acting on $\boldsymbol{\xi}$ and is (Lynden-Bell and Ostriker 1967, see also chapter 11)

$$\mathscr{L}(\boldsymbol{\xi}) = \rho_0(\boldsymbol{u}_0\cdot\nabla)(\boldsymbol{u}_0\cdot\nabla)\boldsymbol{\xi} + \rho_0(\boldsymbol{\xi}_0\cdot\nabla)\nabla\psi_0 + \nabla\Big[(1-\Gamma_1)p_0\,\mathrm{div}\boldsymbol{\xi}\Big]$$
$$-p_0\nabla(\mathrm{div}\boldsymbol{\xi}) - \nabla[(\boldsymbol{\xi}\cdot\nabla)p_0] + (\boldsymbol{\xi}\cdot\nabla)(\nabla p_0). \tag{3.7}$$

The first term on the right-hand side of equation (3.7) is a part of $\rho_0 D_0^2\boldsymbol{\xi}/Dt^2$, the second term is $\rho_0\delta(\nabla\psi)$, and the remaining terms come from $\rho_0\delta[(1/\rho)\nabla p]$. Equation (3.6) is the basic equation describing perturbations in terms of displacement vector, $\boldsymbol{\xi}$.

An important characteristic of the operator, $\mathscr{L}(\boldsymbol{\xi})$, is that it is Hermitian in the sense that for arbitrary $\boldsymbol{\xi}$ and $\boldsymbol{\eta}$ we have (Lynden-Bell and Ostriker 1967)

$$\int \boldsymbol{\eta}\cdot\mathscr{L}(\boldsymbol{\xi})dV = \int \boldsymbol{\xi}\cdot\mathscr{L}(\boldsymbol{\eta})dV, \tag{3.8}$$

where the integration is performed over the whole volume where gases exist, and the surface integration has been assumed to vanish.

If we consider perturbations with frequency ω (taken to be real) and azimuthal wavenumber m, the displacement vector, $\boldsymbol{\xi}$, is written as

$$\boldsymbol{\xi} = \Re[\hat{\boldsymbol{\xi}}\exp(i\omega t)] = \Re\left[\breve{\boldsymbol{\xi}}\exp[(i\omega t - m\varphi)]\right], \tag{3.9}$$

where $\Re$ represents the real part. Then, the wave equation can be expressed as

$$-\omega^2\rho_0\hat{\boldsymbol{\xi}} + 2i\omega\rho_0(\boldsymbol{u}_0\cdot\nabla)\hat{\boldsymbol{\xi}} + \mathscr{L}(\hat{\boldsymbol{\xi}}) = 0 \tag{3.10}$$

and the equation corresponding to equation (3.8) is

$$\int \hat{\boldsymbol{\eta}}^*\cdot\mathscr{L}(\hat{\boldsymbol{\xi}})dV = \int \hat{\boldsymbol{\xi}}^*\cdot\mathscr{L}(\hat{\boldsymbol{\eta}})dV, \tag{3.11}$$

where the asterisk denotes complex conjugate.

Explicitly writing down wave equation (3.10) is not always useful, because wave motions expressed in terms of Eulerian variables are more useful in practical studies than those in terms of Lagrangian variables. For wave motions expressed in terms of Lagrangian variables, see, for example, Kato (2004, 2008).

It is of importance, however, to mention here the relation between $\hat{\xi}_r$ and $\hat{\xi}_\varphi$ in wave motions (speaking more rigorously, the relation between $\breve{\xi}_r$ and $\breve{\xi}_\varphi$). Let us introduce the cylindrical coordinated (r, φ, z) whose center is at the disk center and the z-axis is perpendicular to the disk plane. Then, when we consider wave motions whose azimuthal wavenumber is small, the terms resulting from the pressure force can be neglected in the φ-component of wave equation, and the φ-component of equation of motion (3.10) is

$$-\omega^2\hat{\xi}_\varphi + 2i\omega[(\boldsymbol{u}_0\cdot\nabla)\hat{\boldsymbol{\xi}}]_\varphi + [(\boldsymbol{u}_0\cdot\nabla)(\boldsymbol{u}_0\cdot\nabla)\hat{\boldsymbol{\xi}} + (\hat{\boldsymbol{\xi}}\cdot\nabla)(\nabla\psi_0)]_\varphi = 0. \tag{3.12}$$

Here, we consider oscillations with azimuthal wavenumber m [see equation (3.9)]. Neglecting the pressure force in the radial force balance in unperturbed disks, we can reduce the above equation to[1]

$$i(\omega - m\Omega)\breve{\xi}_\varphi + 2\Omega\breve{\xi}_r = 0, \tag{3.13}$$

where the unperturbed flow is taken to be rotation alone, i.e., $\boldsymbol{u}_0 = (0, r\Omega, 0)$. This relation between $\breve{\xi}_r$ and $\breve{\xi}_\varphi$ is often used hereafter.

3.1.1 Orthogonality of Normal Modes

In relation to latter studies (see Chap. 11), we present here an orthogonal relation of normal mode oscillations. Let us write the set of normal mode oscillations satisfying wave equation (3.6) with some relevant boundary conditions as $\boldsymbol{\xi}_\alpha(\boldsymbol{r}, t) = \Re[\hat{\boldsymbol{\xi}}_\alpha \exp(i\omega_\alpha t)]$ with $\alpha = 1, 2, 3, \ldots$, where ω_α is real. The subscript α attached to $\hat{\boldsymbol{\xi}}$ is to distinguish oscillation modes (not to represent a component of vector). Then, $\hat{\boldsymbol{\xi}}_\alpha(\boldsymbol{r}, t)$ satisfies

$$-\omega_\alpha^2 \rho_0 \hat{\boldsymbol{\xi}}_\alpha + 2i\omega_\alpha \rho_0 (\boldsymbol{u}_0 \cdot \nabla)\hat{\boldsymbol{\xi}}_\alpha + \mathscr{L}(\hat{\boldsymbol{\xi}}_\alpha) = 0. \tag{3.14}$$

Now, this equation is multiplied by $\hat{\boldsymbol{\xi}}_\beta^*(\boldsymbol{r}, t)$ and integrated over the whole volume, where the superscript $*$ denotes the complex conjugate and $\beta \neq \alpha$. The volume integration of $\rho_0 \hat{\boldsymbol{\xi}}_\beta^*(\boldsymbol{r}, t) \cdot \hat{\boldsymbol{\xi}}_\alpha(\boldsymbol{r}, t)$ over the whole volume is written hereafter as $\langle \rho_0 \hat{\boldsymbol{\xi}}_\beta^* \cdot \hat{\boldsymbol{\xi}}_\alpha \rangle$. Then, we have

$$-\omega_\alpha^2 \langle \rho_0 \hat{\boldsymbol{\xi}}_\beta^* \hat{\boldsymbol{\xi}}_\alpha \rangle + 2i\omega_\alpha \langle \rho_0 \hat{\boldsymbol{\xi}}_\beta^* (\boldsymbol{u}_0 \cdot \nabla)\hat{\boldsymbol{\xi}}_\alpha \rangle + \langle \hat{\boldsymbol{\xi}}_\beta^* \cdot \mathscr{L}(\hat{\boldsymbol{\xi}}_\alpha) \rangle = 0. \tag{3.15}$$

Similarly, the linear wave equation with respect to $\hat{\boldsymbol{\xi}}_\beta^*$ is integrated over the whole volume after being multiplied by $\hat{\boldsymbol{\xi}}_\alpha$. The results give

$$-\omega_\beta^2 \langle \rho_0 \hat{\boldsymbol{\xi}}_\alpha \hat{\boldsymbol{\xi}}_\beta^* \rangle - 2i\omega_\beta \langle \rho_0 \hat{\boldsymbol{\xi}}_\alpha (\boldsymbol{u}_0 \cdot \nabla)\hat{\boldsymbol{\xi}}_\beta^* \rangle + \langle \hat{\boldsymbol{\xi}}_\alpha \cdot \mathscr{L}(\hat{\boldsymbol{\xi}}_\beta^*) \rangle = 0. \tag{3.16}$$

[1]The following formula is convenient to write down a vector $(\boldsymbol{A} \cdot \nabla)\boldsymbol{B}$ into components in cylindrical coordinates (r, φ, z):

$$(\boldsymbol{A} \cdot \nabla)\boldsymbol{B} = \left(A_r \frac{\partial}{\partial r} + A_\varphi \frac{\partial}{r\partial \varphi} + A_z \frac{\partial}{\partial z}\right)\boldsymbol{B} + \frac{A_\varphi}{r}(\boldsymbol{i}_z \times \boldsymbol{B}),$$

where $\boldsymbol{A}$ and $\boldsymbol{B}$ are arbitrary vectors, and $\boldsymbol{i}_z$ is the unit vector in the z-direction.

Since the operator $\mathscr{L}$ is Hermitian (Lynden-Bell and Ostriker 1967), we have the relation:

$$\langle \hat{\boldsymbol{\xi}}_\alpha \cdot \mathscr{L}(\hat{\boldsymbol{\xi}}^*_\beta) \rangle = \langle [\mathscr{L}(\hat{\boldsymbol{\xi}}^*_\alpha)]^* \cdot \hat{\boldsymbol{\xi}}^*_\beta \rangle = \langle \mathscr{L}(\hat{\boldsymbol{\xi}}_\alpha) \cdot \hat{\boldsymbol{\xi}}^*_\beta \rangle \tag{3.17}$$

Hence, the difference of the above two equations [equations (3.15) and (3.16)] gives, when $\omega_\beta \neq \omega_\alpha$,

$$(\omega_\alpha + \omega_\beta)\langle \rho_0 \hat{\boldsymbol{\xi}}_\alpha \cdot \hat{\boldsymbol{\xi}}^*_\beta \rangle + 2i \Big\langle \rho_0 \hat{\boldsymbol{\xi}}_\alpha \cdot [(\boldsymbol{u}_0 \cdot \nabla)\hat{\boldsymbol{\xi}}^*_\beta] \Big\rangle = 0 \tag{3.18}$$

or

$$(\omega_\alpha + \omega_\beta)\langle \rho_0 \hat{\boldsymbol{\xi}}_\alpha \cdot \hat{\boldsymbol{\xi}}^*_\beta \rangle - 2i \Big\langle \rho_0 \hat{\boldsymbol{\xi}}^*_\beta \cdot [(\boldsymbol{u}_0 \cdot \nabla)\hat{\boldsymbol{\xi}}_\alpha] \Big\rangle = 0. \tag{3.19}$$

In deriving these equalities, we have used an integration by part, assuming that ρ_0 vanishes on the disk surface. This orthogonality relation [equation (3.18) or (3.19)] is used in Chap. 11 in deriving the wave-wave resonant instability in deformed disks.

Different from the case of non-rotating stars, the eigenfunctions of normal modes of disk oscillations are not orthogonal in the sense that $\langle \rho_0 \hat{\boldsymbol{\xi}}_\alpha \cdot \hat{\boldsymbol{\xi}}^*_\beta \rangle = 0$ when $\alpha \neq \beta$, because of the presence of unperturbed flow, $\boldsymbol{u}_0$.

3.1.2 *Lagrangian Description of Wave Energy and Its Conservation*

Let us derive an expression for wave energy of perturbations. We assume that a weakly growing artificial force is acting on a perturbation, which is $\delta \boldsymbol{f}$ per unit mass. During the initial epoch ($t = -\infty$) when the force began to work on the perturbation, the force and the perturbation had negligible amplitudes. After that, by the effects of the force, the perturbation grows. The work done on the perturbation between $t = -\infty$ and t should be regarded as the wave energy, E, of the perturbation at t. In this sense, the wave energy of perturbation, E, can be written as

$$E(t) = \int \int_{-\infty}^{t} \frac{\partial \boldsymbol{\xi}}{\partial t} \cdot \delta \boldsymbol{f} \, dt dV. \tag{3.20}$$

Here, the volume integration is performed over the whole volume where the disk gas exists.

Since $\delta \boldsymbol{f}$ is an external force acting on unit mass, the integrand of equation (3.20) can be expressed as

$$\frac{\partial \boldsymbol{\xi}}{\partial t} \left(\rho_0 \frac{\partial^2 \boldsymbol{\xi}}{\partial t^2} + 2\rho_0 (\boldsymbol{u}_0 \cdot \nabla) \frac{\partial \boldsymbol{\xi}}{\partial t} + \mathscr{L}(\boldsymbol{\xi}) \right). \tag{3.21}$$

Integration by part with respect to t under the use of the assumption of no perturbation at $t = -\infty$ gives

$$\int_{-\infty}^{t} \rho_0 \frac{\partial \boldsymbol{\xi}}{\partial t} \frac{\partial^2 \boldsymbol{\xi}}{\partial t^2} dt = \frac{1}{2} \rho_0 \left(\frac{\partial \boldsymbol{\xi}}{\partial t} \right)^2. \tag{3.22}$$

The volume integration of $2\rho_0(\partial \boldsymbol{\xi}/\partial t)(\boldsymbol{u}_0 \cdot \nabla)(\partial \boldsymbol{\xi}/\partial t)$ vanishes if the integration is performed by part under consideration of vanishing of the surface integrations and $\nabla(\rho_0 \boldsymbol{u}) = 0$. Next, considering that the operator $\mathscr{L}$ is Hermitian, we have

$$\int_{-\infty}^{t} dt \int \frac{\partial \boldsymbol{\xi}}{\partial t} \mathscr{L}(\boldsymbol{\xi}) dV = \frac{1}{2} \int_{-\infty}^{t} dt \int \frac{\partial}{\partial t} \Big[\boldsymbol{\xi} \cdot \mathscr{L}(\boldsymbol{\xi}) \Big] dv = \frac{1}{2} \int \boldsymbol{\xi} \cdot \mathscr{L}(\boldsymbol{\xi}) dV. \tag{3.23}$$

Combining the above results, we have from equation (3.20)

$$E = \frac{1}{2} \int \left[\rho_0 \left(\frac{\partial \boldsymbol{\xi}}{\partial t} \right)^2 + \boldsymbol{\xi} \cdot \mathscr{L}(\boldsymbol{\xi}) \right] dV. \tag{3.24}$$

This can be regarded as wave energy of perturbations. The first term in the brackets of equation (3.24) can be regarded as kinetic energy, while the second term potential energy.

Conservation of wave energy, E, can be derived directly from equation (3.24) by taking its time derivative. That is, we have

$$\frac{\partial E}{\partial t} = \frac{1}{2} \int \left[2\rho_0 \frac{\partial \boldsymbol{\xi}}{\partial t} \frac{\partial^2 \boldsymbol{\xi}}{\partial t^2} + \frac{\partial \boldsymbol{\xi}}{\partial t} \cdot \mathscr{L}(\boldsymbol{\xi}) + \boldsymbol{\xi} \cdot \mathscr{L}\left(\frac{\partial \boldsymbol{\xi}}{\partial t} \right) \right] dV$$

$$= \int \frac{\partial \boldsymbol{\xi}}{\partial t} \left(\rho_0 \frac{\partial^2 \boldsymbol{\xi}}{\partial t} + \mathscr{L}(\boldsymbol{\xi}) \right) dV = -2 \int \rho_0 \frac{\partial \boldsymbol{\xi}}{\partial t} \left[(\boldsymbol{u}_0 \cdot \nabla) \frac{\partial \boldsymbol{\xi}}{\partial t} \right] dV = 0. \tag{3.25}$$

In deriving the second relation the Hermitian of $\mathscr{L}$ has been used, and in the last relation integration by part has been adopted.

Next, let us express the wave energy, E, in terms of $\hat{\boldsymbol{\xi}}$ when perturbations have a form of normal mode as $\boldsymbol{\xi} = \Re[\hat{\boldsymbol{\xi}} \exp(i\omega t)]$. We have[2]

$$2\left(\frac{\partial \boldsymbol{\xi}}{\partial t} \right)^2 = \omega^2 \hat{\boldsymbol{\xi}} \cdot \hat{\boldsymbol{\xi}}^* + \Re[-\omega^2 \hat{\boldsymbol{\xi}}^2 \exp(i2\omega t)], \tag{3.26}$$

---

[2]The formula:

$$\Re(A)\Re(B) = \frac{1}{2}\Re[AB + AB^*] = \frac{1}{2}[AB + A^*B]$$

is used, where A and B are complex variables and A^* and B^* are the complex conjugates of A and B, respectively.

and

$$2\boldsymbol{\xi} \cdot \mathscr{L}(\boldsymbol{\xi}) = \hat{\boldsymbol{\xi}} \cdot \mathscr{L}(\hat{\boldsymbol{\xi}}^*) + \Re[\boldsymbol{\xi} \cdot \mathscr{L}(\boldsymbol{\xi})\exp(i2\omega t)]. \tag{3.27}$$

Furthermore, we have shown that wave energy is time-independent. Hence, it is obvious that wave energy defined by equation (3.24) is written as

$$E = \frac{1}{2}\int \left[\omega^2 \rho_0 \hat{\boldsymbol{\xi}}^* \hat{\boldsymbol{\xi}} + \hat{\boldsymbol{\xi}}^* \cdot \mathscr{L}(\hat{\boldsymbol{\xi}})\right] dV. \tag{3.28}$$

If $\mathscr{L}(\boldsymbol{\xi})$ is eliminated from equation (3.28) by using wave equation (3.10), we have

$$\begin{aligned} E &= \frac{1}{2}\int \left[\omega^2 \rho_0 \hat{\boldsymbol{\xi}}^* \hat{\boldsymbol{\xi}} - i\omega\rho_0 \hat{\boldsymbol{\xi}}^* (\boldsymbol{u}_0 \cdot \nabla)\hat{\boldsymbol{\xi}}\right] dV \\ &= \frac{1}{2}\int \left[\omega^2 \rho_0 \hat{\boldsymbol{\xi}} \hat{\boldsymbol{\xi}}^* + i\omega\rho_0 \hat{\boldsymbol{\xi}} (\boldsymbol{u}_0 \cdot \nabla)\hat{\boldsymbol{\xi}}^*\right] dV. \end{aligned} \tag{3.29}$$

If the flow in the unperturbed state is rotation alone, i.e., $\boldsymbol{u}_0 = (0, r\Omega, 0)$, and the azimuthal dependence of wavy perturbations is proportional to $\exp(-im\varphi)$, we have

$$\begin{aligned} \breve{\boldsymbol{\xi}}^* (\boldsymbol{u}_0 \cdot \nabla)\breve{\boldsymbol{\xi}} &= -im\Omega \breve{\boldsymbol{\xi}}^* \breve{\boldsymbol{\xi}} - \Omega \breve{\xi}_r^* \breve{\xi}_\varphi + \Omega \breve{\xi}_\varphi^* \breve{\xi}_r \\ &= -im\Omega \breve{\boldsymbol{\xi}}^* \breve{\boldsymbol{\xi}} - i(\omega - m\Omega)\breve{\xi}_\varphi^* \breve{\xi}_\varphi. \end{aligned} \tag{3.30}$$

In deriving the last relation, we have used equation (3.13). If relation (3.30) is used the wave energy given by equation (3.29) is written in the form:

$$E = \frac{1}{2}\int \rho_0 \omega(\omega - m\Omega)(\breve{\xi}_r^* \breve{\xi}_r + \breve{\xi}_z^* \breve{\xi}_z) dV. \tag{3.31}$$

This form of wave energy is important and useful, because this expression shows that the sign of wave energy is directly related to the sign of $\omega - m\Omega$ in the region where the wave predominantly exists. The radius where $\omega = m\Omega$ is the corotation radius, and as will be shown later there are evanescent regions of oscillations around the corotation radius. That is, waves which exist inside the corotation radius ($\omega - m\Omega < 0$) penetrate little into outside the corotation one. Similarly, those outside the corotation radius ($\omega - m\Omega > 0$) penetrate little inside the corotation one. Hence, if we consider waves with $\omega > 0$, the waves inside the corotation radius have negative energy, while those outside the corotation radius have positive energy.

Friedman and Schutz (1978) pointed out that the above wave energy, E, is not invariant under a gauge transformation, and introduced a more general expression for wave energy. See Friedman and Schutz (1978) for details.

3.1.3 Generalization to Magnetized Disks

The Lagrangian expression for linear perturbations, equation (3.6), is formally extended to the cases of adiabatic, small-amplitude MHD perturbations, although detailed expressions for $\mathscr{L}(\boldsymbol{\xi})$ is changed. This generalized form of $\mathscr{L}(\boldsymbol{\xi})$ is still Hermitian.

In the hydromagnetic case the equation corresponding to equation (3.1) is generalized to

$$\frac{D_0^2\boldsymbol{\xi}}{Dt^2}=\delta\biggl(-\nabla\psi-\frac{1}{\rho}\nabla p+\frac{1}{4\pi\rho}\mathrm{curl}\,\boldsymbol{B}\times\boldsymbol{B}\biggr), \tag{3.32}$$

where $\boldsymbol{B}$ is the magnetic flux density. Hence, as the equation corresponding to equation (3.6) we have

$$\rho_0\frac{\partial^2\boldsymbol{\xi}}{\partial t^2}+2\rho_0(\boldsymbol{u}_0\cdot\nabla)\frac{\partial\boldsymbol{\xi}}{\partial t}+\mathscr{L}(\boldsymbol{\xi})+\mathscr{L}_{\rm B}(\boldsymbol{\xi})=0, \tag{3.33}$$

where

$$\begin{aligned}\mathscr{L}_{\rm B}(\boldsymbol{\xi})=&-\frac{1}{4\pi}(\mathrm{curl}\,\delta\boldsymbol{B})\times\boldsymbol{B}_0-\frac{1}{4\pi}(\mathrm{curl}\boldsymbol{B}_0)\times\delta\boldsymbol{B}\\&-\frac{1}{4\pi}(\mathrm{div}\boldsymbol{\xi})(\mathrm{curl}\boldsymbol{B}_0\times\boldsymbol{B}_0)-\frac{1}{4\pi}\delta(\mathrm{curl}\,\boldsymbol{B}_0\times\boldsymbol{B}_0).\end{aligned} \tag{3.34}$$

By use of the linear part of the induction equation, we can express $\delta\boldsymbol{B}$ in terms of $\boldsymbol{\xi}$ as

$$\delta\boldsymbol{B}=\mathrm{curl}(\boldsymbol{\xi}\times\boldsymbol{B}_0). \tag{3.35}$$

Hence, we can explicitly express $\mathscr{L}_{\rm B}(\boldsymbol{\xi})$ in terms of $\boldsymbol{\xi}$ (e.g., Bernstein et al. 1958).[3] The results show that, although detailed procedures are not shown here, $\mathscr{L}_{\rm B}(\boldsymbol{\xi})$ is

[3]By using the cartesian coordinates, we can express the component of $\mathscr{L}_{\rm B}(\boldsymbol{\xi})$ as

$$\begin{aligned}4\pi(\mathscr{L}_{\rm B})_i=&\frac{\partial}{\partial r_i}\biggl[B_{0j}\frac{\partial}{\partial r_k}(B_{0k}\xi_j-B_{0j}\xi_k)\biggr]\\&-\frac{\partial}{\partial r_j}\biggl[B_{0i}\frac{\partial}{\partial r_k}(B_{0k}\xi_j-B_{0j}\xi_k)+B_{0j}\frac{\partial}{\partial r_k}(B_{0k}\xi_i-B_{0i}\xi_k)\biggr]\\&-\biggl(\xi_k\frac{\partial}{\partial r_k}+\mathrm{div}\boldsymbol{\xi}\biggr)\biggl[-\frac{1}{2}\frac{\partial B_0^2}{\partial r_i}+\frac{\partial}{\partial r_j}(B_{0i}B_{0j})\biggr].\end{aligned}$$

also Hermitian as $\mathscr{L}(\boldsymbol{\xi})$ is. That is, for arbitrary $\boldsymbol{\xi}$ and $\boldsymbol{\eta}$ we have

$$\int \boldsymbol{\eta} \cdot \mathscr{L}_{\rm B}(\boldsymbol{\xi}) dV = \int \boldsymbol{\xi} \cdot \mathscr{L}_{\rm B}(\boldsymbol{\eta}) dV. \tag{3.36}$$

Since $\mathscr{L}_{\rm B}$ is a Hermitian operator, we can show, using the same procedures as before, that the wave energy in hydromagnetic perturbations is

$$E = \frac{1}{2} \int \left[\rho_0 \left(\frac{\partial \boldsymbol{\xi}}{\partial t} \right)^2 + \boldsymbol{\xi} \cdot \mathscr{L}(\boldsymbol{\xi}) + \boldsymbol{\xi} \cdot \mathscr{L}_{\rm B}(\boldsymbol{\xi}) \right] dV. \tag{3.37}$$

This wave energy is also conservative, since $\mathscr{L}_{\rm B}$ is Hermitian. We see further that for a normal mode of oscillation the wave energy is expressed again in the form of equation (3.29) even in the case of hydromagnetic oscillations.

3.2 Eulerian Description of Oscillations

We consider small-amplitude adiabatic perturbations on steadily rotating axisymmetric disks with no radial flow. Cylindrical coordinates (r, φ, z) are introduced, where the z-axis is perpendicular to the disk plane and its origin is at the disk center. Then, linear perturbations over the steady state are described by (see Appendix A):

$$\left(\frac{\partial}{\partial t} + \Omega \frac{\partial}{\partial \varphi} \right) \rho_1 + \frac{\partial}{r \partial r}(r \rho_0 u_r) + \frac{\partial}{r \partial \varphi}(\rho_0 u_\varphi) + \frac{\partial}{\partial z}(\rho_0 u_z) = 0, \tag{3.38}$$

$$\left(\frac{\partial}{\partial t} + \Omega \frac{\partial}{\partial \varphi} \right) u_r - 2\Omega u_\varphi = -\frac{\partial h_1}{\partial r}, \tag{3.39}$$

$$\left(\frac{\partial}{\partial t} + \Omega \frac{\partial}{\partial \varphi} \right) u_\varphi + \frac{\kappa^2}{2\Omega} u_r = -\frac{\partial h_1}{r \partial \varphi}, \tag{3.40}$$

$$\left(\frac{\partial}{\partial t} + \Omega \frac{\partial}{\partial \varphi} \right) u_z = -\frac{\partial h_1}{\partial z}, \tag{3.41}$$

$$h_1 = \frac{p_1}{\rho_0} = c_{\rm s} \frac{\rho_1}{\rho_0}. \tag{3.42}$$

Here, (u_r, u_φ, u_z), p_1, and ρ_1 are the Eulerian velocity, pressure, and density perturbations over the unperturbed ones, respectively. Equation (3.38) is the equation of continuity, equations (3.39), (3.40), and (3.41) are in turn the r-, φ-, and z- components of equation of motion. Equation (3.42) is the definition of h_1, and due to the assumption of adiabatic perturbations, we have $-[(1/\rho_0)\nabla p]_1 = -\nabla h_1$, which has been adopted in equations (3.39), (3.40), and (3.41).

When all perturbations are assumed to be proportional to $\exp[i(\omega t - m\varphi)]$ and written, for example, as

$$u_r = \Re\Big[\breve{u}_r \exp[i(\omega t - m\varphi)]\Big], \tag{3.43}$$

the above hydrodynamical equations describing perturbed quantities are

$$i\tilde{\omega}\breve{\rho}_1 + \frac{\partial}{r\partial r}(r\rho_0\breve{u}_r) - i\frac{m}{r}\rho_0\breve{u}_\varphi + \frac{\partial}{\partial z}(\rho_0\breve{u}_z) = 0, \tag{3.44}$$

$$i\tilde{\omega}\breve{u}_r - 2\Omega\breve{u}_\varphi = -\frac{\partial\breve{h}_1}{\partial r}, \tag{3.45}$$

$$i\tilde{\omega}\breve{u}_\varphi + \frac{\kappa^2}{2\Omega}\breve{u}_r = i\frac{m}{r}\breve{h}_1, \tag{3.46}$$

$$i\tilde{\omega}\breve{u}_z = -\frac{\partial\breve{h}_1}{\partial z}, \tag{3.47}$$

where $\tilde{\omega}$ is defined by $\tilde{\omega} = \omega - m\Omega$.

Elimination of $\breve{u}_\varphi$ from equations (3.45) and (3.46) gives a relation between $\breve{h}_1$ and $\breve{u}_r$, which is

$$\frac{\partial\breve{h}_1}{\partial r} - \frac{2m\Omega}{r\tilde{\omega}}\breve{h}_1 = \frac{\tilde{\omega}^2 - \kappa^2}{i\tilde{\omega}}\breve{u}_r. \tag{3.48}$$

Next, another relation between $\breve{h}_1$ and $\breve{u}_r$ is derived. Substituting equations (3.45) and (3.47) into equation (3.44), we have, after some manipulation,

$$\frac{1}{\rho_0}\frac{\partial}{\partial z}\left(\rho_0\frac{\partial\breve{h}_1}{\partial z}\right) + \left(\frac{\tilde{\omega}^2}{c_s^2} - \frac{m^2}{r^2}\right)\breve{h}_1$$
$$= i\tilde{\omega}\frac{\partial\breve{u}_r}{\partial r} + i\tilde{\omega}\left(\frac{\partial \ln r\rho_0}{\partial r} + \frac{m\kappa^2}{2r\tilde{\omega}\Omega}\right)\breve{u}_r. \tag{3.49}$$

Equations (3.48) and (3.49) are a set of two differential equations describing relations between $\breve{u}_r$ and $\breve{h}_1$.

Eliminating $\breve{u}_r$ from equations (3.48) and (3.49), we have a partial differential equation with respect to $\breve{h}_1$, which is written explicitly in the form

$$\frac{1}{\rho_0}\frac{\partial}{\partial z}\left(\rho_0\frac{\partial \breve{h}_1}{\partial z}\right)+\left(\frac{\tilde{\omega}^2}{c_s^2}-\frac{m^2}{r^2}\right)\breve{h}_1+\tilde{\omega}\frac{\partial}{\partial r}\left[\frac{\tilde{\omega}}{\tilde{\omega}^2-\kappa^2}\left(\frac{\partial \breve{h}_1}{\partial r}-\frac{2m\Omega}{r\tilde{\omega}}\breve{h}_1\right)\right]$$
$$+\frac{\tilde{\omega}^2}{\tilde{\omega}^2-\kappa^2}\left(\frac{\partial \ln r\rho_0}{\partial r}+\frac{m\kappa^2}{2r\tilde{\omega}\Omega}\right)\left(\frac{\partial \breve{h}_1}{\partial r}-\frac{2m\Omega}{r\tilde{\omega}}\breve{h}_1\right)=0. \tag{3.50}$$

This is the basic equation to be studied in the following chapters in part I. This equation is a second-order partial differential equation. Solving this equation in a general way is thus difficult. We study characteristics of the wave motions described by equation (3.50) by introducing approximations or idealized situations.

Here, we briefly mention Lindblad and corotation resonances again, although they are mentioned in Chap. 2. At the radii of $\tilde{\omega}^2 = \kappa^2$ (Lindblad resonances) the denominators of some terms in equation (3.50) vanish. As mentioned in Chap. 2, however, detailed examinations show that this does not mean that h_1 becomes singular at the radii of $\tilde{\omega}^2 = \kappa^2$, i.e., the radii are points of apparent singularity. On the other hand, at the radius of $\tilde{\omega} = 0$ (corotation resonance) h_1 becomes singular, if ω is real, i.e., the corotation point is a regular singularity. The frequency ω must be complex there. This leads to amplification or damping of oscillations at the corotation point, which will be discussed in Chap. 10.

Before proceeding to Chap. 4, we mention an Eulerian expression of the wave energy.

3.2.1 Eulerian Description of Wave Energy

Let us now express the fluid velocity for perturbed flows by $\boldsymbol{U}(\boldsymbol{r}, t)$, i.e., $\boldsymbol{U}(\boldsymbol{r}, t) = \boldsymbol{u}_0(\boldsymbol{r}, t) + \boldsymbol{u}_1(\boldsymbol{r}, t)$, and that for unperturbed flows by $\boldsymbol{U}_0$, which is $\boldsymbol{u}_0$. Then, by the definition of the displacement vector, $\boldsymbol{\xi}$, we have (Lynden-Bell and Ostriker 1967)

$$\frac{D_0\boldsymbol{\xi}}{Dt} = \boldsymbol{U}(\boldsymbol{r}+\boldsymbol{\xi}, t) - \boldsymbol{U}_0(\boldsymbol{r}, t), \tag{3.51}$$

which gives in the first order in perturbations

$$\frac{D_0\boldsymbol{\xi}}{Dt} = \boldsymbol{u}_1(\boldsymbol{r}, t) + (\boldsymbol{\xi}\cdot\nabla)\boldsymbol{u}_0. \tag{3.52}$$

In the present problem we consider the case where the unperturbed flow is a pure rotation with angular velocity $\Omega(r)$, i.e., $\boldsymbol{u}_0 = (0, r\Omega(r), 0)$ in the cylindrical coordinates. Then, the above relation (3.52) gives, after a slight arrangement,

$$i(\omega - m\Omega)\breve{\xi}_r = \breve{u}_r, \tag{3.53}$$

$$i(\omega - m\Omega)\breve{\xi}_\varphi = \breve{u}_\varphi + \breve{\xi}_r r\frac{d\Omega}{dr} = \frac{4\Omega^2}{\kappa^2}\breve{u}_\varphi, \tag{3.54}$$

$$i(\omega - m\Omega)\breve{\xi}_z = \breve{u}_z. \tag{3.55}$$

The second equality in equation (3.54) is obtained by using equation (3.13).

The use of the above relations gives us, after some manipulations,

$$\omega\breve{\boldsymbol{\xi}}^* \cdot \breve{\boldsymbol{\xi}} - i\breve{\boldsymbol{\xi}}^*[(\boldsymbol{u}_0 \cdot \nabla)\breve{\boldsymbol{\xi}}] = \frac{1}{2}\frac{1}{\omega - m\Omega}\Bigg[(\breve{u}_r^*\breve{u}_r + \breve{u}_z^*\breve{u}_z)$$
$$+\left(\breve{u}_\varphi + \frac{\kappa^2/2\Omega}{i(\omega - m\Omega)}\breve{u}_r\right)^*\left(\breve{u}_\varphi + \frac{rd\Omega/dr}{i(\omega - m\Omega)}\breve{u}_r\right) + c.c.\Bigg], \tag{3.56}$$

where c.c. denotes the complex conjugate. In geometrically thin disks, the φ-component of equation motion describing perturbations is approximately written as [see equation (3.46)]

$$i(\omega - m\Omega)\breve{u}_\varphi + \frac{\kappa^2}{2\Omega}\breve{u}_r = 0, \tag{3.57}$$

unless the azimuthal wavelength of perturbations is short.[4] Then, the terms in the large brackets in equation (3.56) vanishes, and we have from equation (3.29) that the wave energy, E, is written as

$$E = \frac{1}{2}\int \frac{\omega}{\omega - m\Omega}\rho_0(\breve{u}_r^*\breve{u}_r + \breve{u}_z^*\breve{u}_z)dV. \tag{3.58}$$

This expression for E is consistent with that for E in terms of $\breve{\boldsymbol{\xi}}$ given by equation (3.31), because $\breve{\boldsymbol{u}}$ and $\breve{\boldsymbol{\xi}}$ are related by equations (3.53), (3.54), and (3.55). Equation (3.58) shows that the wave energy diverges if equation (3.58) is formally applied in the region around the corotation resonance. This is unrealistic and implies a limitation of applicability of equation (3.58). This comes from the fact that the assumption that the frequency is real is violated, when waves are in the corotation region. In other words, at the corotation point wave amplification or damping occurs. Amplification or damping of waves at the corotation resonance is discussed in Chap. 10.

[4] If relations (3.53) and (3.54) are substituted into equation (3.57), we obtain a relation between $\breve{\xi}_r$ and $\breve{\xi}_\varphi$, which is equation (3.13).

3.2.2 Wave Energy Density and Energy Flux

It is well known that in sound waves in a non-rotating homogeneous medium wave energy density, $(1/2)(\rho_0 \boldsymbol{u}^2 + c_s^2 \rho_1^2/\rho_0)$, and wave energy flux, $p_1 \boldsymbol{u}$, are related by a conservation relation (e.g., Landau and Lifshitz 1960):

$$\frac{\partial}{\partial t}\left[\frac{1}{2}\left(\rho_0 \boldsymbol{u}^2 + \frac{\rho_0}{c_s^2} h_1^2\right)\right] + \mathrm{div}(p_1 \boldsymbol{u}) = 0. \tag{3.59}$$

Here, we examine how this relation is extended to waves in disks. In the present case, the set of basic equations, equations (3.38), (3.39), (3.40), (3.41), and (3.42), give

$$\left(\frac{\partial}{\partial t} + \Omega\frac{\partial}{\partial \varphi}\right)\left[\frac{1}{2}\left(\rho_0 \boldsymbol{u}^2 + \frac{\rho_0}{c_s^2} h_1^2\right)\right] + \mathrm{div}(p_1 \boldsymbol{u}) = -r\frac{d\Omega}{dr}\rho_0 u_r u_\varphi. \tag{3.60}$$

If we consider perturbations whose radial wavenumber, λ, is sufficiently short compared with radius r, i.e., $\lambda < r$, the right-hand side of equation (3.46) can be neglected, and we have

$$\frac{1}{2}\left(\frac{\partial}{\partial t} + \Omega\frac{\partial}{\partial \varphi}\right)u_\varphi^2 + \frac{\kappa^2}{2\Omega}u_r u_\varphi = 0. \tag{3.61}$$

Then, substituting equation (3.61) into equation (3.60), we have

$$\left(\frac{\partial}{\partial t} + \Omega\frac{\partial}{\partial \varphi}\right)\left[\frac{1}{2}\rho_0\left(u_r^2 + \frac{4\Omega^2}{\kappa^2}u_\varphi^2 + u_z^2\right) + \frac{1}{2}\frac{\rho_0}{c_s^2}h_1^2\right] + \mathrm{div}(p_1 \boldsymbol{u}) = 0, \tag{3.62}$$

which have a form of conservation, and the terms in the large brackets can be regarded as *wave energy density* and $p_1 \boldsymbol{u}$ can be regarded as *wave energy flux*.

3.2.3 Wave Action and Its Implication

Without restricting to short wavelength perturbations, we have an important conservative quantity, which is *wave action*. Let us consider a time periodic perturbations whose motions are, for simplicity, restricted in the equatorial plane, i.e., $u_z = 0$. Then, taking time and azimuthal averages of equation (3.60), we have

$$\frac{\partial}{r\partial r}(r\langle p_1 u_r\rangle) + r\frac{d\Omega}{dr}\rho_0\langle u_r u_\varphi\rangle = 0, \tag{3.63}$$

where $\langle A\rangle$ denotes time and azimuthal averages of A. Equation (3.63) can be rewritten as

$$(\omega - m\Omega)\frac{\partial}{r\partial r}\left[\frac{r\langle p_1 u_r\rangle}{\omega - m\Omega}\right] + \frac{rd\Omega/dr}{\omega - m\Omega}\left[-\frac{m}{r}\langle p_1 u_r\rangle + \rho_0(\omega - m\Omega)\langle u_r u_\varphi\rangle\right] = 0, \tag{3.64}$$

where ω and m are, respectively, the frequency and azimuthal wavenumber of wavy perturbations.

Let us consider that we have $\langle p_1 u_r\rangle = (1/2)\Re(\breve{p}_1\breve{u}_r^*)$, where * denotes the complex conjugate. Similarly, we have $\langle u_r u_\varphi\rangle = (1/2)\Re\langle\breve{u}_r^*\breve{u}_\varphi\rangle$. Hence, multiplying equation (3.46) by $\breve{u}_r^*$, we have

$$\Re(\omega - m\Omega)\breve{u}_\varphi\breve{u}_r^* = \frac{m}{r\rho_0}\Re(\breve{p}_1\breve{u}_r^*), \tag{3.65}$$

which show that the two terms in the large brackets of the second term of equation (3.64) are cancelled out in time-periodic perturbations, and we have

$$\frac{\partial}{r\partial r}\left[\frac{r\langle p_1 u_r\rangle}{\omega - m\Omega}\right] = 0. \tag{3.66}$$

This shows that in time-periodic perturbations, $r\langle p_1 u_r\rangle/(\omega - m\Omega)$, which is called *wave action*, is constant along the radial direction. This is a special example of *wave action conservation* which is known in more general situations (see Bretherton and Garret 1969). The concept of wave action conservation in wavetrains is briefly described in Appendix C.

The corotation radius of $\omega - m\Omega = 0$ is a singular point of wave action. This means that the oscillatory flows passing through a coronation radius cannot be purely time-periodic, i.e., frequency must become complex. This brings about an important result that the corotation radius is the only place where wave action can be created or destroyed in wave motions (e.g., Drury 1985). This issue will be discussed in Chap. 10 in relation to overstability of waves at the corotation resonance.

References

Bernstein, I. B., Frieman, E. A., Kruskal, M. D., & Kulsrud, R. M. 1958, Proc. R. Soc. A-Math. Phys. Eng. Sci., A244, 17
Bretherton, F. P., & Garret, C. J. R. 1969, Proc. R. Soc. A-Math. Phys. Eng. Sci., A302, 529
Drury, L.O'C. 1985, Mon. Not. R. Astron. Soc., 217, 821
Friedman, J. L., & Schutz, B. F. 1978, Astrophys. J., 221, 937
Landau, L. D., & Lifshitz, E. M., 1960, *Fluid Mechanics* (Pergamon Press, Oxford)
Lynden-Bell, D., & Ostriker, J. P. 1967, Mon. Not. R. Astron. Soc., 136, 293
Kato, S. 2004, Publ. Astron. Soc. Japan, 56, 905
Kato, S. 2008, Publ. Astron. Soc. Japan, 60, 111

Chapter 4
Vertical Oscillations

Abstract The basic equation to be solved to study oscillation phenomena in geometrically thin disks is equation (3.50). The equation is, however, very complicated. We shall study the characteristics of equation (3.50) step by step by considering some idealized cases. First, in this chapter, we consider purely vertical oscillations, although such oscillations are not realized in actual disk systems, since the pressure variation associated with these oscillations inevitably induces horizontal motions. It is, however, instructive to study pure vertical oscillations to clarify general characteristics of oscillations in disks. In the next chapter (Chap. 5), we examine characteristics of disk oscillations, based on local approximations in the radial direction, and classify in Chap. 6 oscillation modes. In Chap. 7, frequencies of trapped oscillations are examined by use of the WKB method.

Keywords Toroidal magnetic fields • Truncated isothermal disks • Vertically isothermal disks • Vertical oscillations • Vertically polytropic disks

4.1 Vertical Disk Structure

Two cases are considered where the vertical structure of disks is (i) polytropic and (ii) isothermal. First, we treat the polytropic case.

4.1.1 Vertically Polytropic Disks

A barotropic gas with polytropic index N is considered, i.e., the pressure p and density ρ are related by $p \propto \rho^{(N+1)/N}$. The hydrostatic balance in the vertical direction in the unperturbed state is $-\partial p_0/\partial z = \rho_0 \Omega_\perp^2 z$, which leads to

$$\rho_0(r,z) = \rho_{00}(r)\left(1 - \frac{z^2}{H^2}\right)^N, \tag{4.1}$$

$$p_0(r,z) = p_{00}(r)\left(1 - \frac{z^2}{H^2}\right)^{1+N}, \tag{4.2}$$

S. Kato, *Oscillations of Disks*, Astrophysics and Space Science Library 437,
DOI 10.1007/978-4-431-56208-5_4

where subscript 0 represents the quantities in the unperturbed equilibrium state and 00 are those on the equatorial plane. The acoustic speed defined by $c_s^2 = \Gamma dp_0/d\rho_0$ is also given by

$$c_s^2(r,z) = c_{s0}^2(r)\left(1 - \frac{z^2}{H^2}\right), \tag{4.3}$$

where Γ is related to N by $\Gamma = (N+1)/N$ or $N = 1/(\Gamma - 1)$. The hydrostatic balance in the vertical direction also shows that the half-thickness, $H(r)$, of disks is related to $c_{s0}(r)$ and vertical epicyclic frequency, $\Omega_\perp(r)$, by

$$\Omega_\perp^2 H^2 = 2Nc_{s0}^2. \tag{4.4}$$

4.1.2 *Vertically Isothermal Disks*

Distinct from the case of polytropic disks, the vertically isothermal disks with $p_0 = c_s^2\rho_0$, c_s being constant, extend infinitely in the vertical direction. That is, the vertical hydrostatic balance, $-\partial p_0/\partial z = \rho_0\Omega_\perp^2 z$, gives that in the vertical direction the density $\rho_0(r,z)$ is distributed as

$$\rho_0(r,z) = \rho_{00}(r)\exp\left(-\frac{z^2}{2H^2}\right), \tag{4.5}$$

where the scale height, $H(r)$, is related to the isothermal accoustic speed, $c_s(r)$, and the vertical epicyclic frequency, $\Omega_\perp(r)$, by

$$H^2(r) = \frac{c_s^2}{\Omega_\perp^2(r)}. \tag{4.6}$$

4.2 Purely Vertical Oscillations

When we treat oscillations with node(s) in the vertical direction, the vertical structure of disks has non-negligible effects on their frequencies, because disks are so thin that the disk structure changes appreciably in one vertical wavelength. We consider four cases: (i) pressure and density are distributed in the vertical direction with barotropic relation, (ii) they are distributed isothermally, (iii) they are distributed isothermally but truncated at a certain height, and finally (iv) a case where disks are subject to toroidal magnetic fields.

4.2.1 Vertically Polyrtropic Disks

In the case where oscillations are assumed to be purely vertical, the equation of continuity (3.44) is[1]

$$i\tilde{\omega}\rho_1 + \frac{d}{dz}(\rho_0 u_z) = 0, \tag{4.7}$$

where $\tilde{\omega} = \omega - m\Omega$. The subscript 1 denotes Eulerian perturbations. The vertical component of equation of motion is [equation (3.47)]

$$i\tilde{\omega}u_z = -\frac{dh_1}{dz}. \tag{4.8}$$

Eliminating u_z from the above two equations and using $\rho_1 = \rho_0 h_1/c_s^2$, we have

$$\tilde{\omega}^2 h_1 + \frac{c_s^2}{\rho_0}\frac{d}{dz}\left(\rho_0\frac{dh_1}{dz}\right) = 0. \tag{4.9}$$

In the case of polytropic gases, equation (4.9) is written in the form:

$$(1-\eta^2)\frac{d^2h_1}{d\eta^2} - 2N\eta\frac{dh_1}{d\eta} + 2N\frac{\tilde{\omega}^2}{\Omega_\perp^2}h_1 = 0 \tag{4.10}$$

by changing the independent variable from z to a dimensionless variable η defined by $\eta = z/H$.

Equation (4.10) is solved with a boundary condition that h_1 remains finite at the surface of disks, i.e., $(h_1)_{\eta=\pm 1} < \infty$. Then, the eigenfunction, $h_1(\eta)$, is given by the Gegenbauer polynomial, C_n^λ (where $n = 1, 2, 3\ldots$), (Silbergleit et al. 2001, see also Kato 2005, 2010), i.e.,

$$h_{1,n}(\eta) = C_n^\lambda(\eta) \quad (n = 1, 2, 3, \ldots), \tag{4.11}$$

where λ is related to N by

$$\lambda = N - \frac{1}{2} \tag{4.12}$$

[1] In this chapter and Chap. 5, all perturbed quantities are taken to be proportional to $\exp[i(\omega t - m\varphi)]$, but, for example, u_r, $\hat{u}_r$, and $\breve{u}_r$, are not distinguished in order to avoid complexity, except when the distinction is necessary.

and the subscript n represents the number of node(s) in the vertical direction. The eigenvalue, $\tilde{\omega}_n(\equiv \omega_n - m\Omega)$, corresponding to $h_{1,n}$ is given by

$$\tilde{\omega}_n^2 = \frac{n(n+2N-1)}{2N}\Omega_\perp^2. \tag{4.13}$$

Finally, explicit forms of $h_{1,n}(\eta)$ of fundamental three modes are

$$h_{1,n}(\eta) \propto \begin{cases} \eta & \text{for} \quad n=1 \\ 1-(1+2N)\eta^2 & \text{for} \quad n=2 \\ \eta-(1+2N/3)\eta^3 & \text{for} \quad n=3. \end{cases} \tag{4.14}$$

It is noted that in the fundamental mode ($n=1$), $h_1 \propto \eta$ and $u_z = \text{const}$. This is an up-and-down motion of the disk plane with frequency $\Omega_\perp$.

4.2.2 *Vertically Isothermal Disks*

In the case of vertically isothermal disks, equation (4.9) is reduced to

$$\frac{d^2h_1}{d\eta^2} - \eta\frac{dh_1}{d\eta} + \frac{\tilde{\omega}^2}{\Omega_\perp^2}h_1 = 0, \tag{4.15}$$

where η is defined again by $\eta = z/H$, but the range of variation of η is $-\infty < \eta < \infty$. The boundary conditions that $\rho_0 h_1$ does not become infinite at infinity ($\eta = \pm\infty$) imposes $\tilde{\omega}^2/\Omega_\perp^2$ to be (positive) integer (Okazaki et al. 1987). Then, the above equation becomes the Hermite equation, and the eigenfunctions are expressed by the Hermite polynomials, i.e.,

$$\tilde{\omega}_n^2 \equiv (\omega_n - m\Omega)^2 = n\Omega_\perp^2 \qquad (n = 1, 2, 3\ldots), \tag{4.16}$$

and

$$h_{1,n}(\eta) = \mathscr{H}_n \propto \begin{cases} \eta & \text{for} \quad n=1, \\ \eta^2-1 & \text{for} \quad n=2, \\ \eta^3-3\eta & \text{for} \quad n=3. \end{cases} \tag{4.17}$$

The subscript n is again attached to $\tilde{\omega}$ and h_1 in order to show the number of node(s) of oscillations in the vertical direction. It is noted again that the fundamental mode ($n = 1$) is simply an up-and-down motion and thus its eigenfrequency is $\Omega_\perp$, independent of the disk structure.

4.2.3 Vertically Truncated Isothermal Disks

It is known that in the very high state (the steep power-law state) of X-ray binaries, geometrically thin cool disks are surrounded by hot coronae, and their innermost parts are truncated. The truncation will also occur in the vertical height by disk evaporation due to hot corona. Considering such situations, we examine here isothermal disks where vertical extension is terminated at certain finite height, say $\eta = \pm\eta_s$.

In this case the equation describing the purely vertical oscillations is the same as equation (4.15). Boundary conditions are, however, changed. They are imposed at the truncated surface of $\eta = \pm\eta_s$, where η_s represents the truncation height. The truncation is assumed to be a result of a hot corona. Inside the corona perturbations are transported with much higher speed compared with in the disk because of high temperature. Hence, we adopt the boundary conditions that the Lagrangian variation of pressure, δp, vanishes at the surface of $\eta = \pm\eta_s$. In the present problem $\delta p = p_1 + \xi_z dp_0/dz = \rho_0(h_1 - z\Omega_\perp^2\xi_z)$ and $\tilde{\omega}^2\xi_z = dh_1/dz$. Hence, the boundary condition, $\delta p = 0$, can be written as $\eta dh_1/d\eta - (\tilde{\omega}^2/\Omega_\perp^2)h_1 = 0$, which can be reduced to $d^2h_1/d\eta^2 = 0$ [see equation (4.15)] at $\eta = \pm\eta_s$.

The equation to be solved to find eigenvalue, $K \equiv (\omega - m\Omega)^2/\Omega_\perp^2$, and eigenfunction, h_1, is the same as equation (4.15), but this equation is rewritten here in a form convenient to show orthogonality of eigenfunctions, which is

$$\frac{d}{d\eta}\left[\exp\left(-\frac{\eta^2}{2}\right)\frac{dh_1}{d\eta}\right] + K\exp\left(-\frac{\eta^2}{2}\right)h_1 = 0. \tag{4.18}$$

The boundary condition to be imposed is

$$\frac{d^2h_1}{d\eta^2} = 0 \qquad \text{at } \eta = \pm\eta_s. \tag{4.19}$$

By solving this eigenvalue problem we obtain a discrete set of eigenvalue, K_n ($n = 1, 2, 3\ldots$), and the corresponding eigenfunction $h_{1,n}$ ($n = 1, 2, 3, \ldots$).

First, we show that the eigenfunctions are orthogonal in the sense that

$$\int_{-\eta_s}^{\eta_s} \exp\left(-\frac{\eta^2}{2}\right)\frac{dh_{1,m}}{d\eta}\frac{dh_{1,n}}{d\eta}d\eta = 0, \qquad \text{for } m \neq n. \tag{4.20}$$

To show this orthogonal relation, we take a derivative of equation (4.18) with respect to η to lead to

$$\exp\left(-\frac{\eta^2}{2}\right)\frac{d}{d\eta}\left\{\exp\left(\frac{\eta^2}{2}\right)\frac{d}{d\eta}\left[\exp\left(-\frac{\eta^2}{2}\right)\frac{dh_{1,n}}{d\eta}\right]\right\}$$
$$+\exp\left(-\frac{\eta^2}{2}\right)K_n\frac{dh_{1,n}}{d\eta} = 0. \tag{4.21}$$

Then, let us multiply equation (4.21) by $dh_{1,m}/d\eta$ and integrate from $-\eta_s$ to η_s. By performing the integration by part, we have

$$\frac{dh_{1,m}}{d\eta}\frac{d}{d\eta}\left[\exp\left(-\frac{\eta^2}{2}\right)\frac{dh_{1,n}}{d\eta}\right]\Bigg|_{-\eta_s}^{\eta_s}$$
$$-\int_{-\eta_s}^{\eta_s}\exp\left(\frac{\eta^2}{2}\right)\frac{d}{d\eta}\left[\exp\left(-\frac{\eta^2}{2}\right)\frac{dh_{1,m}}{d\eta}\right]\frac{d}{d\eta}\left[\exp\left(-\frac{\eta^2}{2}\right)\frac{dh_{1,n}}{d\eta}\right]d\eta$$
$$+K_n\int_{-\eta_s}^{\eta_s}\exp\left(-\frac{\eta^2}{2}\right)\frac{dh_{1,m}}{d\eta}\frac{dh_{1,n}}{d\eta}=0. \tag{4.22}$$

It is noted here that the term of surface integral is symmetric with respect to $h_{1,n}$ and $h_{1,m}$, since $d^2h_1/d\eta^2=0$ on the surface. We can also derive a similar equation with equation (4.22), starting from equation of $h_{1,m}$. Taking the difference of both equations, we have the orthogonal relation (4.20) unless K_n and K_m are equal.

Up to the present, we have adopted h_1 as the dependent variable describing oscillations. Instead of h_1, we can adopt u_z as the dependent variable. For example, using $u_z \propto dh_1/d\eta$ [see equation (4.8)], we can derive from equation (4.15) the following equation with respect to u_z[2]:

$$\frac{d^2u_z}{d\eta^2}-\eta\frac{du_z}{d\eta}+\left(\frac{\tilde{\omega}^2}{\Omega_\perp^2}-1\right)u_z=0. \tag{4.23}$$

Equation (4.23) is written in the form

$$\frac{d}{d\eta}\left[\exp\left(-\frac{\eta^2}{2}\right)\frac{du_z}{d\eta}\right]+(K-1)\exp\left(-\frac{\eta^2}{2}\right)u_z=0. \tag{4.24}$$

The boundary condition to be adopted to solve this equation is $du_z/d\eta=0$, since the boundary condition, $dh_1^2/d\eta^2=0$, is equivalent to $du_z/d\eta=0$ [see equation (4.8)]. The orthogonality relation (4.20) can be expressed in terms of u_z as

$$\int_{-\eta_s}^{\eta_s}\exp\left(-\frac{\eta^2}{2}\right)u_{z,m}u_{z,n}d\eta=0, \qquad \text{for } m\neq n. \tag{4.25}$$

Eigenvalues, K_n, and eigenfunctions, $u_{z,n}$ (not $h_{1,n}$), calculated numerically are shown in Figs. 4.1 and 4.2, respectively, for three basic modes ($n=1$, 2, and 3) in the cases where the truncation height, z_s, is a few times the vertical scale height, H, i.e., $\eta_s\equiv z_s/H=2$, 3, and 4. In the limit of $\eta_s=\infty$, $K_n=n$ [see equation (4.16)],

[2]The difference between equations (4.23) and (4.15) is related to the difference of node numbers of $u_z(\eta)$ and $h_1(\eta)$.

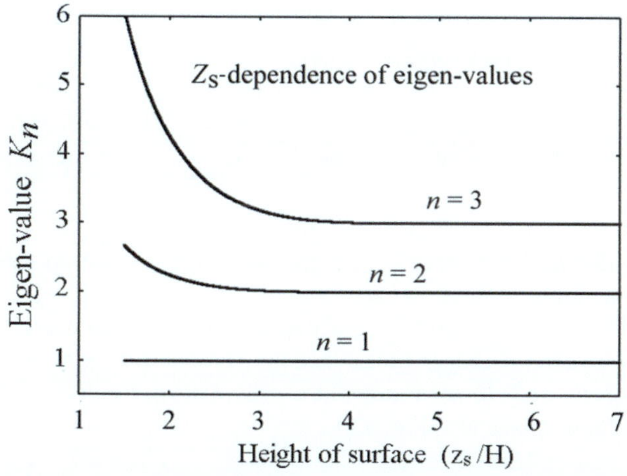

Fig. 4.1 Eigenvalue, K_n, of purely vertical oscillations in vertically isothermal disks as functions of the truncated height, $\eta_s \equiv z_s/H$. Three basic modes of $n = 1$, $n = 2$, and $n = 3$ are shown. In the limit of $\eta_s = \infty$, we have $K_n = n$, but K_n increases with decrease of η_s in oscillations of $n \geq 2$. In the case of $n = 1$, the eigenvalue K_n is unity and independent of η_s, because this oscillation is just a harmonic oscillations of disk plane in the vertical direction under the restoring force of $\Omega_\perp^2 \xi_z$. It is noted that this figure is slightly different from figure 1 of Kato (2012) because of the difference of adopted notations.

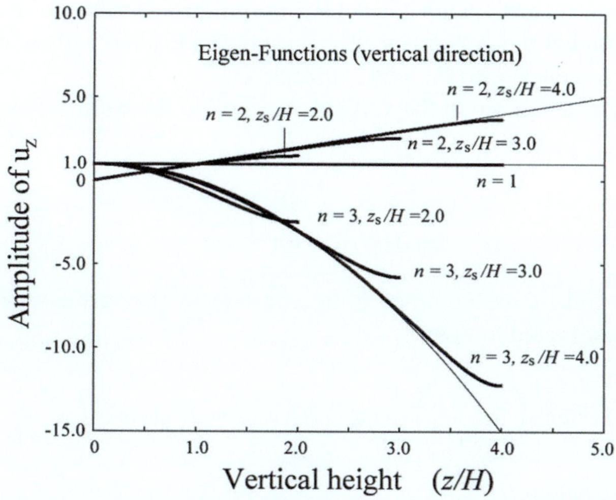

Fig. 4.2 Eigenfunctions, u_z, of purely vertical oscillations in vertically isothermal disks. Eigenfunctions of three basic modes ($n = 1$, $n = 2$, and $n = 3$) are shown for three disks of $\eta_s (\equiv z_s/H) = 2$, 3, and 4. The amplitudes of oscillations are taken to be arbitrary. It is note that in non-truncated disks $u_z = 1$ (i.e., $h_1 \propto \mathscr{H}_1$) for $n = 1$, $u_z = \eta$ (i.e., $h_1 \propto \mathscr{H}_2$ and $u_z \propto 2\eta$) for $n = 2$, and $u_z = 1 - \eta^2$ (i.e., $h_1 \propto \mathscr{H}_3$ and $u_z \propto 3\eta^2 - 3$) for $n = 3$. Eigenfunctions in the non-truncated isothermal disks are shown by *thin curves* (After S. Kato 2012. PASJ ©).

Table 4.1 Eigenvalue $K_{n,s}$ of vertical oscillations.

Node number	Truncation height $\eta_s(\equiv z_s/H)$					
	1.0	1.3	1.5	1.8	2.0	2.5	∞
$n = 1$	1.0	1.0	1.0	1.0	1.0	1.0	1.0
$n = 2$	4.00	3.016	2.670	2.365	2.243	2.083	2.0

but the eigenvalue K_n increases as η_s decreases for the oscillation modes of $n \geq 2$ (see Fig. 4.1 and Table 4.1).

In the mode of $n = 1$, both eigenvalue and eigenfunction are free from the truncation height. This is conceivable, since the vertical motion associated with this mode is a simple up-and-down motion of the disk plane in the vertical direction (i.e., u_z is independent of η, see also Fig. 4.2).

4.2.4 Isothermal Disks with Toroidal Magnetic Fields

Disk rotations are generally differential, and turbulent magnetic fields generated by the magneto-rotational instability (MRI; Balbus and Hawley 1991) are stretched in the azimuthal direction. In this sense, it will be natural to suppose that global magnetic fields in equilibrium disks are mainly toroidal. Here, for simplicity, we consider the case where magnetic fields in equilibrium state are purely toroidal and stratified so that the Alfvén speed, c_A, is constant in the vertical direction, i.e., $(B_0^2/4\pi\rho_0)^{1/2} = \text{const.}$ with $\boldsymbol{B}_0 = [0, B_0(r,z), 0]$.

The hydrostatic balance in the vertical direction in the equilibrium disk is given by

$$-\frac{1}{\rho_0}\frac{d}{dz}\left(p_0 + \frac{B_0^2}{8\pi}\right) = \Omega_\perp^2 z. \tag{4.26}$$

Since both c_s and c_A are assumed to be constant in the vertical direction, equation (4.26) is integrated to give

$$\rho_0(r,z) = \rho_{00}(r)\exp\left(-\frac{z^2}{2H^2}\right) \quad \text{and} \quad B_0(r,z) = B_{00}(r)\exp\left(-\frac{z^2}{4H^2}\right), \tag{4.27}$$

where the scale height, H, is related to c_s, c_A, and $\Omega_\perp$ by

$$H^2(r) = \frac{c_s^2 + c_A^2/2}{\Omega_\perp^2}, \tag{4.28}$$

and the ratio between c_s and c_A is an arbitrary constant, which is a parameter specifying the disk structure.

We consider a small amplitude MHD perturbation over the above equilibrium disks. The velocity perturbations over rotation are denoted by (u_r, u_φ, u_z), and the perturbed part of the magnetic fields over the unperturbed one by (b_r, b_φ, b_z). In order to demonstrate the essential part of vertical oscillations, we assume that the velocity perturbations are mainly vertical, i.e., u_r and u_φ are negligible compared with u_z. The perturbations are not always axisymmetric, but the azimuthal wavenumber m is taken to be not many. Then, as easily understood from the frozen-in relation, b_r and b_z can be neglected in the lowest order of approximations [see the MHD equations in Appendix A]. However, b_φ cannot be neglected, since compression and expansion of gases in the vertical direction induces a change of azimuthal magnetic fields, i.e., $b_\varphi \neq 0$. Since this variation of b_φ is governed by the azimuthal component of induction equation, which is in the framework of the present approximation [see equation (A.47)]

$$i\tilde{\omega}\frac{b_\varphi}{B_0} = -\left(\frac{d}{dz} - \frac{z}{2H^2}\right)u_z. \tag{4.29}$$

The feedback of b_φ on u_z is governed by the equation of motion of u_z. In the present framework of approximations we can adopt [see equation (A.45)]

$$i\tilde{\omega}u_z = -\left(\frac{d}{dz} + \frac{c_{\rm A}^2}{2c_{\rm s}^2}\frac{z}{H^2}\right)h_1 - c_{\rm A}^2\left(\frac{d}{dz} - \frac{z}{H^2}\right)\left(\frac{b_\varphi}{B_0}\right). \tag{4.30}$$

Finally, in the framework of the present approximation, the equation of continuity is written as [see equation (A.42)]

$$i\tilde{\omega}h_1 = -c_{\rm s}\left(\frac{d}{dz} - \frac{z}{H^2}\right)u_z. \tag{4.31}$$

Substituting h_1 and b_φ given by equations (4.31) and (4.29) into equation (4.30) under the use of equation (4.28), we have a wave equation of u_z expressed in terms of the vertical coordinate z. If the coordinate is changed from z to the dimensionless coordinate η defined by $\eta = z/H$, we have finally

$$\left[\frac{d^2}{d\eta^2} - \eta\frac{d}{d\eta} + \frac{\tilde{\omega}^2 - \Omega_\perp^2}{c_{\rm s}^2 + c_{\rm A}^2}H^2\right]u_z = 0, \tag{4.32}$$

where H is related to $\Omega_\perp$ by equation (4.28). Equation (4.32) is an extension of equation (4.23), i.e., in the limit of $c_{\rm A}^2 = 0$, equation (4.32) is reduced to equation (4.23).

Equation (4.32) is a Hermite-type equation. The boundary conditions at infinity ($\eta = \pm\infty$) requires that the last term of equation (4.32) must be zero or a positive integer:

$$\frac{\tilde{\omega}_n^2 - \Omega_\perp^2}{c_{\rm s}^2 + c_{\rm A}^2}H^2 = n - 1, \qquad (n = 1, 2, 3, \ldots) \tag{4.33}$$

and $u_z \propto \mathscr{H}_{n-1}(\eta)$. The order of the Hermite polynomials, $n-1$, represents the number of node of u_z in the vertical direction. The reason why the node number is represented by $n-1$ (not n) is that the node number of u_z is generally less than that of h_1 by unity[3] (see also the next paragraph) and we classify modes by the node number of h_1.

Derivation of an wave equation with respect to h_1 from the set of equations (4.29), (4.30), and (4.31) is somewhat troublesome compared with the derivation of wave equation with respect to u_z, but the results show that

$$\left[\frac{d^2}{d\eta^2} - \eta\frac{d}{d\eta} + \frac{\tilde{\omega}^2 + (c_A^2/2c_s^2)(1 + c_A^2/2c_s^2)^{-1}\Omega_\perp^2}{c_s^2 + c_A^2}H^2\right]h_1 = 0. \tag{4.34}$$

By the reason mentioned in the previous paragraph, we adopt

$$\frac{\tilde{\omega}_n^2 + (c_A^2/2c_s^2)(1 + c_A^2/2c_s^2)^{-1}\Omega_\perp^2}{c_s^2 + c_A^2}H^2 = n. \tag{4.35}$$

We can really show that equations (4.33) and (4.35) are identical, and the frequency of the trapped oscillations is given by

$$\frac{\tilde{\omega}_n^2}{\Omega_\perp^2} = \left[\frac{c_s^2 + c_A^2}{c_s^2 + c_A^2/2}\right](n-1) + 1. \tag{4.36}$$

In the limit of $c_A = 0$, this equation is reduced to equation (4.16).

In the case of $n = 1$, the disk gas moves up and down with no node in the vertical direction (i.e., u_z is independent of z) and $\tilde{\omega}^2 = \Omega_\perp^2$, as equations (4.36) shows. In other cases of $n \geq 2$, $\tilde{\omega}^2$ increases with increase of c_A^2/c_s^2. This can be understood since the vertical oscillations considered here are the fast mode among the three MHD oscillations.

Finally, we should mention possible selections of particular vertical oscillation modes by their being coupled with horizontal motions. The vertical oscillations considered in this chapter are local and present at any radius of disks with different frequencies. In real situations, however, these vertical oscillations cannot be purely vertical: Horizontal motions are inevitably induced in each of these oscillations, because the pressure variation associated with them induces horizontal motions. In some cases these horizontal motions will be trapped in radial region of disks (see Chaps. 6 and 7 for trapping). Hence, we may have nearly vertical normal oscillation modes with discrete frequencies. Such oscillations might be possible candidates of quasi-periodic oscillations observed in LMXBs and others. This possibility will be examined in Sect. 7.3 with no magnetic field and in Sect. 8.1 with toroidal magnetic fields.

---

[3]For example, equation (4.31) shows that when $u_z \propto \mathscr{H}_{n-1}$ we have $h_1 \propto \mathscr{H}_n$.

References

Balbus, S. A., & Hawley, J. F. 1991, Astrophys. J., 376, 214
Silbergleit, A. S., Wagoner, R. V., & Ortega-Rodrígues, M. 2001, Astrophys. J., 548, 335
Kato, S. 2005, Publ. Astron. Soc. Jpn., 57, 699
Kato, S. 2010, Publ. Astron. Soc. Jpn., 62, 635
Kato, S. 2012, Publ. Astron. Soc. Jpn, 64,78
Okazaki, A. T., Kato, S., & Fukue, J. 1987, Publ. Astron. Soc. Jpn., 39, 457

Chapter 5
Disk Oscillations in Radial Direction

Abstract In Chap. 4, we have considered purely vertical oscillations in order to examine some essential characteristics of disk oscillations. Purely vertical oscillations, however, cannot be normal modes of oscillations, because the pressure variations associated with the oscillations inevitably induce horizontal motions. In addition to these nearly vertical oscillations, disk have nearly horizontal oscillations. The latter oscillations also cannot be purely horizontal, because the pressure forces associated with the oscillations induce vertical motions. Disk oscillations thus cannot consist of horizontal nor vertical motions alone. In this chapter, we examine these couplings, emphasizing the resulting behaviors of horizontal motions. In this chapter, however, we are mainly interested in the cases where the coupling is weakly radial-dependent. The cases where the coupling is strongly radial-dependent will be argued in Sect. 8.2.

Keywords Wave equation • Perturbation method • Radial variation of disk thickness

5.1 Approximations for Driving Radial Wave Equations

The basic equation describing disk oscillations is equation (3.50). This is a partial differential equation, and it is hard to solve it rigorously. In solving equation (3.50) we introduce the approximation that the density, $\rho_0(r, z)$, is stratified in the vertical direction as $\rho_0(r, z) = \rho_{00}(r)\exp(-\eta^2/2)$ [isothermal distribution, see equations (4.5) and (4.6)], where $\eta = z/H(r)$. The radial variations of $\rho_{00}(r)$ and $H(r)$ are taken arbitrary. If the disk is truncated at a certain height, z_s, we introduce a dimensionless quantity, η_s, defined by $\eta_s = z_s/H$, and η_s is taken as a parameter. The supposition that the above approximate procedure is acceptable is based on the following considerations.

Examination in Chap. 4 shows that the vertical structure of disks has no strong influences on purely vertical oscillations, as far as the node number n is small, e.g., $n = 0, 1$, and 2.[1] For example, in the case of $n = 0$, there is no motion in the

[1] We are really interested in oscillations with small number of n.

S. Kato, *Oscillations of Disks*, Astrophysics and Space Science Library 437,
DOI 10.1007/978-4-431-56208-5_5

vertical direction in the lowest approximations, and thus the horizontal motions is independent of vertical structure of disks in the lowest order of approximations. In the case of $n = 1$, the frequency of purely vertical oscillations is $\Omega_\perp$ when observed from the corotating frame (i.e., $\tilde{\omega} = \Omega_\perp$), which is independent of vertical disk structure. Furthermore, the corresponding eigenfunction is given by $h_1 \propto z$, which is also independent of vertical disk structure. In the case of $n = 2$, eigenfrequency and eigenvalue depend a little on vertical disk structure, but not much.[2] Really, the final results of analyses based on the above approximation show that the most important factors for determining wave trapping in the radial direction are the radial distributions of rotation and epicyclic frequencies. The vertical disk structure is less important for wave trapping in the radial direction.

The independent variables (r, z) are now changed to the set of (r, η). Then, $\partial/\partial r$ and $\partial/\partial z$ are changed to

$$\frac{\partial}{\partial r} \to \frac{\partial}{\partial r} - \eta \frac{d\ln H}{dr}\frac{\partial}{\partial \eta}, \quad \frac{\partial}{\partial z} \to \frac{1}{H}\frac{\partial}{\partial \eta}. \tag{5.1}$$

We now rewrite the basic equation (3.50) to, using $\rho_0(r, z) = \rho_{00}(r)\exp(-\eta^2/2)$,

$$\left(\frac{\partial^2}{\partial \eta^2} - \eta\frac{\partial}{\partial \eta}\right)h_1 + H^2\mathscr{L}_0(h_1) + \frac{d\ln H}{d\ln r}H^2\mathscr{L}_1(h_1) + \left(\frac{d\ln H}{d\ln r}\right)^2 H^2\mathscr{L}_2(h_1) = 0, \tag{5.2}$$

where operators, $\mathscr{L}_0$, $\mathscr{L}_1$, and $\mathscr{L}_2$, are defined, respectively, by

$$\begin{aligned}\mathscr{L}_0(h_1) = &\left(\frac{\tilde{\omega}^2}{c_s^2} - \frac{m^2}{r^2}\right)h_1 + \tilde{\omega}\frac{\partial}{\partial r}\left[\frac{\tilde{\omega}}{\tilde{\omega}^2 - \kappa^2}\left(\frac{\partial}{\partial r} - \frac{2m\Omega}{r\tilde{\omega}}\right)\right]h_1 \\ &+ \frac{\tilde{\omega}^2}{\tilde{\omega}^2 - \kappa^2}\left[\frac{d\ln r\rho_{00}}{dr} + \frac{m\kappa^2}{2r\tilde{\omega}\Omega}\right]\left(\frac{\partial}{\partial r} - \frac{2m\Omega}{r\tilde{\omega}}\right)h_1,\end{aligned} \tag{5.3}$$

$$\mathscr{L}_1(h_1) = \frac{\tilde{\omega}^2}{\tilde{\omega}^2 - \kappa^2}\left(\eta\frac{\partial}{\partial \eta}\mathscr{L}_{10}(h_1) + \eta^2\mathscr{L}_{12}(h_1)\right), \tag{5.4}$$

and

$$\mathscr{L}_2(h_1) = \frac{\tilde{\omega}^2}{\tilde{\omega}^2 - \kappa^2}\frac{1}{r^2}\left(\eta^2\frac{\partial^2}{\partial \eta^2} - \eta^3\frac{\partial}{\partial \eta}\right)h_1. \tag{5.5}$$

[2]Eigenvalues of vertical oscillations in vertically polytropic disks are similar to those in vertically truncated isothermal disks, if a relevant one to one correspondence is made by introducing η_s. Thus, the above-mentioned approximate procedure (the approximation of vertically isothermal disks) will be valid even if the vertical density distribution is not exactly isothermal.

Hereafter, m^2/r^2 in the first parentheses on the right-hand side of equation (5.3) is neglected compared with $\tilde{\omega}^2/c_s^2$.

In equation (5.4) operators $\mathscr{L}_{10}$ and $\mathscr{L}_{12}$ are introduced, which are defined, respectively, by

$$\mathscr{L}_{10}(h_1) = -\frac{2}{r}\frac{\partial h_1}{\partial r} + \frac{1}{r}\left[\frac{d}{dr}\ln\left(\frac{\tilde{\omega}^2 - \kappa^2}{\rho_{00}}\right)\right]h_1, \tag{5.6}$$

$$\mathscr{L}_{12}(h_1) = \frac{1}{r}\left(\frac{\partial}{\partial r} - \frac{2m\Omega}{r\tilde{\omega}}\right)h_1. \tag{5.7}$$

We solve equation (5.2) by two approximate methods.

5.1.1 Perturbation Method

The first is a perturbation method. In the limiting case where the disk thickness, H, is constant, i.e., $d\ln H/d\ln r = 0$, equation (5.2) is reduced to

$$\left(\frac{\partial^2}{\partial \eta^2} - \eta\frac{\partial}{\partial \eta}\right)h_1 + H^2\mathscr{L}_0(h_1) = 0. \tag{5.8}$$

This equation can be solved by separating $h_1(r, \eta)$ as

$$h(r, \eta) = f(r)g(\eta). \tag{5.9}$$

This separation shows that equation (5.2) is reduced to

$$\frac{1}{g(\eta)}\left(\frac{d^2}{d\eta^2} - \eta\frac{d}{d\eta}\right)g(\eta) = -K, \tag{5.10}$$

and

$$\frac{1}{f}H^2\mathscr{L}_0(f) = K, \tag{5.11}$$

where K is a separation constant, and should be determined later from boundary conditions. Equations (5.10) and (5.11) are ordinaryy differential equations, and thus we can solve them easily.

Starting from the set of equations (5.10) and (5.11), we can solve equation (5.2) by a perturbation method, taking $d\ln H/d\ln r$ as a small expansion parameter. This process is presented in Sect. 5.2. A further development of the method will be given in Sect. 8.2.3.

5.1.2 Galerkin's Method

Another method we adopt here to solve equation (5.2) is a kind of Galerkin's method. The variable $h_1(r,\eta)$ is approximated by a finite sum of expansion functions, $g_i(\eta)$ $(i = 1, 2, \ldots .N)$:

$$h_1(r,\eta) = \sum_{i=0}^{i=N} f_i(r) g_i(\eta), \tag{5.12}$$

where $g_i(\eta)$'s $(i = 1, 2, \ldots N)$ are a set of functions which are chosen arbitrary but under the expectation that equation (5.12) becomes a good approximation of the real solution of $h_1(r,\eta)$. The coefficients $f_i(r)$'s are determined later so that equation (5.12) can well approximate the real $h_1(r,\eta)$.

The above expression for $h_1(r,\eta)$ is substituted into equation (5.2). The resulting equation is integrated over the vertical direction after being multiplied by some given trial functions $\tilde{g}_i(\eta)(i = 1, 2, \ldots M)$. We have then a set of simultaneous ordinary differential equations with respect to $f_i(r)$ $(i = 1, 2, \ldots N)$. The number of equations is M. If the number M is taken to be equal to N, the number of equations and that of functions $f_i(r)$'s become the same, and thus we can have an approximate form of $h_1(r,\eta)$. If $g_i(\eta)$ and $\tilde{g}_i(\eta)$ are taken properly, the resulting $h_1(r,\eta)$ will be a good approximation of the real $h_1(r,\eta)$.

This is the essence of Galerkin's method. An approximate solution derived by this method will be presented in Sect. 5.3. A further development of the method will be presented in Sect. 8.2.4.

5.2 Wave Equation Derived by Perturbation Method

First, we consider wave equations expressing the radial variation of h_1 by a perturbation method, where $d\ln H/d\ln r$ is taken to be a small expansion parameter.

5.2.1 Wave Equation in the Limit of $d\ln H/d\ln r = 0$

In the limit of $d\ln H/d\ln r = 0$, we have a set of equations (5.10) and (5.11), as mentioned before. Equation (5.10) is solved by imposing boundary conditions at $\eta = \pm\infty$, which leads to

$$g(\eta) = \mathscr{H}_n(\eta) \quad \text{and} \quad K = n, \tag{5.13}$$

where $\mathscr{H}_n$ is the Hermite polynomial and n is zero or positive integer, giving the node number of $g(\eta)$ in the vertical direction.[3] Then, putting $K = n$ we can derive from equation (5.11) a wave equation describing radial motions of oscillations, which is[4]

$$\frac{\tilde{\omega}^2-\kappa^2}{\tilde{\omega}}\frac{d}{dr}\biggl[\frac{\tilde{\omega}}{\tilde{\omega}^2-\kappa^2}\biggl(\frac{df}{dr}-\frac{2m\Omega}{r\tilde{\omega}}f\biggr)\biggr]$$

$$\biggl(\frac{d\ln r\rho_{00}}{dr}+\frac{m\kappa^2}{2r\tilde{\omega}\Omega}\biggr)\biggl(\frac{df}{dr}-\frac{2m\Omega}{r\tilde{\omega}}f\biggr)+\frac{(\tilde{\omega}^2-\kappa^2)(\tilde{\omega}^2-n\Omega_\perp^2)}{\tilde{\omega}^2c_s^2}f=0. \qquad (5.14)$$

In deriving this equation, m^2/r^2 in equation (5.3) has been neglected compared with $\tilde{\omega}^2/c_s^2$. Equation (5.14) is a second order ordinary differential equation and describes radial behavior of $f(r)$.

If we consider oscillations whose radial wavelength is so short that radial variations of all unperturbed disk quantities (except for the radial variations of Ω and c_s) are neglected, equation (5.14) is reduced to

$$\frac{\tilde{\omega}^2-\kappa^2}{\tilde{\omega}}\frac{d}{dr}\biggl(\frac{\tilde{\omega}}{\tilde{\omega}^2-\kappa^2}\frac{df}{dr}\biggr)+\frac{(\tilde{\omega}^2-\kappa^2)(\tilde{\omega}^2-n\Omega_\perp^2)}{\tilde{\omega}^2c_s^2}f=0. \qquad (5.15)$$

This equation is the most simplified wave equation describing the radial variation of $f(r)$.

5.2.2 *Wave Equation Till the Order of* ($d\ln H/d\ln r$)

Next, we proceed to the case where $d\ln H/d\ln r \neq 0$, but it can be taken to be a small expansion parameter. In the case of $d\ln H/d\ln r \neq 0$, by the presence of $\mathscr{L}_1$ and $\mathscr{L}_2$, equation (5.2) cannot be separated into two ordinary differential equations. This means that the radial-dependent couplings between vertical and horizontal motions occur through the term of $d\ln H/d\ln r$. The purpose here is to derive ordinary differential equations under the approximation that the radial-dependent couplings are weak.

Assuming the effects of $d\ln H/d\ln r \neq 0$ on wave equation is weak, we approximately separate $h_1(r,\eta)$ as

$$h_1(r,\eta) = f(r)g(\eta,r) \qquad (5.16)$$

---

[3] In the cases of vertically truncated disks, K is different from n (see Chap. 4).

[4] In the disks we are considering, there is no distinction among Ω, Ω_K, and $\Omega_\perp$. However, in order to emphasize the origin, K/H^2 is written hereafter as $n\Omega_\perp^2/c_s^2$ (not $n\Omega^2/c_s^2$).

with a weak *radial*-dependence of g. This means that the separation constant, K, has now a weak radial dependence.

First, we divide equation (5.2) by fg to lead to

$$\begin{aligned}
&\frac{1}{g}\left(\frac{\partial^2 g}{\partial\eta^2}-\eta\frac{\partial g}{\partial\eta}\right)\\
&\quad+\frac{\tilde{\omega}^2}{\tilde{\omega}^2-\kappa^2}\frac{d\ln H}{d\ln r}\frac{H^2}{r^2}\left[\eta\frac{\partial g}{g\partial\eta}\frac{r^2}{f}\mathscr{L}_{10}(f)+\eta^2\frac{r^2}{f}\mathscr{L}_{12}(f)\right]\\
&\quad+\frac{\tilde{\omega}^2}{\tilde{\omega}^2-\kappa^2}\frac{H^2}{r^2}\left[\frac{r^2}{g}\frac{\partial^2 g}{\partial r^2}+\frac{r^2}{g}\frac{\partial g}{\partial r}\frac{d}{dr}\ln\left(\frac{r\rho_{00}}{\tilde{\omega}^2-\kappa^2}f^2\right)\right]\\
&\quad+\frac{\tilde{\omega}^2}{\tilde{\omega}^2-\kappa^2}\frac{d\ln H}{d\ln r}\frac{H^2}{r^2}\left(-2\eta\frac{r}{g}\frac{\partial g}{\partial r}+\eta^2\frac{r}{g}\frac{\partial g}{\partial r}\right)\\
&\quad+\frac{\tilde{\omega}^2}{\tilde{\omega}^2-\kappa^2}\left(\frac{d\ln H}{d\ln r}\right)^2\frac{H^2}{r^2}\left(\eta^2\frac{\partial^2 g}{g\partial\eta^2}-\eta^3\frac{\partial g}{g\partial\eta}\right)=-K(r),
\end{aligned}\tag{5.17}$$

and

$$H^2\frac{1}{f}\mathscr{L}_0(f)=K(r).\tag{5.18}$$

It is noted here that in the limit of $d\ln H/d\ln r=0$, the separation constant K is really a constant. In disks with $d\ln H/d\ln r\neq 0$, however, no exact separation is possible, but the r-dependence of K can be taken to be weak when $d\ln H/d\ln r$ is small.

To emphasize that we are now interested in the n-th mode of oscillations in the vertical direction, h_1 is expressed by attaching n as

$$h_1=f_n(r)g_n(\eta,r),\tag{5.19}$$

and $g_n(\eta,r)$ and K are expanded as

$$g_n(\eta,r)=g_n^{(0)}(\eta)+g_n^{(1)}(r,\eta)+\dots\quad\text{and}\quad K_n(r)=K_n^{(0)}+K_n^{(1)}(r)+\dots\tag{5.20}$$

The zeroth order solutions are already given in Sect. 5.2.1:

$$g_n^{(0)}(\eta)=\mathscr{H}_n(\eta)\quad\text{and}\quad K_n^{(0)}=n.\tag{5.21}$$

Let us proceed to the next order approximation. From equation (5.17) we have

$$\frac{\partial^2 g_n^{(1)}}{\partial \eta^2} - \eta \frac{\partial g_n^{(1)}}{\partial \eta} + n g_n^{(1)}$$
$$= -\frac{\tilde{\omega}^2}{\tilde{\omega}^2 - \kappa^2} \frac{d\ln H}{d\ln r} \frac{H^2}{r^2} \left(\frac{r^2}{f} \mathscr{L}_{10}(f) \eta \frac{\partial g_n^{(0)}}{\partial \eta} + \frac{r^2}{f} \mathscr{L}_{12}(f) \eta^2 g_n^{(0)} \right)$$
$$- \frac{\tilde{\omega}^2}{\tilde{\omega}^2 - \kappa^2} \frac{H^2}{r^2} \left[r^2 \frac{\partial^2 g_n^{(1)}}{\partial r^2} + r^2 \frac{\partial g_n^{(1)}}{\partial r} \frac{d}{dr} \ln\left(\frac{r\rho_{00}}{\tilde{\omega}^2 - \kappa^2} f^2 \right) \right]$$
$$- K_n^{(1)}(r) g_n^{(0)}. \tag{5.22}$$

In deriving the right-hand side of equation (5.22), we have used $\partial g_n^{(0)}/\partial r = 0$. Equation (5.22) is solved by expanding $g_n^{(1)}(r, \eta)$ as

$$g_n^{(1)}(r, \eta) = \sum_{m \neq n} a_m^{(1)}(r) g_m^{(0)}(\eta), \tag{5.23}$$

where $g_m^{(0)}(\eta)$ is the Hermite polynomial, i.e., $g_m^{(0)}(\eta) = \mathscr{H}_m(\eta)$.

From the orthogonal relation of the Hermite polynomials we have

$$\int_{-\infty}^{+\infty} \exp\left(-\frac{\eta^2}{2} \right) g_n^{(0)}(\eta) g_m^{(0)}(\eta) d\eta = n!(2\pi)^{1/2} \delta_{nm}. \tag{5.24}$$

Furthermore, we have the recurrent formulae:

$$g_{m+1}^{(0)}(\eta) - \eta g_m^{(0)}(\eta) + m g_{m-1}^{(0)}(\eta) = 0 \tag{5.25}$$

and

$$\frac{d}{d\eta} g_m^{(0)}(\eta) = m g_{m-1}^{(0)}(\eta). \tag{5.26}$$

The solvability condition of equation (5.22) is that the right-hand side of the equation has no component proportional to $g_n^{(0)}$. Using the orthogonal relation (5.24), we find from equation (5.22) that the condition is

$$K_n^{(1)} f_n + \frac{\tilde{\omega}^2}{\tilde{\omega}^2 - \kappa^2} \frac{d\ln H}{d\ln r} \frac{H^2}{r^2} \left(n r^2 \mathscr{L}_{10}(f_n) + (2n+1) r^2 \mathscr{L}_{12}(f_n) \right)$$
$$+ \frac{\tilde{\omega}^2}{\tilde{\omega}^2 - \kappa^2} \frac{H^2}{r^2} \left[r^2 \frac{d^2 a_n^{(1)}}{dr^2} + \left\{ \frac{d}{dr} \ln\left(\frac{r\rho_{00} f_n^2}{\tilde{\omega}^2 - \kappa^2} \right) \right\} r \frac{d a_n^{(1)}}{dr} \right] f_n = 0. \tag{5.27}$$

In this equation $a_n^{(1)}$ appears, but there is no restriction on $a_n^{(1)}$ in the perturbation method. It can be taken arbitrary. Here, we take it to be zero as is done in usual perturbation methods, since the term of $a_n^{(1)}g_n^{(0)}$ has the same η-dependence as $g_n^{(0)}$ and can be included in the zero-th order solution, $g_n^{(0)}$. Then, substituting $K_n^{(1)}$ obtained from equation (5.27) into equation (5.18), we have an equation for f_n, which is

$$\frac{\tilde{\omega}^2-\kappa^2}{\tilde{\omega}}\frac{d}{dr}\biggl[\frac{\tilde{\omega}}{\tilde{\omega}-\kappa^2}\biggl(\frac{d}{dr}-\frac{2m\Omega}{r\tilde{\omega}}\biggr)\biggr]f_n+\frac{(\tilde{\omega}^2-\kappa^2)(\tilde{\omega}^2-n\Omega_\perp^2)}{c_s^2\tilde{\omega}^2}f_n$$
$$+\biggl(\frac{d\ln r\rho_{00}}{dr}+\frac{m\kappa^2}{2r\tilde{\omega}\Omega}\biggr)\biggl(\frac{d}{dr}-\frac{2m\Omega}{r\tilde{\omega}}\biggr)f_n$$
$$+\frac{d\ln H}{d\ln r}\biggl(n\mathscr{L}_{10}+(2n+1)\mathscr{L}_{12}\biggr)f_n=0. \tag{5.28}$$

This is the wave equation describing the radial behavior of oscillations, when the effecters of $d\ln \mathrm{H}/d\ln r$ are taken into account as a small quantity. Equation (5.28) is an extension of equation (5.14). The last term on the right-hand side of equation (5.28) is the term resulting from the effects of $d\ln H/d\ln r\neq 0$.

The remaining problem is to determine the coefficients $a_m^{(1)}(r)$ $(m\neq n)$ of expansion of $g_n^{(1)}$ [see equation (5.23)]. This is done by integrating equation (5.22) times $\exp(-\eta^2/2)g_m^{(0)}$ over $\eta=-\infty$ to $\eta=\infty$. The results are

$$(n-m)a_m^{(1)}+\frac{\tilde{\omega}^2}{\tilde{\omega}^2-\kappa^2}\frac{d\ln H}{d\ln r}\frac{H^2}{r^2}\biggl[(m+1)(m+2)\frac{r^2}{f_n}\mathscr{L}_{10}(f)\delta_{m,n-2}$$
$$+\frac{r^2}{f_n}\mathscr{L}_{12}(f)\biggl(\delta_{m,n+2}+(m+1)(m+2)\delta_{m,n-2}\biggr)\biggr]$$
$$+\frac{\tilde{\omega}^2}{\tilde{\omega}^2-\kappa^2}\frac{H^2}{r^2}\biggl[r^2\frac{d^2a_m^{(1)}}{dr^2}+\biggl\{\frac{d}{dr}\ln\biggl(\frac{r\rho_{00}f_n^2}{\tilde{\omega}^2-\kappa^2}\biggr)\biggr\}r\frac{da_m^{(1)}}{dr}\biggr]=0. \tag{5.29}$$

This is an second order ordinary differential equation for $a_m^{(1)}$, where $m=n\pm 2$. The expansion coefficients, $a_m^{(1)}$, is obtained by solving this differential equation for $a_m^{(1)}$ after f_n derived from equation (5.28) is substituted into this equation. Equation (5.29) for $a_m^{(1)}$ $(m=n\pm 2)$ is, however, unnecessary for solving the differential equation for f_n till the order of $(d\ln H/d\ln r)^1$ [i.e., equation (5.28)], but it becomes necessary if we want to solve one more higher order equation with respect to $d\ln H/d\ln r$ (see Sect. 8.2).

5.2.3 Wave Equation Expressed in Terms of u_r

In Sects. 5.2.1and 5.2.2, we have derived wave equations with respect to h_1, i.e., equations (5.14) and (5.28). In the case of p-mode oscillations of $n = 0$, for example, one of boundaries of the trapped region is at $\tilde{\omega}^2 - \kappa^2 = 0$ (Lindblad resonances) or close to it, as mentioned before. In equation (5.28) the radius of the Lindblad resonances is the points of apparent singularity. Hence, it will be relevant to use another variable to describe oscillations so that the apparent singularity does not appear in wave equation. For this purpose it is convenient to use u_r as an independent variable instead of h_1.[5]

Variables h_1 and u_r are related by [e.g., equation (3.48)]

$$\frac{\partial h_1}{\partial r} - \frac{2m\Omega}{r\tilde{\omega}}h_1 = \frac{\tilde{\omega}^2 - \kappa^2}{i\tilde{\omega}}u_r. \tag{5.30}$$

Hence, as a function representing the radial variation of u_r, we introduce f_u defined by

$$u_r(r,\eta) = f_u(r)g(\eta, r). \tag{5.31}$$

Equation (5.30) then shows that f and f_u are related by

$$f_u = \frac{i\tilde{\omega}}{\tilde{\omega}^2 - \kappa^2}\left(\frac{df}{dr} - \frac{2m\Omega}{r\tilde{\omega}}f\right). \tag{5.32}$$

Using equation (5.32) we can write equation (5.28) in the form

$$iA(r)f = \tilde{\omega}\frac{df_u}{dr} + \tilde{\omega}\left[\frac{d}{dr}\mathrm{ln}\left(r\rho_{00}H^{n+1}\right) + \frac{m\kappa^2}{2r\tilde{\omega}\Omega}\right]f_u = 0, \tag{5.33}$$

where $A(r)$ is defined by

$$A(r) = \frac{\tilde{\omega}^2 - n\Omega_\perp^2}{c_s^2} + \frac{n\tilde{\omega}^2}{\tilde{\omega}^2 - \kappa^2}\frac{d\mathrm{ln}H}{d\mathrm{ln}r}\frac{1}{r^2}\left[r\frac{d}{dr}\mathrm{ln}\left(\frac{\tilde{\omega}^2 - \kappa^2}{\rho_{00}H^{2n+1}}\right) - \frac{2m\Omega}{\tilde{\omega}}\right]. \tag{5.34}$$

If equation (5.33) is multiplied by $[\tilde{\omega}/(\tilde{\omega}^2 - \kappa^2)](d/dr - 2m\Omega/r\tilde{\omega})$ after divided by A in order to eliminate f and to obtain an equation of f_u, we have, after some

[5] In this subsection, the subscript n to be attached to f and f_u is neglected in order to avoid complexity.

manipulation,

$$\frac{d^2f_u}{dr^2}+\frac{d}{dr}\ln\left(\frac{r\rho_{00}H^{n+1}}{A}\right)\frac{df_u}{dr}$$
$$+B\left[\frac{d}{dr}\ln\left(\frac{B}{A}\right)-\frac{m\kappa^2}{2r\tilde{\omega}\Omega}\right]f_u+A\frac{\tilde{\omega}^2-\kappa^2}{\tilde{\omega}^2}f_u=0, \qquad (5.35)$$

where

$$B(r)=\frac{d}{dr}\ln\left(r\rho_{00}H^{n+1}\right)+\frac{m\kappa^2}{2r\tilde{\omega}\Omega}. \qquad (5.36)$$

Equation (5.35) is the equation corresponding to equation (5.28), and a basic equation describing oscillations in terms of f_u, i.e., u_r.

It is noted that in the limit of $d\ln H/d\ln=0$, equation (5.35) is reduced to

$$\left(\frac{d}{dr}-\frac{2m\Omega}{r\tilde{\omega}}\right)\left[\frac{\tilde{\omega}c_{\rm s}^2}{\tilde{\omega}^2-n\Omega_\perp^2}\left(\frac{d}{dr}+\frac{d\ln r\rho_0}{dr}+\frac{m\kappa^2}{2r\tilde{\omega}\Omega}\right)f_u\right]+\frac{\tilde{\omega}^2-\kappa^2}{\tilde{\omega}}f_u=0. \qquad (5.37)$$

This is the equation corresponding to equation (5.14). In equation (5.14) the wave equation is written in terms of f, while it is written in equation (5.37) in terms of f_u.

If the radial wavelength of oscillations is so short that radial variations of all unperturbed quantities except for those related to rotation and $c_{\rm s}$ are neglected, equation (5.37) is further reduced to

$$\frac{1}{\tilde{\omega}}\frac{d}{dr}\left(\frac{\tilde{\omega}c_{\rm s}^2}{\tilde{\omega}^2-n\Omega_\perp^2}\frac{df_u}{dr}\right)+\frac{\tilde{\omega}^2-\kappa^2}{\tilde{\omega}^2}f_u=0. \qquad (5.38)$$

This is the equation corresponding to equation (5.15), expressing the wave equation in terms of f_u.

5.3 Wave Equation Derived by Galerkin's Method

As mentioned in Sect. 5.1, we approximate $h_1(r,\eta)$ in the form of equation (5.12). As the expansion functions, $g_i(\eta)$, we adopt here a series of the Hermite polynomials, $\mathscr{H}_i(\eta)$, $(i=1,2\ldots N)$. Furthermore, as trial functions, we adopt $\exp(-\eta^2/2)\mathscr{H}_i(\eta)(i=1,2,\ldots M)$ by the following reasons. The vertical density distribution in the unperturbed disks has been approximated in the form of $\rho_0(r,\eta)=\rho_{00}(r)\exp(-\eta^2/2)$. Related to this, adoption of Hermite polynomials is convenent because their orthogonal relation can be used with simple results.

Let us now consider the radial behavior of oscillations which have n node(s) in the vertical direction, where $n = 0, 1, 2, \ldots$. As the first step of application of Galerkin's method, we approximate the mode by a single term as (i.e., $N = 1$)[6]

$$h_1(r, \eta) = f(r)\mathscr{H}_n(\eta). \tag{5.39}$$

Equation (5.39) is substituted into equation (5.2) and the resulting equation is integrated from $\eta = -\infty$ to $\eta = \infty$ after being multiplied by $\exp(-\eta^2/2)\mathscr{H}_n(\eta)$. Then, we have, after manipulations,

$$\begin{aligned}&\frac{\tilde{\omega}^2-\kappa^2}{\tilde{\omega}}\frac{d}{dr}\left[\frac{\tilde{\omega}}{\tilde{\omega}-\kappa^2}\left(\frac{d}{dr}-\frac{2m\Omega}{r\tilde{\omega}}\right)\right]f+\frac{(\tilde{\omega}^2-\kappa^2)(\tilde{\omega}^2-n\Omega_\perp^2)}{c_s^2\tilde{\omega}^2}f\\&+\left(\frac{d\mathrm{ln}r\rho_{00}}{dr}+\frac{m\kappa^2}{2r\tilde{\omega}\Omega}\right)\left(\frac{d}{dr}-\frac{2m\Omega}{r\tilde{\omega}}\right)f\\&+\frac{d\mathrm{ln}H}{d\mathrm{ln}r}\left(n\mathscr{L}_{10}+(2n+1)\mathscr{L}_{12}\right)f-\frac{n(n+2)}{r^2}\left(\frac{d\mathrm{ln}H}{d\mathrm{ln}r}\right)^2f=0.\end{aligned} \tag{5.40}$$

It is of interest to notice that the wave equation for f derived here [equation (5.40)] is almost the same as that derived by the perturbation method in Sect. 5.2.2 [equation (5.28)]. The difference is the appearance of the term proportional to $(d\mathrm{ln}H/d\mathrm{ln}r)^2$ in the present treatment. The term was absent in equation (5.28) derived by the perturbation method, because it is a higher order term with respect to $(d\mathrm{ln}H/d\mathrm{ln}r)$.

It is noted that Galerkin's method is powerful. The method can be applied, in principle, to cases where the vertical distribution of density, $\rho_0(r, z)$, is arbitrary. However, it is difficult to know the accuracy of the results obtained, unless many terms are adopted in expansion (5.12).

[6] A case of $N = 2$ is considered in Sect. 8.2.4.

Chapter 6
Classification of Oscillations and Their Characteristics

Abstract A rough but instructive way for understanding the basic characteristics of the wave motions described by equation (3.50) is to separate $h_1(r, \eta)$ into r- and η-dependent terms as $h_1(r, z) = f(r)g(\eta)$. Based on such approximations, we have derived in Chap. 5 wave equations describing radial behavior of oscillations. In this chapter, in the framework of the same approximations, we classify the wave motions and examine their characteristics. The oscillations in geometrically thin disks are classified into (i) p-modes, (ii) c-modes, (iii) vertical p-modes, and (iv) g-modes.

Keywords c-modes • g-modes • One-armed precession mode • p-modes • Tilt mode • Trapping of oscillations • Vertical p-modes

6.1 Classification by Local Approximations

The results in Chap. 5 show that the most simplified wave equation is equation (5.15), which is written here as

$$\tilde{\omega}\frac{d}{dr}\left(\frac{\tilde{\omega}}{\tilde{\omega}^2-\kappa^2}\frac{df}{dr}\right)+\frac{\tilde{\omega}^2-K_n\Omega_\perp^2}{c_s^2}f=0. \tag{6.1}$$

In deriving this equation, $h_1(\equiv p_1/\rho_0)$ associated with oscillations is separated as $h_1(r,\eta) = g(\eta)f(r)$. In equation (5.15), K_n is written as n, where n is the node number of oscillations in the vertical direction, i.e., $n = 0, 1, 2, \ldots$. Here, however, in order to emphasize applicability to cases where the disk gases are polytropic or the disks are terminated at a certain height, η_s, K_n has been retained without reducing to n (for K_n see Chap. 4).[1]

[1]In the case of polytropic disks, K_n is $n + n(n-1)/2N$ (see equation (4.13)), and in the case of truncated isothermal disks, K_n is given in Sect. 4.2.3 (see Table 4.1). In the followings, for simplicity, we restrict our attention mainly to the cases of isothermal disks which infinitely extend in the vertical direction, i.e., $K_n = n$.

S. Kato, *Oscillations of Disks*, Astrophysics and Space Science Library 437,
DOI 10.1007/978-4-431-56208-5_6

If wave perturbations are local in the radial direction with radial wavenumber, k_r, equation (6.1) becomes

$$(\tilde{\omega}^2 - \kappa^2)(\tilde{\omega}^2 - n\Omega_\perp^2) = k_r^2\tilde{\omega}^2 c_s^2. \tag{6.2}$$

This dispersion relation (6.2) for disk oscillations was first derived by Okazaki et al. (1987). The purpose of this section is to classify disk oscillations by using dispersion relation (6.2). In the Keplerian disks, we have $\kappa = \Omega_\perp = \Omega_K$, where Ω_K is the Keplerian angular velocity of disk rotation. In many cases, however, we are interested in relativistic disks. The general relativistic formulation of disk oscillations is very complicated. In order to pick up only the essential part of general relativity on oscillations, we adopt a conventional way that the Newtonian formulation is adopted under the use of the general relativistic expressions for κ and $\Omega_\perp$. That is, we use equation (6.2) even in the case of relativistic disks with relativistic expressions for κ and $\Omega_\perp$. Detailed general relativistic expressions for κ and $\Omega_\perp$ are given in Chap. 2 and their derivations are presented in Appendix B. We only note here that in the general relativity we have $\kappa < \Omega_\perp < \Omega_K$.

We are also interested in oscillations in tidally deformed disks. In tidally deformed disks, the disks of the primary is time-periodically deformed by tidal force of a secondary. When we consider disk oscillations of the primary star, the time-averaged disks are concerned. In time-averaged tidally deformed disks we have usually $\kappa < \Omega_K < \Omega_\perp$, when the orbital plane coincides with the disk plane (see Sect. 2.2).

In the limit of pressureless disks, $c_s = 0$, the dispersion relation (6.2) is decomposed into two relations:

$$\tilde{\omega}^2 - n\Omega_\perp^2 = 0 \qquad \text{and} \qquad \tilde{\omega}^2 - \kappa^2 = 0. \tag{6.3}$$

The former represents the purely vertical oscillations discussed in Chap. 4. In pressureless disks, however, only the fundamental mode of $n = 1$ is possible, since the overtones ($n = 2, 3, 4, \ldots$) are results of presence of pressure force. This oscillation of $\tilde{\omega}^2 = \Omega_\perp^2$ is the vertical epicyclic oscillation. The latter of equation (6.3) shows the horizontal epicyclic oscillation whose frequency is κ, i.e., inertial oscillations.

In real situations with pressure, the vertical oscillations inevitably involve pressure variations, which leads to horizontal motions. Even in the case of horizontal oscillations of $\tilde{\omega}^2 = \kappa^2$, situations are the same as in the case of vertical oscillations, i.e., vertical motions are inevitably associated with the horizontal motions. In other words, in both cases vertical oscillations and horizontal ones are essentially coupled through pressure variation, which leads to the dispersion relation (6.2).

Let us proceed to classification of oscillations. We consider three cases of $n = 0$, $n = 1$, and $n \geq 2$. In addition, we are particularly interested in low-frequency modes in Sect. 6.3

6.1.1 Oscillations with $n = 0$ (Inertial-Acoustic Mode or p-Mode)

In the case of $n = 0$, we have oscillation modes of $\tilde{\omega}^2 = \kappa^2 + k_r^2 c_s^2$ alone (see equation (6.2)). These modes are an extension of inertial oscillations ($\tilde{\omega}^2 = \kappa^2$) to cases where pressure effects are taken into account. That is, they are *inertial-acoustic oscillations*. We call hereafter, for simplicity, the modes *p-modes* (*pressure modes*).

6.1.2 Oscillations with $n = 1$ (Corrugation Mode and g-Mode)

In the case of $n = 1$, we have two kinds of oscillation modes. In one of them, $\tilde{\omega}^2$ is larger than $\Omega_\perp^2$, while in the other modes $\tilde{\omega}^2$ is smaller than κ^2. In the former modes, the equatorial plane of disks moves up and down in the vertical direction with radial wavelength k_r. The restoring force acting on the oscillations is mainly the *gravitational force* which acts so as to return fluid elements to the original equatorial plane. In this sense, the naming of p-modes to these modes will be irrelevant. We call hereafter these modes *c-modes* (*corrugation modes*), because this naming is often adopted to similar oscillations in galactic disks. The other modes which have $\tilde{\omega}^2 < \kappa^2$ are called hereafter *g-modes*.

It is important to note that in a rough sense, the corrugation mode is a up-down motion of the disk plane under gravitational restoring force, and thus the motions are approximately incompressible. Furthermore, in the case of one-armed corrugation waves the frequency is extremely low in the Keplerian disks, which will be mentioned in Sect. 6.1.4.

The fact that the corrugation mode is a roughly incompressible motion can be shown by the following considerations. The Lagrangian time change of density, $d\rho/dt$, is given by

$$\frac{d\rho}{dt} = i\tilde{\omega}\rho_1 + \left(u_r\frac{\partial\rho_0}{\partial r} + u_z\frac{\partial\rho_0}{\partial z}\right) \sim i\tilde{\omega}\rho_1 - \frac{z}{H^2}\rho_0 u_z. \tag{6.4}$$

In the case of $n = 1$ the z-component of equation of motion gives

$$i\tilde{\omega}u_z = -\frac{\partial h_1}{\partial z} \sim -\frac{h_1}{z} = -\frac{c_s^2}{z}\frac{\rho_1}{\rho_0}, \tag{6.5}$$

because $h_1 \propto z$. Combination of the above two equations leads to $d\rho/dt \sim i\tilde{\omega}\rho_1[1 - c_s^2/\tilde{\omega}^2 H^2)] \sim 0$, because $\tilde{\omega}^2 \simeq \Omega_\perp^2 = c_s^2/H^2$ in the modes of $n = 1$.

It is noted here that the terminology "g-mode" for oscillation modes with $\tilde{\omega}^2 < \kappa^2$ might be confusing or misleading, since the modes are quite different from the g-modes in the stellar pulsation. The g-modes in stellar pulsation represent the oscillation modes resulting from the restoring force of the Brundt-Väisärä frequency. If the square of the Brundt-Väisärä frequency is negative, the g-mode oscillations in stellar pulsation changes to the convection modes. In the present issues of disk oscillations, the Brundt-Väisärä frequency is assumed to be zero (i.e., we are considering adiabatic perturbations in convectively neutral media). In spite of this situation, we call the modes of $\tilde{\omega}^2 < \kappa^2$ g-modes by usual convention.

If the Brundt-Väisärä frequency is not zero, the dispersion relation (6.2) is modified so that additional modes are included (see e.g., Hirotani and Kato 1995, and also Kato et al. 1998).

6.1.3 Oscillations with $n \geq 2$ (Vertical p-Mode and g-Mode)

In the case of $n \geq 2$ we have again two kinds of oscillation modes. One of $\tilde{\omega}^2$ is larger the $n\Omega_\perp^2$, while the other one is smaller than κ^2. In the former modes, $\tilde{\omega}^2$ is larger than $\Omega_\perp^2$ by the presence of the pressure restoring force acting in the vertical direction. In this sense, it will be relevant to call the modes *vertical p-modes*. The modes with $\tilde{\omega}^2 < \kappa^2$ are called g-modes.

The above classification of oscillations is summarized in Table 6.1.

6.1.4 Comments on One-Armed Low-Frequency Global Oscillations

Some of one-armed modes are interesting and important because they have very low frequencies (ω, not $\tilde{\omega}$, is small) with global patterns. Let us first consider one-armed p-modes ($n = 0$ and $m = 1$). Their propagation regions are specified by $(\omega - \Omega)^2 > \kappa^2$. One of their propagation region is thus $\omega < \Omega - \kappa$. Since Ω and κ are close in Newtonial disks, we can expect low frequency global oscillations. This

Table 6.1 Classification of disk oscillations by vertical node number n.

	Node numbers in the vertical direction		
	$n = 0$	$n = 1$	$n \geq 2$
Higher $\tilde{\omega}^2$	**p-modes** $(\tilde{\omega}^2 > \kappa^2)$	**c-modes** $(\tilde{\omega}^2 > \Omega_\perp^2)$	**Vertical p-modes** $(\tilde{\omega}^2 > n\Omega_\perp^2)$
(Special naming) $m = 1$	.... **Precession mode**	.... **Tilt mode**	....
Lower $\tilde{\omega}^2$	Absence	**g-modes** $(\tilde{\omega}^2 < \kappa^2)$	**g-modes** $(\tilde{\omega}^2 < \kappa^2)$

Note: In the case of higher $\tilde{\omega}^2$ oscillations, one-armed ($m = 1$) modes with $n = 0$ and $n = 1$ have special naming by their characteristics of oscillations

Table 6.2 Low-frequency global oscillations.

(m, n)	Mode	Propagation region	Alternative name
$(1, 0)$	p-mode	$\omega < \Omega - \kappa$	Eccentric precession mode
$(1, 1)$	c-mode	$\omega < \Omega - \Omega_\perp$	Tilt mode
$(2, 4)$	Vertical p-mode	$\omega < 2(\Omega - \Omega_\perp)$	
$(2, 3)$	Vertical p-mode in Polytropic disks with $N = 3$	$\omega < 2(\Omega - \Omega_\perp)$	

issue will be discussed in Sect. 6.3.1, and will be applied to superhumps in dwarf novae in Sects. 7.4.1 and 12.2, and V/R variations in Be stars in Sect. 8.2.

The one-armed ($m = 1$) c-mode oscillations ($n = 1$) are also of interest, since their propagation regions are specified by $(\omega - \Omega)^2 > \Omega_\perp^2$, and one of their regions is $\omega < \Omega - \Omega_\perp$. This propagation region shows the presence of low frequency global oscillation mode, which is nothing but the tilt. This low frequency global corrugation mode will be discussed also in Sect. 6.3.2, and will be applied to negative superhumps of dwarf novae in Sects. 7.4.2 and 12.3. The oscillation modes which can become global, low-frequency ones are summarized in Table 6.2.

It should be noted that in relativistic disks with a high spin of the central source, the difference between Ω and $\Omega_\perp$ is large. In such cases the one-armed corrugation wave is no longer a low-frequency global one.

It is noted here that realization of low frequency oscillations is not restricted only to one-armed oscillations. To demonstrate this, we extend our arguments to cases of polytropic disks and vertically truncated ones. To do so, the eigenvalue of vertical oscillations, K, is retained here without equating it to n. Then, in vertical p-mode oscillations, their propagation regions are $(\omega - m\Omega)^2 > K\Omega_\perp^2$, which consists of $\omega < m\Omega - K^{1/2}\Omega_\perp$ and $\omega > m\Omega + K^{1/2}\Omega_\perp$. Hence, if $m\Omega \sim K^{1/2}\Omega_\perp$ is realized in a wide region, we can expect low frequency oscillations. In the case of vertically extended isothermal disks, for example, the condition, $m\Omega \sim K^{1/2}\Omega_\perp$, is realized for $m = 2$ and $n = 4$. In the case of polytropic disks, the condition is realized when $m^2 \sim n(n + 2N - 1)/2N$ (see equation (4.13)), which is $N \sim n(n - 1)/2(m^2 - n)$. If we take, for example, $m = 2$ and $n = 3$, this relation is realized for $N = 3$. That is, in the vertically polytropic disks with $N = 3$, the oscillations with $m = 2$ and $n = 3$ have low frequencies.

Finally, a difference between oscillations in gaseous disks and those in stellar disks is briefly mentioned. In density waves, $(\omega - m\Omega)^2$ is always smaller than κ^2, i.e., $(\omega - m\Omega)^2 < \kappa^2$. This is because the stellar disks are collisionless systems and there is no pressure restoring force there. Rather, the self-gravity acts so as to decrease the restoring force resulting from inertial force, leading to $(\omega - m\Omega)^2 < \kappa^2$. In this book we are considering nonself-gravitating gaseous disks with pressure. Hence, the p-mode oscillations have $(\omega - m\Omega)^2 > \kappa^2$. In other words, the wave propagation region of the p-mode oscillations considered here (i.e., $(\omega - m\Omega)^2 > \kappa^2$) is the evanescent region of density waves, while the propagation region of density waves (i.e., $(\omega - m\Omega)^2 < \kappa^2$) is the evanescent region of the p-mode oscillations.

6.2 Trapping of Oscillations in Relativistic Disks

In accretion disks of low-mass X-ray binaries, quasi-periodic oscillations are observed (for review, see van der Klis 2000; McClintock and Remillard 2004; Remillard and McClintock 2006). For example, twin high-frequency quasi-periodic oscillations are known, whose frequencies are roughly comparable with the disk's angular velocity of rotation in their innermost relativistic region (see Chap. 1). The presence of discrete set of oscillations will suggest the wave trapping process in disks.

The issue to be addressed here is whether wave trapping processes are expected in accretion disks. In binary systems, for example, their disks are terminated at a radius by tidal force (tidal truncation, see Paczynski 1977), and outgoing waves will be reflected back there inward. In the innermost region of relativistic disks around compact objects (neutron stars or black holes), however, there is practically no outer boundary in disks, because they extend far into the outer region. Two possible processes of reflection of waves are, however, conceivable even in such disks. One is wave reflection by sharp change of geometrical structure of disks. For example, if disk structure changes at a certain radius from an inner hot part to an outer geometrically thin part (i.e., the presence of inner torus), oscillations in the inner part of disks will be trapped in the inner hot region. Another and more practical possibility is wave trapping due to radial changes of angular velocity of disk rotation and of epicyclic frequencies. Especially, in general relativistic disks the radial epicyclic frequency, $\kappa(r)$, is drastically changed from that of the Newtonian disks (see Chap. 2). In this section we examine the wave trapping due to radial distributions of epicyclic frequencies in relativistic disks. Trapping of oscillations in disks of binary systems will be discussed in Sect. 6.3.

6.2.1 Trapping of Relativistic p-Mode Oscillations n = 0)

The propagation regions of p-mode oscillations ($n = 0$) are specified by $\tilde{\omega}^2 > \kappa^2$, as mentioned before. Hence, the propagation regions of axisymmetric ($m = 0$) p-mode oscillations are specified by $\omega > \kappa$ (when $\omega > 0$). In relativistic disks, the radial epicyclic frequency, $\kappa(r)$, has a maximum, $\kappa_{\max}$, at a radius close to the innermost stable circular orbit (ISCO), and decreases both inward and outward (see Chap. 2). This shows that the propagation region of p-mode oscillations is separated into two regions when $\omega < \kappa_{\max}$, as shown in Fig. 6.1. If the inner edge of disks works as a reflecting boundary of the oscillations, the oscillations in the inner propagation region are trapped there. That is, in the inner propagation region, trapping of discrete set of oscillations is expected (see Fig. 6.1). Oscillations in the outer propagation region with $\omega > \kappa_{\max}$, on the other hand, are not trapped, because the propagation region extends far outside and no particular frequency oscillations are selected there.

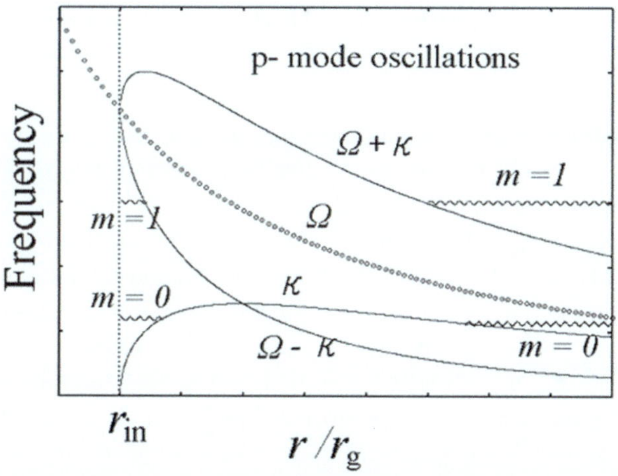

Fig. 6.1 Schematic diagram showing propagation regions of p-mode oscillations ($n = 0$) in relativistic disks. Cases of $m = 0$ and $m = 1$ are shown. The radius, $r_{\rm in}$, is the inner edge of the disks, which will be the radius of ISCO (inner stable circular orbit) in the cases of geometricaly thin disks. $r_{\rm g}$ is the Schwarzscild radius defined by $r_{\rm g} = 2GM/c^2$, where M is the mass of the central object.

Similar argument can be made for non-axisymmetric ($m \neq 0$) oscillations. That is, their propagation regions are $(\omega - m\Omega)^2 - \kappa^2 > 0$, and separated into two regions of $\omega < -\kappa + m\Omega$ and $\omega > \kappa + m\Omega$, as also shown in Fig. 6.1 for $m = 1$. This result means that in the inner propagation region there will be trapped oscillations if the inner edge of disks works as a reflection boundary.

It is of important to note here that the frequencies of trapped oscillations are on the order of the angular frequency of disk rotation in the innermost region, and are comparable with the frequencies of high-frequency quasi-periodic oscillations (HFQPOs and kHz QPOs) observed in low-mass X-ray binaries (see Chap. 1). To know the discrete frequencies of trapped oscillations quantitatively, we must solve the eigenvalue problems, which is the subjects in Chap. 7.

The non-axisymmetric ($m \neq 0$) trapped p-mode oscillations are one of the prominent candidates of HFQPOs and kHz QPOs, since they will be excited by corotation resonance (Lai and Tsang (2009), see also Chap. 10) in addition to the fact that their frequencies are comparable with those of HFQPOs and kHz QPOs. It is noted that the radius of corotation resonance ($\omega = m\Omega$) exists outside the propagation region (evanescent region) of trapped oscillations. In spite of this, amplification of oscillations occurs, because of a part of the oscillations penetrate there.

6.2.2 Trapping of Relativistic c-Mode ($n = 1$) and Vertical p-Mode Oscillations ($n \geq 2$)

In these oscillations, their propagation regions are specified by $\tilde{\omega}^2 > K_n \Omega^2$, where K_n is n ($= 1, 2, 3, \ldots$) if the disks are vertically isothermal and extend infinity. The propagation regions are again separated into two ones. One is $\omega < -K_n^{1/2}\Omega_\perp + m\Omega$ and the other is $\omega > K_n^{1/2} + m\Omega$, where $m = 0, 1, 2, 3, \ldots$. The propagation region for $\omega > K_n^{1/2}\Omega_\perp + m\Omega$ extends outside without boundary, and is not interesting here. The regions of $\omega < -K_n^{1/2}\Omega_\perp + m\Omega$ for some m and n are bounded if the inner edge of the disks works as a reflection boundary. To demonstrate this, the boundary curves of $\omega = -K_n^{1/2}\Omega_\perp + m\Omega$ for $m = 1$ and $m = 2$, and the corresponding trapped regions are schematically shown in Fig. 6.2 in the case where $K_n = 2$. Figure 6.2 shows that the one-armed ($m = 1$) oscillations are not trapped, but the two-armed ($m = 2$) ones are trapped in the inner region of disks.

It is of importance to note that for trapping to occur, $m \geq 2$ is necessary when $K_n = 2$. In the case of c-mode oscillations ($K_n = 1$), $m \geq 1$ is necessary for trapping to occur. The case of $n = 1$ and $m = 1$, however, has a special position, which will be mentioned in Sect. 6.2.4.

The frequencies of trapped oscillations are again on the order of the angular velocity of the disk rotation in their innermost region, and thus these trapped oscillations will also one of possible candidates of HFQPOs and kHz QPOs. Compared with the p-mode oscillations considered in Sect. 6.2.1, the frequencies of the present trapped oscillations are sensitive to the disk structure. This is because

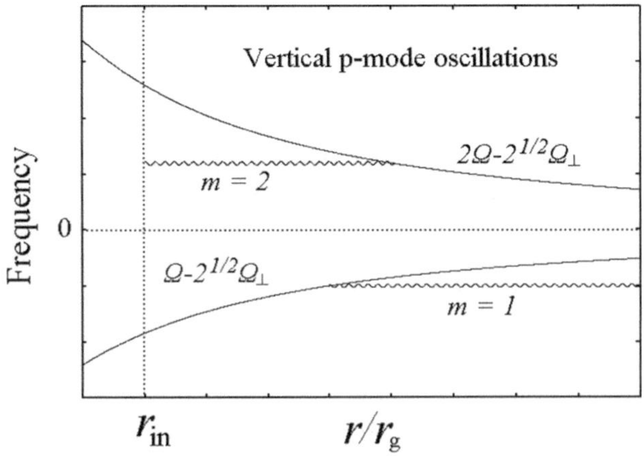

Fig. 6.2 Schematic propagation diagram showing propagation regions of vertical p-modes ($n \geq 2$) oscillations in relativistic disks. Two cases of $m = 1$ and $m = 2$ with $K_n(= n) = 2$ are shown. In the case of $n = 2$ oscillations, one-armed ($m = 1$) oscillations are not trapped, but two-armed ($m = 2$) ones can be trapped in the innermost region of disks. Similar arguments can be made for $n \geq 3$ oscillations. Oscillations with a large n, however, will not be of interest in observational viewpoints.

K_n depends on the vertical structure of disks. That is, for example, in the case of vertically infinite isothermal disks the value of K_n is 2 for $n = 2$, while in truncated isothermal disks, K_n for $n = 2$ becomes smaller than 2, depending on the truncation height (see Sect. 4.2.3). Furthermore, in vertically polytropic disks the value of K_n depends on the polytropic index (see Sect. 4.2.1). In addition, if disks are subject to magnetic fields, the frequencies of vertical oscillations also depend on the structure and strength of the fields (see Sect. 4.2.4). These dependences of the trapped oscillations on the disk structure might be favorable for describing the facts that the frequencies of twin kHz QPOs observed in neutron star LMXBs vary with time.

One of problems related to these c-mode and vertical p-mode oscillations is their excitation. They are not excited (rather damped) by the corotation resonance (see Chap. 10). They may be excited by wave-wave resonant process (Chaps. 11 and 12), if disks are deformed. This issue will be discussed in Chap. 11. In addition, stochastic processes of turbulence (see Goldreich and Keeley 1977) will be one of candidates of excitation processes (see Sect. 13.2), but there is no detailed studies on this direction up to the present time.

6.2.3 *Trapping of Relativistic g-Mode Oscillations ($n \geq 1$)*

The propagation region of the g-mode oscillations is given by $\tilde{\omega}^2 < \kappa^2$, as mentioned before, which is written as $-\kappa + m\Omega < \omega < \kappa + m\Omega$. This propagation region is schematically shown in Fig. 6.3 in the case of relativistic disks for $m = 0$ and $m = 1$ oscillations.

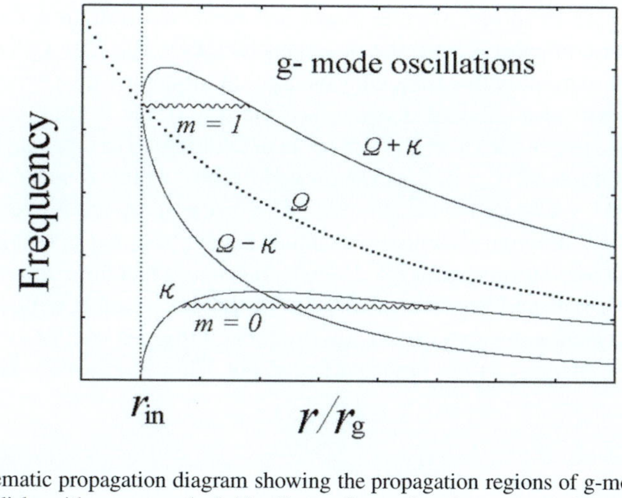

Fig. 6.3 Schematic propagation diagram showing the propagation regions of g-mode oscillations in relativistic disks with no magnetic fields. Cases of $m = 0$ and $m = 1$ are shown with $n = 1$. In the case of non-axisymmetric ($m \neq 0$) oscillations, the radius of corotation resonance ($\omega = m\Omega$) appears in their propagation region, unless ω is in the range of $\Omega_{\rm in} < \omega < (\Omega + \kappa)_{\max}$.

In the case of axisymmetric ($m = 0$) g-mode oscillations, they can be trapped below the curve of $\kappa(r)$, if $\omega < \kappa_{\max}$. This trapping of axisymmetric g-mode oscillations is due to the very nature of radial distribution of relativistic $\kappa(r)$, and free from the wave reflection at the inner edge of disks. This trapping was first pointed out by Okazaki et al. (1987), and suggested by Nowak et al. (1997) to be a cause of 67 Hz oscillations observed in the black-hole candidate GRS 1915 + 105.

Recently, an important work was done by Fu and Lai (2009). This is that if the disks are subject to poloidal magnetic fields, the self-trapping of the axisymmetric g-mode oscillations is strongly affected and will be destroyed. As their results are important, further quantitative studies are required so that the effects of vertical disk stratification on the oscillations are taken into account more in detail.

In the case of $m \geq 1$, oscillations are trapped between two curves of $m\Omega \pm \kappa$, as schematically shown in Fig. 6.3 for $m = 1$. In the cases of non-axisymmetric ($m \neq 0$) modes, however, the point (radius) of corotation resonance appears in their propagation region, except when the frequency is very close to the maximum value of $m\Omega + \kappa$, as shown in Fig. 6.3. It is known that at the corotation point the g-mode oscillations are damped (see Chap. 10).

In summary, non-axisymmetric g-mode oscillations will be less interesting compared with axisymmetric ones, because they will be damped by corotation resonance except for special cases where $\omega \sim (\Omega + m\kappa)_{\max}$.

6.2.4 *Trapping of Relativistic One-Armed (m = 1) c-Mode Oscillations (n = 1)*

Up to the present, we have considered high frequency oscillations in relativistic disks in order to examine whether there are trapped oscillation modes whose frequencies are comparable with the observed HFQPOs and kHz QPOs. Here, we emphasize that extremely low frequency oscillation modes are also expected even in the innermost part of relativistic disks, unless the spin of the central source is high.

Let us consider c-mode ($n = 1$ and $m = 1$) oscillations. Their propagation region is specified by $(\omega - \Omega)^2 > \Omega_\perp^2$ as mentioned before, which is separated into two regions; $\omega > \Omega + \Omega_\perp$ and $\omega < \Omega - \Omega_\perp$. The former propagation region extends outwards without boundary. Hence, we cannot expect trapping of oscillations. Our concern here is oscillations with $\omega < \Omega - \Omega_\perp$. It is noted that the difference between Ω and $\Omega_\perp$ is small and positive in disks with the Kerr metric with a small spin parameter, a_*. Hence, we can expect low frequency trapped oscillations with $\omega > 0$ (prograde oscillation) in the inner part of disks. This is schematically shown in Fig. 6.4.

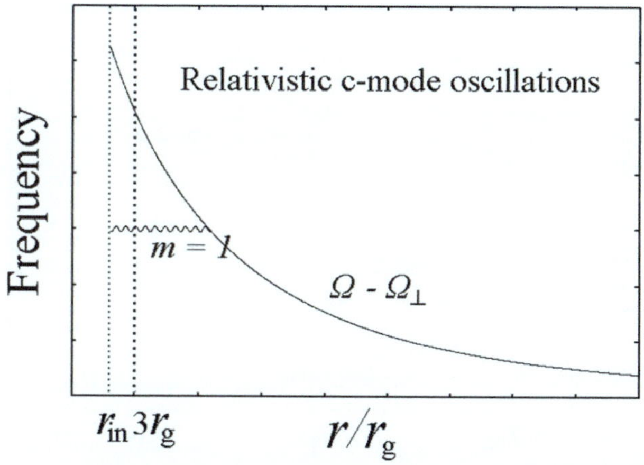

Fig. 6.4 Schematic propagation diagram showing a propagation region of relativistic c-mode oscillations with $m = 1$ and $n = 1$. The inner edge of disks, $r_{\rm in}$, is shown to be smaller than $3r_{\rm g}$, since the cases where the central source has spin are considered. The spin, however, must be small for the trapped oscillations to have low frequencies. Otherwise, the difference between Ω and $\Omega_\perp$ is large and frequencies of trapped oscillations become high.

6.3 Trapping of Low-Frequency Oscillations in Newtonian Disks

Up to the present, we have focused our attention on possible trapping of oscillations in relativistic disks. Here, cases of non-relativistic (Newtonian) disks are considered. In Newtonian disks we can also expect many trapped oscillations, if there are relevant reflection boundaries of waves. Among them, we mention here one-armed, low-frequency oscillations. Since in Newtonian disks both of $\Omega_\perp$ and κ are close to Ω ($\sim \Omega_{\rm K}$), various one-armed low-frequency trapped oscillations are expected. Here, we focus, in particular, on binary systems, since extensive observational evidences of low-frequency oscillations are accumulated in these systems (see Chap. 1), and thus detailed comparison between observations and disk oscillation models is possible.

In the followings, for simplicity, the orbit of the secondary star is assumed to be circular around the primary and the orbital plane coincides with the disk plane.

6.3.1 One-Armed Eccentric Precession Mode (m = 1, n = 0) in Binary System

The propagation region of one-armed ($m = 1$) p-mode ($n = 0$) is specified by $(\omega - \Omega)^2 > \kappa^2$. This region consists of two parts. One of them is $\omega < -\kappa + \Omega$. In tidally deformed disks, $\kappa(r)$ is slightly smaller than $\Omega(r)$ in general (see equations (2.15)

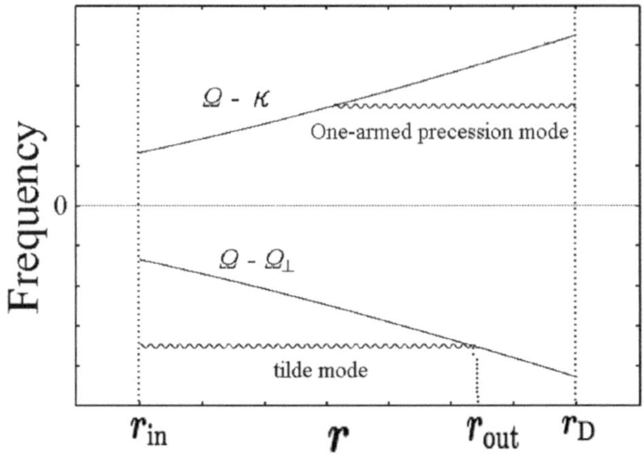

Fig. 6.5 Schematic propagation diagram showing the propagation regions of one-armed precession mode ($m = 1$ and $n = 0$) and tilt mode ($m = 1$ and $n = 1$) in non-relativistic disks of a binary system. The former oscillation is prograde, while the latter retrograde. The radii of the inner and outer edges of disks are shown by $r_{\rm in}$ and $r_{\rm D}$, respectively.

and (2.16)). Hence, $-\kappa(r) + \Omega(r) > 0$ and increases with increase of r, as shown in Fig. 6.5. In tidally deformed disks, the disk size is limited by the tidal force resulting from the secondary (tidal truncation of disks Paczynski 1977). Hence, we can expect low-frequency oscillations trapped between the radius where $\omega = \Omega - \kappa$ is realized and the disk outer edge, $r_{\rm D}$ (see Fig. 6.5). These trapped oscillations are prograde. This low-frequency trapped mode is known to be the cause of superhumps of dwarf novae (Osaki 1985). The excitation of this mode is discussed in Chaps. 11 and 12.

6.3.2 Tilt Mode ($m = 1$, $n = 1$) in Binary System

One-armed corrugation mode (tilt mode) has a propagation region specified by $\omega < \Omega - \Omega_\perp$, as mentioned before. In Newtonian disks the difference between Ω and $\Omega_\perp$ is always small. The angular velocity of disk rotation, $\Omega(r)$, is smaller than $\Omega_{\rm K}$, since the presence of tidal force of the secondary and the outward pressure force on the disk, i.e., $\Omega < \Omega_{\rm K}$. In binary systems, the vertical epicyclic frequency, $\Omega_\perp(r)$, is usually larger than $\Omega_{\rm K}$, because an additional force coming from the secondary works so as to return the disk gases to the equator, i.e., $\Omega_\perp > \Omega_{\rm K}$. From the above two relations, we see that $\Omega - \Omega_\perp$ is negative and its negative value increases with increase of r (see equations (2.15) and (2.25)). This is also schematically shown in Fig. 6.5. Since the propagation region is specified by $\omega < \Omega - \Omega_\perp$, we can expect the presence of low-frequency trapped oscillations with $\omega < 0$, as is shown in Fig. 6.5. These trapped oscillations are retrograde.

This kind of retrograde oscillations are considered to be one of possible causes of negative superhumps in dwarf novae (Kato 2014; Lubow 1992). Excitation of these oscillations in tidally deformed disks will be discussed in Chaps. 11 and 12.

References

Fu, W., & Lai, D. 2009, Astrophys. J., 690. 1386
Goldreich, P., & Keeley, D. A. 1977, Astrophys. J., 212, 243
Hirotani, K., & Kato, S. 1995, Publ. Astron. Soc. Jpn., 47, 647
Kato, S. 2014, Publ. Astron. Soc. Jpn., 66, 21
Kato, S., Fukue, J., & Mineshige, S. 1998, *Black Hole Accretion Disks* (Kyoto University Press)
Lai, D., & Tsang, D. 2009, Mon. Not. R. Astron. Soc., 393, 979
Lubow, S. H. 1992, Astrophys. J., 398, 525
McClintock, J. E., & Remillard, R. A. 2004, in *Compact Stellar X-ray Sources*, eds. W. H. G. Lewin & M. van der Klis (Cambridge University Press), 157
Nowak, M. A., Wagoner, R. V., Begelman, M. C., & Lehr, D. E. 1997, Astrophys. J., 477, L91
Okazaki, A. T., Kato, S., & Fukue, J. 1987, Publ. Astron. Soc. Jpn., 39, 457
Osaki, Y. 1985, Astron. & Astrophys., 144, 369
Paczyński, B. 1977, Astrophys. J., 216, 822
Remillard, R. A., & McClintock, J. E. 2006, Annu. Rev. Astron. Astrophys., 44, 49
van der Klis, M. 2000, Annu. Rev. Astron. Astrophys., 38, 717

Chapter 7
Frequencies of Trapped Oscillations and Application

Abstract In this chapter, we examine the frequencies of trapped oscillations and their parameter dependences for various oscillation modes. The purpose is to examine whether they can describe periodic or quasi-periodic oscillations observed in X-ray binaries and dwarf novae. First, we examine frequencies of p-mode, c-mode, and vertical p-mode oscillations trapped in the innermost regions of relativistic disks. The results are compared with the high-frequency quasi-periodic oscillations observed in neutron-star or black-hole X-ray binaries. Second, frequencies of low-frequency trapped oscillations in Newtonian disks are examined. The oscillation modes examined are one-armed p-mode (one-armed precession mode) and one-armed vertical p-mode (tilt mode). The results are examined whether they can describe positive and negative superhumps in dwarf novae.

Keywords Frequencies of c-mode • Frequencies of p-modes • Frequencies of vertical p-modes • High-frequency QPOs • Relativistic disks • Superhumps

7.1 Trapped Oscillations and Their Frequencies by WKB Method

In this chapter we solve equation (6.1) (or equation (5.15) or equation (5.38)) by the WKB method. The purpose is to have rough pictures how frequencies of trapped oscillations depend on oscillation modes and to know whether they can qualitatively describe quasi-periodic oscillations observed in low-mass X-ray binaries, dwarf novae, and so on.

It is emphasized here that in deriving equation (6.1) (or equation (5.15) or equation (5.38)), the local approximations of neglecting $|d\ln H/d\ln r|$ and $|d\ln\rho_{00}/d\ln r|$ compared with $1/\lambda$ have been adopted, where λ is the radial wavelength of oscillations. This approximations is allowed as the first step to know the frequencies of trapped oscillations, because the radial width where oscillations are trapped is

S. Kato, *Oscillations of Disks*, Astrophysics and Space Science Library 437,
DOI 10.1007/978-4-431-56208-5_7

usually not wide, except for special cases.[1] The radial variations of $\Omega(r)$, $\kappa(r)$, and $\Omega_\perp(r)$, however, have been taken into account. Under these approximations we solve the wave equations by using WKB method.

We have two equations from which our examination is to be started: equations (5.15) and (5.38). Which equation should be adopted depends on what wave modes are considered. Hence, we treat a few cases separately.

It is noted that we adopt vertically isothermal disks. Eigenfrequencies of trapped oscillations in polytropic disks have been examined extensively by Wagoner's group (e.g., Nowak and Wagoner 1991, 1992; Perez et al. 1997; Wagoner 1999; Wagoner et al. 2001; Silbergleit et al. 2001; Ortega-Rodríguez et al. 2002, 2008, and references therein).

7.1.1 p-Mode Oscillations in Relativistic Disks ($n = 0$)

We first consider p-mode ($n = 0$) oscillations in relativistic disks. As discussed in Chap. 6, a propagation region of p-mode oscillations with frequency ω is inside the radius $r_{\rm out}$ defined by $\omega = m\Omega - \kappa$ (for the case of $m = 1$, see Fig. 6.1). The inner edge of the propagation region, $r_{\rm in}$, is taken at $r_{\rm isco}$ (the radius of the innermost stable circular orbit). The radius of $\omega = m\Omega - \kappa$ is the point of the inner Lindblad resonance (see Chap. 2), and an apparent singularity appears there in equation (5.15). In equation (5.38), however, the radius is a regular point. In equation (5.38), the radius of $\tilde{\omega}^2 = K_n\Omega_\perp^2$ appears as an apparent singularity, but the radius is outside the propagation region of p-mode oscillations (this is obvious because in the case of $n = 0$, $\tilde{\omega}^2 - K_n\Omega_\perp^2$ is $\tilde{\omega}^2$ and always positive). Hence, in the present case of p-mode oscillations, it will be relevant to start from equation (5.38) in order to avoid an apparent singularity in the propagation region.

In order to change equation (5.38) to a standard form, we introduce a new variable $\tau_u(r)$ defined by[2]

$$\tau_u(r) = \int_{r_{\rm in}}^{r} \frac{|\tilde{\omega}^2 - K_n\Omega_\perp^2|}{c_{\rm s}^2}dr = \int_{r_{\rm in}}^{r} \frac{\tilde{\omega}^2}{c_{\rm s}^2}dr. \tag{7.1}$$

Then, equation (5.38) is written as

$$\frac{d^2 f_u}{d\tau_u^2} + Q_u f_u = 0, \tag{7.2}$$

[1] In the cases where the V/R variations of emission lines in Be stars are considered, the local approximations are irrelevant, since the trapped region of oscillations describing the V/R variations is wide (see Chap. 8).

[2] In this chapter, n in equations (5.15) and (5.38) are written as K_n so that truncated disks can be also taken into account.

where Q_u is given by

$$Q_u = \frac{c_s^2}{|\tilde{\omega}^2 - K_n\Omega_\perp^2|}\frac{\tilde{\omega}^2-\kappa^2}{\tilde{\omega}^2} = \frac{c_s^2}{\tilde{\omega}^2}\frac{\tilde{\omega}^2-\kappa^2}{\tilde{\omega}^2}. \tag{7.3}$$

Equation (7.2) shows that the propagation region of oscillations in the radial direction is the region of $Q_u > 0$. In p-mode oscillations the outer boundary of the trapped region is at $r_{\rm out}$ defined by $\omega = m\Omega - \kappa$ (i.e., $\tilde{\omega}^2 = \kappa^2$) as mentioned before. In terms of τ_u defined by equation (7.1), the outer boundary is denoted $\tau_{u,\rm out}$ where $Q_u = 0$. The outer boundary $\tau_{u,\rm out}$ is a turning point of Q_u, and near $\tau_{u,\rm out}$, we can write Q_u as $Q_u = a^2(\tau_{u,\rm out} - \tau_u)$, where $a^2(>0)$ is the expansion coefficient. If $Q_u = a^2(\tau_{u,\rm out} - \tau_u)$ is adopted, equation (7.2) may be solved in terms of Bessel functions of the order of $1/3$. Their two independent solutions are

$$(\tau_{u,\rm out}-\tau)^{1/2}J_{1/3}\left[\frac{3}{2}a(\tau_{u,\rm out}-\tau_u)^{3/2}\right] \quad \text{and} \quad (\tau_{u,\rm out}-\tau_u)^{1/2}J_{-1/3}\left[\frac{3}{2}a(\tau_{u,\rm out}-\tau_u)^{3/2}\right]. \tag{7.4}$$

In the propagation region far inside of $\tau_{u,\rm out}$, the WKB solution is well-known to be[3]

$$Q_u^{-1/4}\exp\left[\pm i\int^r Q_u^{1/2}d\tau_u\right]. \tag{7.5}$$

Hence, we must look for functions which tend to equation (7.4) near to $\tau_{u,\rm out}$ and does to equation (7.5) far inside of $\tau_{u,\rm out}$. As such solutions we have (Morse and Feshbach 1953)

$$f_u \simeq \left(\frac{w}{Q_u}\right)^{1/2}[AJ_{1/3}(w) + BJ_{-1/3}(w)], \tag{7.6}$$

where

$$w(\tau_u) = \int_{\tau_u}^{\tau_{u,\rm out}} Q_u(\tau_u')^{1/2}d\tau_u', \tag{7.7}$$

and A and B are integration constants. The relation between A and B is determined later.

[3]The WKB method is applicable to equation (7.2), although Q_u is a small quantity. This is because the range of τ_u where oscillations are trapped is rather wide.

For $\tau_u \ll \tau_{u,\text{out}}$ where w is large, from the asymptotic behavior of Bessel functions, we find that equation (7.6) tends to (Morse and Feshbach 1953)

$$f_u \simeq \left(\frac{2}{\pi Q_u}\right)^{1/2}\left[A\cos\left(w-\frac{5}{12}\pi\right)+B\cos\left(w-\frac{1}{12}\pi\right)\right]. \tag{7.8}$$

Outside of the trapping radius $\tau_{u,\text{out}}$, the eigenfunctions must spatially damp to tend to zero. For $\tau_u \gg \tau_{u,\text{out}}$, Q_u is negative, so that both $Q_u^{1/2}$ and w are imaginary, having a branch point at $\tau_{u,\text{out}}$. Continuation of the expression (7.6) into the region of $\tau_u > \tau_{u,\text{out}}$ requires $A = B$ for spatial damping of eigenfunctions (Morse and Feshbach 1953). Hence, we have

$$f_u \simeq 2A\left(\frac{2}{\pi Q_u}\right)^{1/2}\cos\left(\frac{\pi}{6}\right)\cos\left(w-\frac{1}{4}\pi\right). \tag{7.9}$$

Next, an inner boundary condition is considered. The inner boundary, $\tau_{u,\text{in}}$, is taken at the inner edge of disks (i.e., the radius of the innermost stable circular orbit (ISCO), r_{ISCO}, or the radius of sonic point), inside which the gas falls with supersonic speed and the gas density decreases there sharply inwards. Hence, a relevant boundary condition to be imposed there will be vanishing of the Lagrangian pressure variation, i.e., $\delta p = 0$ or $\delta \rho = 0$.[4] The equation of continuity then shows that the condition is roughly written as $\partial\xi_r/\partial r = 0$, since $\xi_z \sim 0$ in p-mode oscillation ($n = 0$), where the ξ_r and ξ_z are the r- and z-components of the Lagrangian displacement vector of perturbations, respectively. Although ξ_r and u_r are related by $u_r = (\omega - m\Omega)\xi_r$, $\partial\xi_r/\partial r = 0$ will be approximated by $df_u/dr \sim 0$ if the radial wavelength of oscillations is short. Hence, for simplicity, we adopt here $df_u/dr = 0$ as the boundary condition at r_{in}. Then, we find from equation (7.9) that the condition, $df_u/dr = 0$, is $w - \pi/4 = n_r\pi$ at $\tau_u = \tau_{u,\text{in}}$, which is

$$\int_{\tau_{u,\text{in}}}^{\tau_{u,\text{out}}} Q_u^{1/2} d\tau_u = \left(\frac{1}{4}+n_r\right)\pi \quad (n_r = 0, 1, 2, \ldots), \tag{7.10}$$

where n_r denotes the node number of f_u in the radial direction. Returning variable from τ_u to r, we find the trapping condition (7.10) is written as

$$\int_{r_{u,\text{in}}}^{r_{u,\text{out}}} \frac{(\tilde{\omega}^2-\kappa^2)^{1/2}}{c_s} dr = \left(\frac{1}{4}+n_r\right)\pi \quad (n_r = 0, 1, 2, \ldots). \tag{7.11}$$

Equation (7.11) is the trapping condition for the p-mode ($n = 0$) oscillations.

[4] There is no strong support for validity of this boundary condition. We think, however, a partial reflection of inward waves to outgoing waves will occur there, because strong mode couplings due to inhomogeneity of disk structure produce there outgoing waves (e.g., Honma et al. 1992; Kato et al. 1988; Matsumoto et al. 1998).

It is noted that if we adopt $f_u = 0$ instead of $df_u/dr = 0$ as the inner boundary condition, the term $(1/4 + n_r)\pi$ in equation (7.11) is changed to $(3/4 + n_r)\pi$.

7.1.2 c-Mode ($n = 1$) and Vertical p-Mode ($n \geq 2$) Oscillations in Relativistic Disks

In c-mode ($n = 1$) and vertical p-mode ($n \geq 2$) oscillations in relativistic disks, the trapping region is between $r_{\rm in}(\sim r_{\rm ISCO})$ and $r_{\rm out}$, the later being the radius where $\omega = m\Omega - (K_n)^{1/2}\Omega_\perp$ is realized (see Chap. 6, and for the case of $n = 2$ and $m = 2$, see Fig. 6.2). In equation (5.38) the point of $\omega = m\Omega - (K_n)^{1/2}\Omega_\perp$ is an apparent singularity, while in equation (5.15) the point is not so. In this sense it is proper to start from equation (5.15) in the present problem, distinct from the case of p-mode oscillations. As in the case of p-mode oscillations, we introduce a new variable defined by

$$\tau(r) = \int_{r_{\rm in}}^{r} \frac{\tilde{\omega}^2 - \kappa^2}{|\tilde{\omega}|} dr. \tag{7.12}$$

Then, equation (5.15) is written in a standard form as

$$\frac{d^2 f}{d\tau^2} + Q(\tau) f = 0, \tag{7.13}$$

where

$$Q(\tau) = \frac{\tilde{\omega}^2 - K_n\Omega_\perp^2}{c_{\rm s}^2(\tilde{\omega}^2 - \kappa^2)}. \tag{7.14}$$

The outer boundary of the propagation region is the turning point of Q, and the same WKB procedures as those adopted in the case of p-mode oscillations can be applied. The inner boundary condition is again taken as $\delta p = 0$, which is reduced to $f = 0$ in case of $n \geq 2$ and to $d^2f/dr^2 = 0$ in the cases of $n = 1$.[5] Applying the

---

[5]The boundary condition $\delta p (= c_{\rm s}^2 \delta\rho) = 0$ can be written as

$$\frac{\partial (r\xi_r)}{r\partial r} + \frac{\partial \xi_z}{\partial z} \sim 0, \tag{*}$$

by using the equation of continuity, where ξ_r and ξ_z are, respectively, the r- and z-components of displacement vector associated with oscillations. In vertical p-mode oscillations with $n = 1$, ξ_z is roughly independent of z. Hence, the boundary condition is approximately reduced to $\partial\xi_r/\partial r \sim 0$. This boundary condition is the same as that of p-mode oscillations, and is $df_u/dr \sim 0$. In terms of f, this is written as $d^2f/dr^2 \sim 0$ (see, e.g., equation (5.32)). In cases of $n \geq 2$, the second term of equation (*) predominates over the first one, and the boundary condition is written as $\partial\xi_z/\partial z \sim$

above inner boundary condition to a similar equation with equation (7.8), we have as the trapping condition for $n \geq 1$

$$\int_{\tau_{\rm in}}^{\tau_{\rm out}} Q^{1/2} d\tau = \left(\frac{3}{4} + n_r\right)\pi \quad (n_r = 0, 1, 2, \ldots), \tag{7.15}$$

Returning the variable from τ to r, we have finally the trapping condition:

$$\int_{r_{\rm in}}^{r_{\rm out}} \left[\frac{(\tilde{\omega}^2 - \kappa^2)(\tilde{\omega}^2 - K_n\Omega_\perp^2)}{c_{\rm s}^2\tilde{\omega}^2}\right]^{1/2} dr = \left(\frac{3}{4} + n_r\right)\pi \quad (n_r = 0, 1, 2, \ldots). \tag{7.16}$$

It is noted that the right-hand side of this equation is $(3/4 + n_r)\pi$, not $(1/4 + n_r)\pi$.

If $df/dr = 0$ is adopted as the inner boundary condition, $(3/4 + n_r)\pi$ on the right-hand side of equation (7.16) is changed to $(1/4 + n_r)\pi$.

7.1.3 g-Mode Oscillations in Relativistic Disks ($n \geq 1$)

In the case of g-mode oscillations in relativistic disks, the propagation region is specified by $m\Omega - \kappa < \omega < m\Omega + \kappa$ (Chap. 6, and for $m = 0$ and $m = 1$ modes, see Fig. 6.3). That is, the both ends of the propagation region are turning points of Q. In this case, solutions around $r_{\rm in}$ and $r_{\rm out}$ are both obtained by using the turning point solutions discussed in Sect. 7.1.1 for the p-mode oscillations. Then, both solutions are fitted in a region between $r_{\rm in}$ and $r_{\rm out}$. The results show that the condition of the trapping is (Morse and Feshbach 1953)

$$\int_{r_{\rm in}}^{r_{\rm out}} \left[\frac{(\tilde{\omega}^2 - \kappa^2)(\tilde{\omega}^2 - K_n\Omega_\perp^2)}{c_{\rm s}^2\tilde{\omega}^2}\right]^{1/2} dr = \left(\frac{1}{2} + n_r\right)\pi \quad (n_r = 0, 1, 2, \ldots). \tag{7.17}$$

It is noted here that except for axisymmetric mode ($m = 0$) and for special cases of $m \neq 0$ oscillations (see Fig. 6.3), the g-mode oscillations are damped by the presence of the corotation resonance in their trapped region (Chap. 10).

---

0. The z-component of equation of motion reduces this boundary condition to $\partial^2 h_1/\partial z^2 \sim 0$. Since $h_1(r, \eta)$ is written as $h_1 = f(r)\mathcal{H}_n(\eta)$, $\partial^2 h_1/\partial z^2 = n(n-1)f(r)\mathcal{H}_{n-2}(\eta)$ and the boundary condition is finally reduced to $f = 0$.

7.1.4 *One-Armed, Low-Frequency Oscillations in Binary Systems*

As mentioned before, there are two oscillation modes which are interesting in relation to oscillatory phenomena observed in binary systems. One is one-armed eccentric precession mode (one-armed p-mode, i.e., $m = 1$ and $n = 0$), and the other is tilt mode (one-armed vertical p-mode, i.e., $m = 1$, and $n = 1$).

(i) one-armed eccentric precession mode ($m = 1, n = 0$)

Oscillations of this mode are trapped in the region between the inner radius where $\omega = \Omega - \kappa = 0$ is realized and the outer edge of disks (see Fig. 6.5). In wave equation (5.15) the radius of $\omega = \Omega - \kappa$ (inner Lindblad resonance) is an apparent singular point, while it is not so in wave equation (5.38). Hence, we start from wave equation (5.38), as in the case of p-mode oscillations in relativistic disks.

As in the case of p-mode oscillations in relativistic disks, we introduce a new variable defined by

$$\tau_u(r) = \int_{r_{\rm in}}^{r} \frac{\tilde{\omega}^2}{c_{\rm s}^2} dr. \tag{7.18}$$

Then, equation (5.38) is written as

$$\frac{d^2 f_u}{d\tau_u^2} + Q_u f_u = 0, \tag{7.19}$$

where Q_u is given by

$$Q_u = \frac{c_{\rm s}^2}{\tilde{\omega}^2} \frac{\tilde{\omega}^2 - \kappa^2}{\tilde{\omega}^2}. \tag{7.20}$$

In the present case, the difference from the case of p-mode oscillations in relativistic disks is boundary conditions. In the present problem, the inner boundary is a turning point of Q_u and the outer boundary is a free surface where boundary condition $\delta p = 0$ will be relevant. That is, boundary conditions at $r_{\rm in}$ and $r_{\rm out}$ are exchanged from those in the case of p-mode oscillations in relativistic disks. Except this, there is no essential difference. Hence, the trapping condition is the same as that in the case of p-mode oscillations in relativistic disks:

$$\int_{r_{\rm in}}^{r_{\rm out}} Q_u^{1/2} d\tau_u = \left(\frac{1}{4} + n_r\right)\pi \quad (n_r = 0, 1, 2, \ldots). \tag{7.21}$$

Returning the variable from τ_u to r, we can write the trapping condition as

$$\int_{r_{\rm in}}^{r_{\rm out}} \frac{[(\omega-\Omega)^2-\kappa^2]^{1/2}}{c_{\rm s}}dr = \left(\frac{1}{4}+n_r\right)\pi \quad (n_r=0,1,2,\ldots). \tag{7.22}$$

(ii) Tilt mode in binary systems ($m = 1, n = 1$)

In the case of tilt mode in binary systems, it is relevant to start from equation (5.15), as in the case of vertical p-mode oscillations in relativistic disks. Different from the vertical p-modes in relativistic disks, $u_r = 0$ will be relevant as the boundary condition at the inner edge, because the inner edge of the disk will be the surface of the primary star (see Fig. 6.5). Since equation (5.15) is a differential equation with respect to f, the boundary condition $u_r = 0$ need to be expressed in terms of f. As discussed before, we use here $df/dr = 0$ as a rough inner boundary condition. The outer edge of the propagation region is a turning point which represent the outer boundary of the propagation region (see Fig. 6.5). The above consideration about boundary conditions leads to the trapping condition:

$$\int_{r_{\rm in}}^{r_{\rm out}} \left[\frac{(\tilde{\omega}^2-\kappa^2)(\tilde{\omega}^2-\Omega_\perp^2)}{\tilde{\omega}^2c_{\rm s}^2}\right]^{1/2} dr = \left(\frac{1}{4}+n_r\right)\pi \quad (n_r=0,1,2,\ldots), \tag{7.23}$$

where $\tilde{\omega} = \omega - \Omega$. It is noted that this trapping condition is different from equation (7.16).

7.2 Frequencies of Trapped p-Mode ($n = 0$) Oscillations and QPOs

Frequencies of trapped p-mode ($n = 0$) oscillations are examined. Results of numerical calculations obtained by using equation (7.11) are presented in Fig. 7.1. In order to compare the results of numerical calculations with high frequency quasi-periodic oscillations observed in black-hole LMXBs, we consider relativistic disks surrounding a black hole of $10\,M_\odot$. The spin of the central source is taken to be zero, i.e., $a_* = 0$. The adopted angular velocity of disk rotation, Ω, and the epicyclic frequency, κ, are those of relativistic Keplerian ones. It is noted that in the present calculations the disk parameter involved is the radial distribution of square of acoustic speed, $c_{\rm s}^2(r)$, alone. It is noted again that radial distributions of $\rho_{00}(r)$ and $H(r)$ are neglected in deriving equation (7.11) under the assumption that the trapped region is narrow.

In the standard Shakura-Sunyaev disks, the square of acoustic speed, i.e., $c_{\rm s0}^2$, is described by (e.g., Kato 2008)

$$c_{\rm s0}^2 = 1.83\times10^{16}\Gamma\alpha^{-1/5}\dot{m}^{2/5}(M/M_\odot)^{-1/5}(r/r_{\rm g})^{-9/10}\ {\rm cm}^2\ {\rm s}^{-2}, \tag{7.24}$$

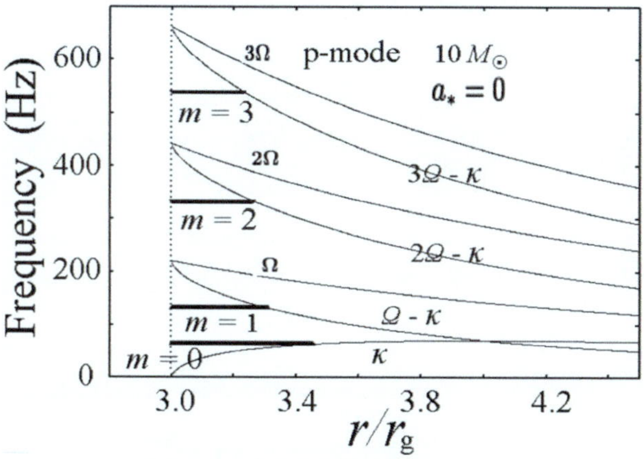

Fig. 7.1 Frequencies and captured regions of trapped p-mode ($n = 0$) oscillations. Four *thick horizontal lines* represent the trapped regions of four oscillation modes. From bottom to upper, they are axisymmetric ($m = 0$), one-armed ($m = 1$), two-armed ($m = 2$), and ($m = 3$) oscillations. The central star is a $10\,M_\odot$ black hole with no spin ($a_* = 0$). The inner boundary of the trapped region is taken at $3r_g$, r_g being the Schwarzschild radius defined by $r_g = 2GM/c^2$. The radial distribution of acoustic speed, $c_s(r)$ has been taken as equation (7.24).

in the case where the gas pressure dominates over the radiation pressure and the opacity mainly comes from the free-free processes. Here, the subscript 0 has been attached to c_s^2, since we adopt this value of c_s^2 as a standard one, and cases where c_s^2 is larger or smaller than this value by a factor are considered sometimes. In our present calculations we adopt the above expression for c_s^2 with $\Gamma = 1, \alpha = 0.3$, and $\dot{m} = 0.3$.

In Fig. 7.1, four oscillation modes of $m = 0, 1, 2,$ and 3 with $n_r = 0$ are shown. The case of $n_r = 0$ means that we consider oscillations which have no node in the radial direction, i.e., the fundamental mode in the radial direction. As expected, trapped oscillations occur in a narrow region around the inner edge of disks. The overtones ($n_r \geq 1$) in the radial direction have lower frequencies compared with those of the fundamental modes, although they are not shown in Fig. 7.1. Figure 7.2 shows how the frequency ratio of $m = 2$ and $m = 3$ oscillations depends on disk parameters, c_s.

Frequencies and trapped regions of two oscillation modes of $m = 2$ and $m = 3$ seem to good candidates of HFQPOs by the following reasons. In HFQPOs observed in black-hole LMXBs, the time change of their frequencies is little. In the present trapped oscillations, the frequency of trapped oscillations is also insensible to parameter changes. Second, in some HFQPOs in black hole LMXBs the QPOs often appear in pairs, and their frequency ratio is close to 3:2 (see Chap. 1). In the present trapped p-mode oscillations, the frequency ratio of $m = 2$ and $m = 3$ oscillations are close to 3:2 as emphasized by Lai and Tsang (2009). In order to demonstrate this, the frequency ratio of $m = 3$ and $m = 2$ oscillations is shown in Fig. 7.2 as functions of the frequency of the $m = 2$ oscillation, when the disk temperature is

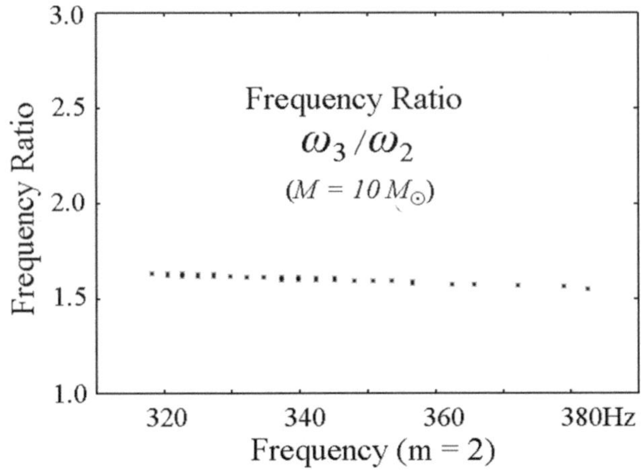

Fig. 7.2 Dependence of frequency ratio ω_3/ω_2 on ω_2, where ω_2 and ω_3 are frequencies of trapped $m = 2$, and $m = 3$ oscillations, respectively. Frequency changes due to change of disk temperature are shown. For moderate changes of trapped frequencies due to changes of disk temperature, the frequency ratio is almost unchanged and close to 3:2 observed in black-hole LMXBs.

changed as a parameter. The third point favor for the p-mode oscillations is that the p-mode oscillations can be really excited by corotation resonant process (Lai and Tsang 2009; Horák and Lai 2013; and see Chap. 10).

7.3 Frequencies of Trapped c- and Vertical p-Modes and QPOs

Trapped frequencies and trapped regions of c-mode ($n = 1$) and vertical p-mode ($n = 2$) oscillations are calculated by using equation (7.16). The results are shown in Fig. 7.3 for two-armed ($m = 2$) and three-armed ($m = 3$) oscillations with the vertical node number of $n = 1$ and $n = 2$. The central star is taken to be a black hole of $10\,M_\odot$ with no spin ($a_* = 0$). It is noted that the node number n denotes that of h_1 in the vertical direction, and u_z has one less node number. That is, in the mode of $n = 1$ the disk plane oscillates in the vertical direction with no node. In the mode of $n = 2$ the equatorial plane of the disks is just the node of oscillations, i.e., above and below the equator the disk oscillates in the opposite phase in the vertical direction. Comparison of Figs. 7.1 and 7.3 shows that the trapped regions are slightly wider in the present case than in the case of p-mode oscillations.

Figures 7.4 and 7.5 show the propagation diagram for $n = 1$ and $n = 2$ oscillations with $m = 2$, in order to show how the frequencies and trapped regions depend on the node number, n_r, in the radial direction. As n_r increases, the frequency of the trapped oscillations decreases and the width of the trapped region becomes wide. It should be noted that in the vertical p-mode oscillations with $n \geq 2$

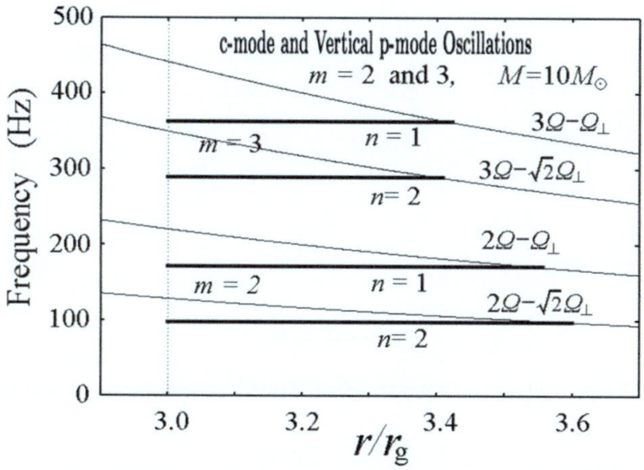

Fig. 7.3 Frequecies and captured regions of trapped c-mode ($n = 1$) and trapped vertical p-mode ($n \geq 2$) oscillations in vertically extended isothermal disks ($\eta_s \equiv z_s/H = \infty$). Four *thick horizontal lines* represent the trapped regions of four oscillation modes. Upper two *horizontal lines* are for three-armed ($m = 3$) oscillations with two different n (i.e., $n = 1$ and $n = 2$). Lower two lines are for two-armed ($m = 2$) oscillations with two different n ($n = 1$ and $n = 2$). The central star is a $10\,M_\odot$ black hole with no spin ($a_* = 0$). The inner boundary of the trapped region is at $3r_g$ and free boundary condition has been adopted there, where r_g is the Schwarzschild radius defined by $r_g = 2GM/c^2$.

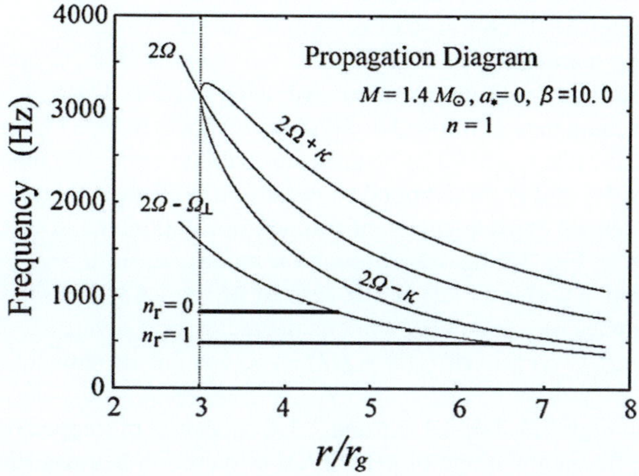

Fig. 7.4 Propagation diagram for two oscillations of (n, n_r) = (1,0) and (1,1). The azimuthal arm number, m, is taken to be $m = 2$. The *horizontal lines* show the radial range where oscillations are trapped. $\beta \equiv c_s^2/(c_s^2)_0 = 10.0$ has been adopted. $M = 1.4\,M_\odot$ and $a_* = 0$ are taken.

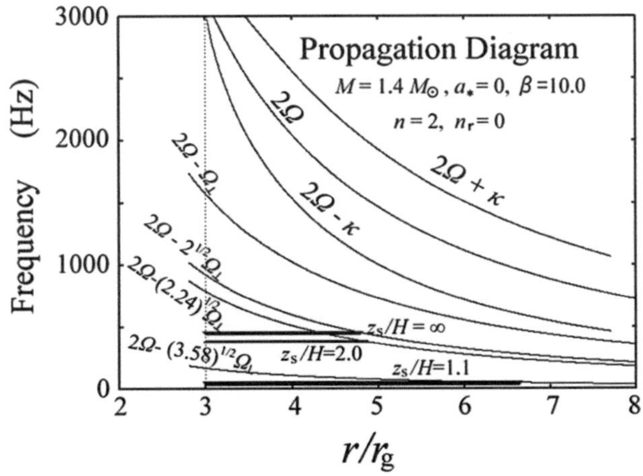

Fig. 7.5 Propagation diagram for vertical p-mode oscillations of $(n, n_r) = (2, 0)$ for three cases of $\eta_s = \infty$, 2.0, and 1.1. Other parameters are the same as Fig. 7.4 (After Kato 2012. PASJ ©).

the truncation of disk thickness in a finite height has non-negligible effects on frequencies of oscillations. In order to demonstrate this, cases of $n = 2$ are shown in Fig. 7.5 for three values of η_s, i.e., $\eta_s = \infty$, 2.0 and 1.1. It should be noted that in the disks whose vertical thickness is thin by truncation by hot corona, frequencies are low. For example, the frequencies are as low as the low frequency quasi periodic oscillations (LHQPOs) in the case of $\eta_s \equiv z_s/H = 1.1$.

A remaining parameter related to disk structure is disk temperature. As a parameter describing disk temperature we adopt again $\beta \equiv c_s^2/c_{s0}^2$. The β-dependence of frequency of trapped oscillations is shown in Fig. 7.6 for some values of η_s. The central star is a $1.4M_\odot$ neutron star with no spin. The inner edge of the disk is taken at $3r_g$ and the free boundary condition is adopted there.

The dependences of frequencies of trapped oscillations on spin of the central star are shown in Fig. 7.7 for four oscillation modes ($n_r = 0$ and 1 with $n = 1$, and $n_r = 0$ and 1 with $n = 2$) in the case of $M = 1.4M_\odot$. In the cases where $n \geq 2$, frequencies of trapped oscillations decreases as the vertical disk thickness decreases. Hence, two cases of $\eta_s(\equiv z_s/H) = \infty$ and 2.0 are shown for oscillations of $n = 2$.

As shown in Figs. 7.3, 7.4, 7.5, 7.6 and 7.7, frequencies of trapped oscillations are roughly in the frequency range of kHz QPOs observed in neutron star LMXBs. In these objects, kHz QPOs are often observed in pairs, but different from the cases of black-hole LMXBs, their frequencies are neither fixed, nor their ratio is fixed to 3:2 (see Chap. 1). In order to examine whether c-mode and vertical p-mode oscillations can describe these characteristics of kHz QPOs, the frequency-frequency relation of two oscillations of $n_r = 0$ and $n_r = 1$ is shown in Figs. 7.8 and 7.9 by changing $\beta(\equiv c_s^2/c_{s0}^2)$ for two cases of $a_* = 0$ and $a_* = 0.2$. The oscillation mode adopted is the c-mode ($n = 1$) with $m = 2$. Figure 7.8 is for $M = 1.4M_\odot$ and Fig. 7.9 is for

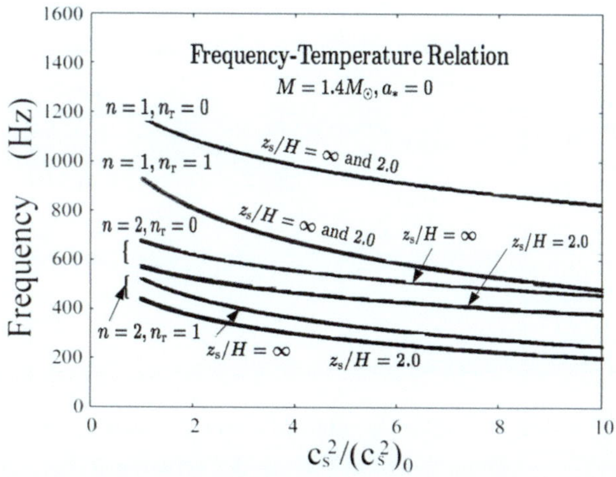

Fig. 7.6 Frequency – temperature relation of trapped two-armed ($m = 2$) c-mode ($n = 1$) and vertical p-mode ($n = 2$) oscillations in vertically isothermal disks. Two cases are shown where the disks extend infinitely in the vertical direction ($z_s/H = \infty$) and are terminated at the height of $2H$, i.e., $\eta_s \equiv z_s/H = 2.0$. Four oscillation modes are shown, i.e., the set of (n, n_r) are (1,0), (1,1), (2,0) and (2,1). In oscillations of $n = 1$, the frequencies are independent of η_s. The mass M and the spin parameter a_* adopted are $M = 1.4\,M_\odot$ and $a_* = 0$.

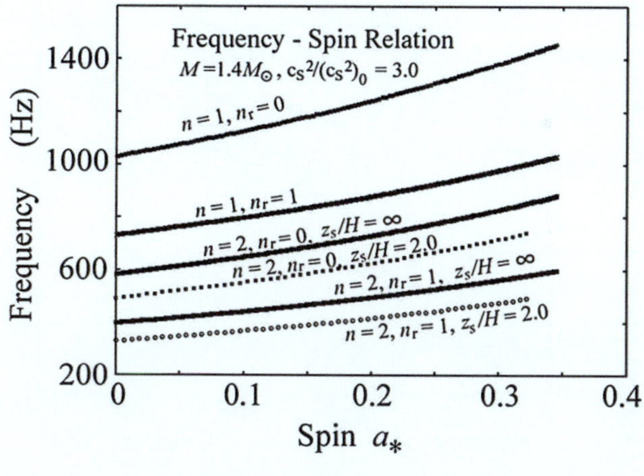

Fig. 7.7 Frequency – spin relation for four oscillation modes of (n, n_r) = (1,0), (1,1), (2,0), and (2,1). In the case of $n = 2$, the relation depends on η_s. Two cases of $\eta_s = \infty$ and 2.0 are shown for the oscillation of $n = 2$. M and β adopted are $M = 1.4\,M_\odot$ and $\beta(\equiv c_s^2/(c_s^2)_0 = 3.0$ (After Kato 2012. PASJ ©).

$M = 1.8\,M_\odot$. These figures show that the observed time variation of frequencies of pair QPOs seems to be well described by assuming that the pair QPOs are a set of two-armed ($m = 2$) $n = 1$ oscillations: One is the fundamental mode ($n_r = 0$) in the radial direction and the other is its first overtone ($n_r = 1$) in the radial direction. If

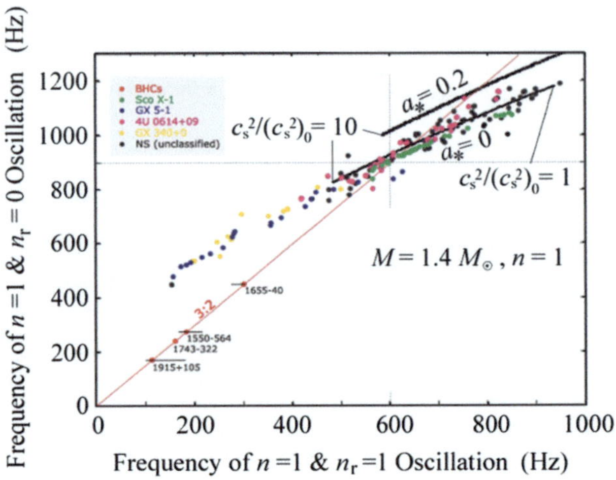

Fig. 7.8 The frequency correlation between (n, n_r)= (1,0) and (1,1) oscillations for two cases of $a_* = 0$ and 0.2. The value of β is changed from 1.0 to 10.0 along the curves. $M = 1.4 M_\odot$ is adopted. The plot of observational data of some typical Z-sources, taken from the figure of Abramowicz (2005), are overlapped on this figure. The straight line labelled by 3:2 is the line on which frequency ratio is 3:2 (After Kato 2012. PASJ ©).

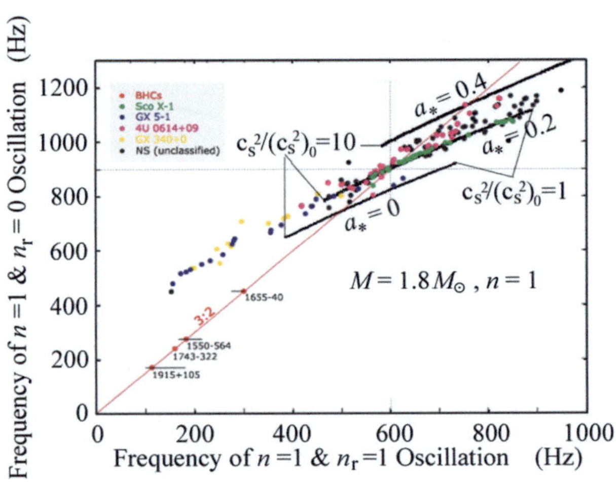

Fig. 7.9 The same as Fig. 7.8, except that $M = 1.8 M_\odot$ and three cases of $a_* = 0$, 0.2, and 0.4 are considered (After Kato 2012. PASJ ©).

twin kHz QPOs are really $n_r = 0$ and $n_r = 1$ oscillations of $n = 1$, Figs. 7.8 and 7.9 suggest that the spin of neutron-star LMXBs is $a_* = 0 \sim 0.1$ if their masses are around $1.4 M_\odot$ and $a_* = 0.2 \sim 0.3$ if their masses are around $1.8 M_\odot$. Furthermore, the figures suggest that c_s^2 changes in these objects in the range of $(1 - 10) \times c_{s0}^2$.

It is noted that the above results are based on the assumption that magnetic fields in neutron-star LMXBs are so weak that they have no essential effects on wave motions. Even if magnetic fields are present, however, qualitative results obtained here are qualitatively unchanged, if the global magnetic fields are toroidal. This will be presented in Sect. 8.1.

One of important problems remained to be clarified is whether the oscillation modes considered here (mainly c-mode oscillations) are excited in disks. This issue will be discussed in Chaps. 11 and 12.

7.4 Frequencies of Trapped One-Armed Oscillations in Binary Systems

Up to the present, we have not considered two kinds of oscillations of $(m, n) = (1, 0)$ and $(m, n) = (1, 1)$. This is because these oscillations have low frequencies, and may be related to other types of oscillatory phenomena than those in LMXBs. As mentioned in Chap. 1, in dwarf novae superoutbursts are observed in addition to normal outbursts. The origin of these superoutbursts is now understood by the thermal-tidal instability model (TTI model) proposed by Osaki (1996). During a superoutburst, a periodic photometric hump with amplitude 0.2–0.3 mag appears, whose period is very close to the orbital period of the system but longer than that by a few percent. This is called the "positive" superhump, because another periodic humps with a period shorter than the orbital period are found in some SU UMa and nova-like variable stars and the latter called the "negative" superhump. The origin of the positive superhump is understood as due to a deformation of accretion disk into an eccentric form (excitation of one-armed ($m = 1$) oscillation with $n = 0$), while the negative superhuman is thought to be produced by a tilt (excitation of one-armed ($m = 1$) oscillation with $n = 1$). We examine the trapping of the one-armed eccentric precession mode ($m = 1$ and $n = 0$) in Sect. 7.4.1 and the trapping of the tilt mode ($m = 1$ and $n = 1$) in Sect. 7.4.2.

7.4.1 Eccentric Precession Mode and Superhumps of Dwarf Novae

The condition of trapping of one-armed precession mode (one-armed p-mode, i.e., $m = 1$ and $n = 0$) is given by equation (7.22). The disk size of primary star is limited by tidal effect (tidal truncation) of secondary star. The disk size is now denoted $r_{\rm D}$. As discussed in Sect. 6.3.1, the one-armed precession mode is trapped between the radius where $\omega = \Omega - \kappa$ is realized, say $r_{\rm c}$, and the disk outer boundary, $r_{\rm D}$ (see Fig. 6.5).

In dwarf novae the disk of the primary star is deformed by the tidal force resulting from secondary star. In the case where the orbit of the secondary star around the primary is circular, the time-averaged part of the tidal potential, $\bar{\psi}_{\rm T}$, is given by

$$\bar{\psi}_{\rm T}(r) = -\frac{GM_{\rm s}}{4a^3}r^2, \tag{7.25}$$

where $M_{\rm s}$ is the mass of the secondary star, and a is the separation distance between primary and secondary stars.

If the pressure force is neglected, the angular velocity of rotation of the disk gas, $\Omega(r)$, is given by (see equation (2.14) in Chap. 2)

$$\Omega^2 r = \frac{GM}{r^2}\left(1 - \frac{1}{2}q\frac{r^3}{a^3}\right), \tag{7.26}$$

where $q = M_{\rm s}/M$ is the mass ratio, M being the mass of the primary. Since we are interested in the case where q is smaller than unity and in the radial region of $r/a < 1$, we have approximately

$$\Omega = \Omega_{\rm K}\left(1 - \frac{1}{4}q\frac{r^3}{a^3}\right), \tag{7.27}$$

where $\Omega_{\rm K}$ is the Keplerian angular velocity of rotation, given by $\Omega_{\rm K} = (GM/r^3)^{1/2}$. The epicyclic frequency defined by $\kappa^2 = 2\Omega(2\Omega + rd\Omega/dr)$ is then approximately given by

$$\kappa = \Omega_{\rm K}\left(1 - q\frac{r^3}{a^3}\right). \tag{7.28}$$

The above two equations lead to

$$\Omega - \kappa = \frac{3}{4}q\Omega_{\rm K}\left(\frac{r}{a}\right)^3. \tag{7.29}$$

Since the frequency of the trapped oscillations is low compared with Ω, we can approximately write the integrand of equation (7.22) in the form:

$$\frac{(2\Omega)^{1/2}}{c_{\rm s}}\left[-\omega + (\Omega - \kappa)\right]^{1/2}. \tag{7.30}$$

Considering that

$$\omega = (\Omega - \kappa)_{\rm c}, \tag{7.31}$$

and $\Omega - \kappa$ is given by equation (7.29), we can reduce equation (7.30) to

$$\left(\frac{3}{2}q\frac{GM}{ac_{\rm s}^2}\right)^{1/2}\left[1-\left(\frac{r_{\rm in}/a}{r/a}\right)^{3/2}\right]^{1/2}\frac{1}{a}. \tag{7.32}$$

Here, we consider that $(r_{\rm D} - r_{\rm c})/r_{\rm c} \ll 1$ (the WKB method is still valid, since $GM/ac_{\rm s}^2$ is a large quantity). Then, we can approximately perform the integration of equation (7.22) to obtain

$$\frac{(r_{\rm D}/a)^{3/2} - (r_{\rm c}/a)^{3/2}}{(r_{\rm c}/a)^{1/2}}\left(\frac{GM}{ac_{\rm s}^2}q\right)^{1/2} = \left(n_{\rm r} + \frac{1}{4}\right)\pi. \tag{7.33}$$

We are interested in the fundamental mode of oscillations in the radial direction, i.e., $n_{\rm r} = 0$. Equation (7.33) gives then an approximate expression for the width of the trapped region:

$$\frac{r_{\rm D} - r_{\rm c}}{a} = \frac{\pi}{6q^{1/2}}\frac{c_{\rm s}}{(GM/a)^{1/2}}. \tag{7.34}$$

It is noted that the normalized width of the trapped region, $(r_{\rm D} - r_{\rm c})/a$, is roughly determined only by parameters q and $c_{\rm s}^2/(GM/a)$, independent of $r_{\rm D}$.

Considering that the frequency of the trapped oscillations, ω, is given by $\omega = (\Omega - \kappa)_{\rm c}$ and $\Omega - \kappa$ is given by equation (7.29), we can express ω normalized by the observed binary orbital frequency, $\Omega_{\rm orb}$, in the form:

$$\frac{\omega}{\Omega_{\rm orb}} = \frac{3}{4}\frac{q}{(1+q)^{1/2}}\left(\frac{r_{\rm c}}{a}\right)^{3/2}. \tag{7.35}$$

It is noted that $\Omega_{\rm orb}$ and the orbital frequency of the secondary star seen from the primary one, $\Omega^*_{\rm orb}$, are related by

$$\Omega_{\rm orb} = \Omega^*_{\rm orb}(1+q)^{1/2}. \tag{7.36}$$

The above consideration shows that the dimensionless parameters describing $\omega/\Omega_{\rm orb}$ are q, $c_{\rm s}/(GM/a)^{1/2}$, and $r_{\rm D}/a$. Figure 7.10 shows $\omega/\Omega_{\rm orb}$ as functions of $r_{\rm D}/a$ in two cases of $q = 0.1$ and 0.3 with $c_{\rm s}/(GM/a)^{1/2} = 0.1$ and 0.01. It is noticed that disk-temperature dependence of frequency is weak. For a change of disk temperature of an order, the frequency change is less than a factor.

In the framework of the present treatment, $r_{\rm D}/a$ is a free parameter. If we want to compare the calculated frequencies with the observed ones of superhump in dwarf novae, we must know what determines the disk radius, $r_{\rm D}$, in actual binary systems. This issue will be discussed in Chaps. 11 and 12, but it is noted here that the truncation radius, $r_{\rm D}$, is determined by the tidal instability. The tidal instability is found numerical simulations by Whitehurst (1988a,b) and later theoretically

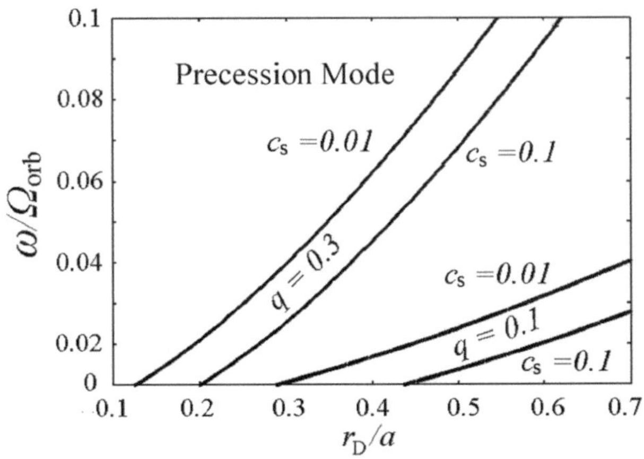

Fig. 7.10 Frequencies of trapped one-armed p-mode oscillations ($m = 1$ and $n = 0$) (in another terminology, eccentric precession mode) as functions of disk radius, $r_{\rm D}$. Parameters are $c_{\rm s}/(GM/a)^{1/2}$ and q. Two cases of $c_{\rm s}/(GM/a)^{1/2} = 0.1$ and 0.01 are shown for $q = 0.1$ and 0.3. It is noticed that the frequencies are insensitive to $c_{\rm s}$.

by Hirose and Osaki (1990, 1993) and (Lubow 1991a,b). Lubow's mode-mode coupling model is further developed to a wave-wave coupling model by Kato (2013, 2014), see also Kato et al. (2011), which will be described in Chaps. 11 and 12.

7.4.2 Tilt Mode and Negative Superhumps of Dwarf Novae

The condition of trapping of one-armed tilt mode ($m = 1$ and $n = 1$) is given by equation (7.23). We consider again the case where the secondary star has a circular orbit around the primary star with separation distance a. The vertical epicyclic frequency, $\Omega_\perp$, is then given by equation (2.25) in Chap. 2, and we have

$$\Omega - \Omega_\perp = -\frac{3}{4}q\Omega\left(\frac{r}{a}\right)^3. \tag{7.37}$$

Then, considering that the frequency of trapped oscillations is low, i.e., $|\omega| \ll \Omega$, and Ω and κ are close, we see that the integrand of equation (7.23) is written as

$$\frac{2}{c_{\rm s}}\Big[(-\omega+\Omega-\kappa)(-\omega+\Omega-\Omega_\perp)\Big]^{1/2} = \frac{2}{c_{\rm s}}\Big[(\Omega-\Omega_\perp)^2_{\rm out}-(\Omega-\Omega_\perp)^2\Big]^{1/2}$$
$$= \frac{3}{2}q\frac{(GM/D)^{1/2}}{c_{\rm s}}\left[\left(\frac{r_{\rm out}}{a}\right)^3-\left(\frac{r}{a}\right)^3\right]^{1/2}\frac{1}{a}, \tag{7.38}$$

where $\Omega - \kappa = -(\Omega - \Omega_\perp)$ has been used.

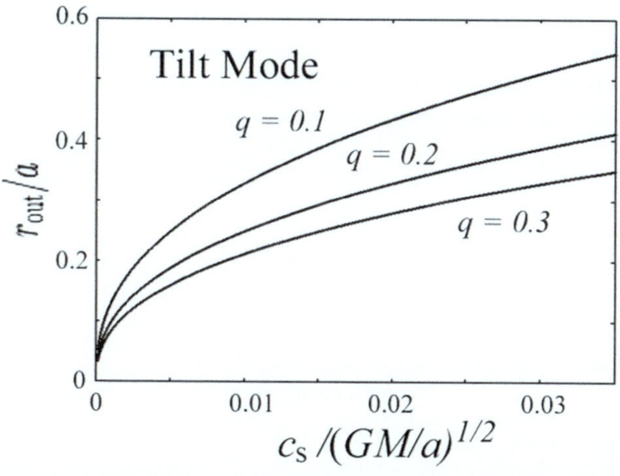

Fig. 7.11 The radius of turning point, $r_{\rm out}$ (the outer capture radius), as functions of disk temperature. In typical dwarf novae disks, $c_{\rm s}$ normalized by orbital velocity, $(GM/a)^{1/2}$, is around 0.02.

Using this expression for the integrand of equation (7.23), from the trapping condition (7.23) we can derive $r_{\rm in}/a$ as functions of $r_{\rm out}/a$, $c_{\rm s}^2/(GM/a)^{1/2}$, and q, like in the case of one-armed precession mode. The radius, $r_{\rm out}$, is the place where the tilt mode is trapped, i.e., the radius where $\omega = \Omega - \Omega_\perp$ (see Fig. 6.5). If we take, for simplicity, the inner edge of disk, $r_{\rm in}$, to be zero, i.e., $r_{\rm in} = 0$, the trapping condition is a relation among $r_{\rm out}/a$, q, and $c_{\rm s}/(GM/a)^{1/2}$.

The relation is shown in Fig. 7.11 as a relation between $r_{\rm out}/a$ and $c_{\rm s}/(GM/a)^{1/2}$ with a parameter q. It is noted that because the frequency of the trapped oscillations is given by $\omega = (\Omega - \Omega_\perp)_{\rm out}$, we have

$$\frac{\omega}{\Omega_{\rm orb}} = -\frac{3}{4}\frac{q}{(1+q)^{1/2}}\left(\frac{r_{\rm out}}{a}\right)^{3/2}. \tag{7.39}$$

It should be noted that if $r_{\rm out}$ derived from the trapping condition (7.23) becomes larger than the disk radius, $r_{\rm D}$, it means that the result is inconsistent with the outer boundary adopted in calculations. A relevant outer boundary condition is then a free surface, which will be close to $df_u/dr \sim 0$.

Another issue to be addressed is whether the tilt mode is really excited on disks. This issue will be discussed in Chaps. 11 and 12.

Finally, it is important to remember that in the case of tilt modes, their propagation region is rather wide. In some cases it will extend from the inner edge of the disks to the outer edge of the disk (cf., Fig. 6.5). In such cases, the assumption that the radial wavelength of perturbations is shorter than the characteristic radial lengths of disk structure may be irrelevant. That is, the effects of radial variations of $d\ln\rho_{00}/d\ln r$, $d\ln H/d\ln r$ and others should be taken into account in wave equation to obtain more reliable results. This can be made by solving equation (5.28), which will be a subject in the near future.

References

Abramowicz, M. A. 2005, Astro. Nachr. 326, 782

Hirose, M., & Osaki, Y. 1990, Publ. Astron. Soc. Jpn., 42, 135

Hirose, M., & Osaki, Y. 1993, Publ. Astron. Soc. Jpn., 45, 595

Honma, F., Matsumoto, R., & Kato, S. 1992, Publ. Astron. Soc. Jpn., 44, 529

Horák, J., & Lai, D. 2013, Mon. Not. R. Astro. Soc., 434, 276

Kato, S. 2008, Publ. Astron. Soc. Jpn., 60, 111

Kato, S. 2012, Publ. Astron. Soc. Jpn., 64, 129

Kato, S. 2013, Publ. Astron. Soc. Jpn., 65, 75

Kato, S. 2014, Publ. Astron. Soc. Jpn., 66, 25

Kato, S., Honma, F., & Matsumoto, R., 1988, Mon. Not. R. Astron. Soc., 231, 37

Kato, S., Okazaki, A. T., & Oktariani, F. 2011, Publ. Astron. Soc. Jpn., 63, 363

Lai, D., & Tsang, D. 2009, Mon. Not. R. Astron. Soc., 400, 470

Lubow, S. H. 1991a, Astrophys. J., 381, 259

Lubow, S. H. 1991b, Astrophys. J., 381, 268

Matsumoto, R., Kato, S., & Honma, F. 1998, in *Physics of Neutron Stars and Black Holes*, ed. Y. Tanaka (Universal Academic Press, Tokyo), p155

Morse, P. M., & Feshbach, H. 1953, *Methods of Theoretical Physics, Part II* (McGraw-Hill, Inc., New York), p.1092

Nowak, M. A., & Wagoner, R.V. 1991, Astrophys. J., 378, 656

Nowak, M. A., & Wagoner, R.V. 1992, Astrophys. J., 393, 697

Ortega-Rodríguez, M., Silbergleit, A. S., & Wagoner, R. V. 2002, Astrophys. J., 567, 1043

Ortega-Rodríguez, M., Silbergleit, A. S., & Wagoner, R. V. 2008, Geophys. Astrophys. Fluid Dyn., 102, 75

Osaki, Y. 1996, Publ. Astron. Soc. Pac., 108, 390

Perez, C. A., Silbergleit, A. S., Wagoner, R. V., & Lehr, D. E. 1997, Astrophys. J., 476, 589

Silbergleit, A. S., Wagoner, R. V., & Ortega-Rodrígues, M. 2001, Astrophys. J., 548, 335

Wagoner, R. V. 1999, Phys. Rep., 311, 259

Wagoner, R. V., Silbergleit, A. S., & Ortega-Rodríguez, M. 2001, Astrophys. J., 559, L25

Whitehurst, R. 1988a, Mon. Not. R. Astron. Soc., 232, 35

Whitehurst, R. 1988b, Mon. Not. R. Astron. Soc., 233, 529

Chapter 8
Two Examples of Further Studies on Trapped Oscillations and Application

Abstract In Chaps. 6 and 7 effects of magnetic fields on oscillations are neglected. In real disks, however, global toroidal magnetic fields will present, because the magnetic fields generated by magneto-rotational instability (MRI) will be generally stretched in the azimuthal direction by the shear of differential rotation. In Sect. 8.1, we examine effects of toroidal magnetic fields on c-mode oscillations. The purpose is to know whether toroidal magnetic fields affect the interpretation that kHz QPOs comes from the c-mode oscillations.

In Chaps. 6 and 7 we had neglected the effects of radial variations of disk thickness, disk density and others on oscillations. This was allowed as the first step, since we are mainly interested in trapped oscillations and their trapped regions are not wide. In some important cases, however, the propagation regions of such oscillations as one-armed precession modes and tilt modes are as wide as the whole disk size. Hence, it will be necessary to relax the above approximations to study, for example, precession of the V/R variations in Be stars (see Chap. 1). This issue is discussed in Sect. 8.2.

The main parts in this chapter are special and technical. In particular, Sect. 8.2 treats a special issue. Thus, the readers who are not interested in mathematical procedures can skip this chapter, except for Sect. 8.1.3.

Keywords Disk thickness • Global oscillations • Toroidal magnetic fields • c-modes oscillations • V/R variations

8.1 Trapped c- and Vertical p-Mode Oscillations in Disks with Toroidal Magnetic Fields

We consider vertically isothermal disks with axisymmetric toroidal magnetic fields. The fields are assumed to be purely toroidal with no poloidal component;

$$\boldsymbol{B}_0(r,z) = [0, B_0(r,z), 0]. \tag{8.1}$$

The fields are further assumed to be distributed so that the Alfvén velocity, $c_{\rm A}$, defined by $(B_0^2/4\pi\rho_0)^{1/2}$ is constant in the vertical direction. Then, as mentioned in

S. Kato, *Oscillations of Disks*, Astrophysics and Space Science Library 437,
DOI 10.1007/978-4-431-56208-5_8

Chap. 4, the hydrostatic balance in the vertical direction leads to

$$\rho_0(r,z) = \rho_{00}(r)\exp\left(-\frac{z^2}{2H^2}\right) \quad \text{and} \quad B_0(r,z) = B_{00}(r)\exp\left(-\frac{z^2}{4H^2}\right), \tag{8.2}$$

where the scale height, H, is related to c_s, c_A, and $\Omega_\perp$ by

$$H^2(r) = \frac{c_\mathrm{s}^2 + c_\mathrm{A}^2/2}{\Omega_\perp^2}, \tag{8.3}$$

and the ratio between c_s and c_A is an arbitrary constant.

On such disks small-amplitude MHD perturbations are superposed. The velocity perturbation over rotation is denoted by (u_r, u_φ, u_z), and the perturbed part of the magnetic field over the unperturbed one by (b_r, b_φ, b_z). Then, the r-, φ-, and z-components of the equation of motion are written as equations (A.43), (A.44), and (A.45), respectively. Similarly, the r-, φ-, and z-components of the induction equation give, respectively, equations (A.46), (A.47), and (A.48). Furthermore, the equation of continuity is written as equation (A.42).

Here, the azimuthal and time dependences of the perturbed quantities are taken to be proportional to $\exp[i(\omega t - m\varphi)]$, where ω and m are frequency and azimuthal wavenumber of the perturbations, respectively. The perturbations are assumed to be local in the sense that their characteristic radial wavelength, λ, is shorter than the characteristic radial scale of disks, λ_D, i.e., $\lambda < \lambda_\mathrm{D}$, where λ_D is on the order of r. By using this approximation, we neglect such quantities as $d\ln\rho_{00}/d\ln r$, $d\ln B_{00}/d\ln r$, and $d\ln H/d\ln r$, compared with terms of the order of r/λ, but the radial variation of Ω is retained to be taken into account. Then, the r-, φ-, and z-components of equation of motion, equations (A.43), (A.44) and (A.45), are reduced to

$$i\tilde{\omega}u_r - 2\Omega u_\varphi = -\frac{\partial h_1}{\partial r} - c_\mathrm{A}^2\frac{\partial}{\partial r}\left(\frac{b_\varphi}{B_0}\right), \tag{8.4}$$

$$i\tilde{\omega}u_\varphi + \frac{\kappa^2}{2\Omega}u_r = 0, \tag{8.5}$$

$$i\tilde{\omega}u_z = -\left(\frac{\partial}{\partial z} + \frac{c_\mathrm{A}^2}{2c_\mathrm{s}^2}\frac{z}{H^2}\right)h_1 - c_\mathrm{A}^2\left(\frac{\partial}{\partial z} - \frac{z}{H^2}\right)\left(\frac{b_\varphi}{B_0}\right) - i\frac{m}{r}c_\mathrm{A}^2\left(\frac{b_z}{B_0}\right), \tag{8.6}$$

where $\tilde{\omega} = \omega - m\Omega$, and h_1 defined by $h_1 = p_1/\rho_0 = c_\mathrm{s}^2\rho_1/\rho_0$ has been introduced instead of p_1. Similarly, the r-, φ-, and z-components of induction equation, equations (A.46), (A.47) and (A.48), are reduced to

$$i\tilde{\omega}\frac{b_r}{B_0} = -i\frac{m}{r}u_r, \tag{8.7}$$

$$i\tilde{\omega}\frac{b_\varphi}{B_0} = r\frac{d\Omega}{dr}\frac{b_r}{B_0} - \frac{\partial u_r}{\partial r} - \left(\frac{\partial}{\partial z} - \frac{z}{2H^2}\right)u_z, \tag{8.8}$$

$$i\tilde{\omega}\frac{b_z}{B_0} = -i\frac{m}{r}u_z. \tag{8.9}$$

Finally, the equation of continuity, equation (A.42), is reduced to

$$i\tilde{\omega}h_1 = -c_s^2\left[\frac{\partial u_r}{\partial r}+\left(\frac{\partial}{\partial z}-\frac{z}{H^2}\right)u_z\right]. \tag{8.10}$$

Now, we further simplify equations (8.6) and (8.8). The last term of equation (8.6), $-i(m/r)c_A^2(b_z/B_0)$, can be expressed in terms of u_z by using equation (8.9). The result shows that the term of $-i(m/r)c_A^2(b_z/B_0)$ is smaller than the left-hand term, $i\tilde{\omega}u_z$, of equation (8.6) by a factor of $c_A^2/(r\Omega)^2$. Considering this, we neglect the last term on the right-hand side of equation (8.6). Next, we consider equation (8.8). The first term on the right-hand side, $r(d\Omega/dr)(b_r/B_0)$, is smaller than the second term, $-\partial u_r/\partial r$, by a factor of λ/r, which can be shown by expressing b_r in terms of u_r by using equation (8.7). Hence, we neglect the term in the following analyses.

After introducing the above approximations into equations (8.6) and (8.8), we multiply $i\tilde{\omega}$ to equation (8.6) in order to express h_1 and b_φ/B_0 in equation (8.6) in terms of u_z and u_r by using equation (8.10) and (8.8). Then, after changing independent variables from (r, z) to (r, η), where η is defined by $\eta = z/H$, we have

$$\left[\frac{\partial^2}{\partial\eta^2} - \eta\frac{\partial}{\partial\eta} + \frac{\tilde{\omega}^2 - \Omega_\perp^2}{c_s^2 + c_A^2}H^2\right]u_z + H\left[\frac{\partial}{\partial\eta} - \frac{c_A^2/2}{c_s^2 + c_A^2}\eta\right]\frac{\partial u_r}{\partial r} = 0. \tag{8.11}$$

This is the basic wave equation to be solved in this section (Kato 2011).

Let us compare this equation with wave equations used in Chaps. 6 and 7. First, if u_r is neglected, the oscillations are purely vertical and equation (8.11) becomes equation (4.32) as expected. Second, if magnetic fields are neglected, equation (8.11) becomes

$$\left[\frac{\partial^2}{\partial\eta^2} - \eta\frac{\partial}{\partial\eta} + \frac{\tilde{\omega}^2 - \Omega_\perp^2}{c_s^2}H^2\right]u_z + H\frac{\partial^2 u_r}{\partial\eta\partial r} = 0. \tag{8.12}$$

This equation is identical with equations used in Chap. 6 for classification of oscillations, as shown below.

In the case of no magnetic fields, the z-component of equation of motion is $i\tilde{\omega}u_z = -\partial h_1/\partial z$. Hence, equation (8.12) can be written in the form:

$$\left[\frac{\partial^2}{\partial\eta^2} - \eta\frac{\partial}{\partial\eta} + \frac{\tilde{\omega}^2 - \Omega_\perp^2}{c_s^2}H^2\right]\frac{\partial h_1}{\partial\eta} = i\tilde{\omega}H^2\frac{\partial^2 u_r}{\partial\eta\partial r}. \tag{8.13}$$

Eliminating u_φ from the r- and φ-components of equation of motion, we have $(\tilde{\omega}^2 - \kappa^2)u_r = i\tilde{\omega}\partial h_1/\partial r$. Substitution of this equation into equation (8.13) to eliminate u_r

gives us an equation of h_1, which is

$$\left(\frac{\partial^2}{\partial\eta^2}-\eta\frac{\partial}{\partial\eta}+\frac{\tilde{\omega}^2-\Omega_\perp^2}{c_s^2}H^2\right)\left(\frac{\partial h_1}{\partial\eta}\right)+\tilde{\omega}H^2\frac{\partial}{\partial r}\left(\frac{\tilde{\omega}}{\tilde{\omega}^2-\kappa^2}\frac{\partial}{\partial r}\right)\left(\frac{\partial h_1}{\partial\eta}\right)=0. \tag{8.14}$$

If we write $h_1(r,\eta)$ in a separable form: $h_1(r,\eta) = g(\eta)f(r)$, and consider that $[d^2/d\eta^2 - \eta d/d\eta + (n-1)](dg/d\eta) = 0$, we have

$$\tilde{\omega}\frac{d}{dr}\left[\frac{\tilde{\omega}}{\tilde{\omega}^2-\kappa^2}\frac{d}{dr}\right]f+\frac{\tilde{\omega}^2-n\Omega_\perp^2}{c_s^2}f=0. \tag{8.15}$$

This is the wave equation considered in Chap. 6 (see equation (6.1)).

8.1.1 Derivation of Wave Equation Describing Radial Behavior

We are interested here in c-mode ($n = 1$) and vertical p-mode ($n \geq 2$) oscillations in the presence of toroidal magnetic fields. In these oscillations, motions are nearly vertical. In this sense, we introduce here an approximation that the main terms in equation (8.11) are those of the first brackets and the terms of the second ones are small perturbed quantities.[1] Although the terms of the second brackets are small, they are of importance to determine the wave trapping in the radial direction, as shown below.

First, we should notice that the third term on the first large bracket in equation (8.11) is a function of r, but its radial change in the trapped region is not large because the trapped region is generally narrow as the final results show. Hence, its weak r-dependence is regarded as a small perturbed quantity when we solve equation (8.11) by a perturbation method. To do so we introduce a small dimensionless quantity $\epsilon(r)$ defined by

$$\frac{\tilde{\omega}^2-\Omega_\perp^2}{c_s^2+c_A^2}H^2=\left[\frac{\tilde{\omega}^2-\Omega_\perp^2}{c_s^2+c_A^2}H^2\right]_c+\epsilon(r), \tag{8.16}$$

where the subscript c represents the value at capture radius (trapped radius), r_c.[2] The capture radius, r_c, is the radius where $\epsilon(r)$ vanishes there and represents the outer boundary of the propagation region of oscillations, as will be shown later. Inside

[1]In nearly vertical oscillations, the vertical component of equation motion gives approximately $u_z \sim h_1/c_s$ (e.g., see equation (8.6) in the limit of $c_A = 0$). Furthermore, the radial component of equation of motion (equation (8.4)) gives $u_r \sim (1/\Omega\lambda)h_1$. Combining these two relations, we have $u_r \sim (H/\lambda)u_z$, i.e., u_r is smaller than u_z by a factor of H/λ. Hence, the terms of the second brackets of equation (8.11) is smaller than the terms of the first brackets by a factor of $(H/\lambda)^2$.

[2]In Chap. 7, r_{out} was used to denote r_c. Both of them have the same meanings.

the radius r_c, $\epsilon(r)$ is a small positive quantity depending on r. The magnitude of $\epsilon(r)$ can be found from equation (8.16) when r_c and ω are determined later by an eigenvalue problem in the radial direction.

By introducing small dimensionless quantity ϵ, we can write equation (8.11) in the lowest order of approximations with respect to ϵ as

$$\frac{\partial^2}{\partial\eta^2}u_z^{(0)} - \eta\frac{\partial}{\partial\eta}u_z^{(0)} + \left[\frac{\tilde{\omega}^2 - \Omega_\perp^2}{c_s^2 + c_A^2}H^2\right]_c u_z^{(0)} = 0, \tag{8.17}$$

where the superscript (0) is attached to u_z in order to emphasize that it is the quantity of the lowest order of approximations. By imposing the boundary condition that $u_z^{(0)}$ does not grow exponentially at $z = \pm\infty$, we find that the z-dependence of $u_z^{(0)}$ can be expressed by a Hermite polynomial, $\mathscr{H}_n(\eta)$, and the term in the large brackets in equation (8.17) is determined as the eigenvalue (see Chap. 4). That is, we have

$$u_z^{(0)} = f_z(r)g_z^{(0)}(\eta), \tag{8.18}$$

where

$$g_z^{(0)}(\eta) = \mathscr{H}_{n-1}(\eta), \quad n = 1, 2, 3\ldots \tag{8.19}$$

and

$$\left[\frac{\tilde{\omega}^2 - \Omega_\perp^2}{c_s^2 + c_A^2}H^2\right]_c = n - 1. \tag{8.20}$$

Here, it is noted that the eigenfunction is taken to be $\mathscr{H}_{n-1}$, not $\mathscr{H}_n$. The reason is that in many previous studies h_1 is adopted as the dependent variable (not u_z) and the z-dependence of h_1 is taken to be proportional to $\mathscr{H}_n$, where n is the node number of h_1 in the vertical direction. In this case u_z is proportional to $\mathscr{H}_{n-1}(\eta)$ and the node number of u_z in the vertical direction is smaller than that of h_1 by unity. As shown in equation (8.18), in the lowest order of approximations, $u_z^{(0)}(r, \eta)$ is expressed in a separable form with respect to r and η. The r-dependence of $u_z^{(0)}$ is free at this stage, which is denoted by $f_z(r)$ in equation (8.18). It will be determined later by solving an eigenvalue problem in the radial direction.

Equation (8.20) can be rewritten in the form

$$\left(\frac{\omega - m\Omega}{\Omega_\perp}\right)_c^2 = \left[\frac{c_s^2 + c_A^2}{c_s^2 + c_A^2/2}\right]_c (n-1) + 1. \tag{8.21}$$

In the limit of $c_A^2 = 0$, this equation is reduced to $\tilde{\omega}_c^2 = n\Omega_{\perp c}^2$, which is the expected result (see Chap. 4). The local dispersion relation in non-magnetized isothermal disks shows that the propagation region of c-mode and vertical p-mode oscillations is described by $\tilde{\omega}^2 > n\Omega_\perp^2$. This means that for oscillations with ω, one of their propagation region is $\omega < m\Omega - n^{1/2}\Omega_\perp$. That is, the outer boundary of this propagation region on the $\omega - r$ plane is given by $\omega = m\Omega - n^{1/2}\Omega_\perp$. Equation (8.21) suggests that in the present magnetized disks, the outer boundary of the propagation region is given by

$$\omega = m\Omega - \left[\frac{c_s^2 + c_A^2}{c_s^2 + c_A^2/2}(n-1) + 1\right]^{1/2}\Omega_\perp. \tag{8.22}$$

Now, we proceed to take into account the deviation of the oscillations from purely vertical ones by a perturbation method. We see soon that separation of u_z into two functions of r and η is no longer valid. Hence, we consider the effects of small perturbed quantities by introducing a weak r-dependence in g. That is, u_z is now written as

$$u_z(r,\eta) = f_z(r)[g_z^{(0)}(\eta) + g_z^{(1)}(r,\eta) + \ldots]. \tag{8.23}$$

Then, from equation (8.11), using equations (8.16) and (8.20) we obtain, as an equation describing $f_z g_z^{(1)}$,

$$f_z(r)\left(\frac{\partial^2}{\partial\eta^2} - \eta\frac{\partial}{\partial\eta} + n - 1\right)g_z^{(1)}(r,\eta) = -\epsilon(r)f_z(r)g_z^{(0)}(\eta) - H\left[\frac{\partial}{\partial\eta} - \frac{c_A^2/2}{c_s^2 + c_A^2}\right]\frac{\partial u_r^{(0)}}{\partial r}, \tag{8.24}$$

where $u_r^{(0)}$ is the lowest order expression for $u_r(r,\eta)$.

The next subject is to solve equation (8.24). To do so, $u_r^{(0)}$ is expressed in terms of $u_z^{(0)}[= f_z(r)g_z^{(0)}(\eta)]$. First, eliminating u_φ from equations (8.4) and (8.5), we have

$$(-\tilde{\omega}^2 + \kappa^2)u_r^{(0)} = -i\tilde{\omega}\left[\frac{\partial h_1^{(0)}}{\partial r} + c_A^2\frac{\partial}{\partial r}\left(\frac{b_\varphi^{(0)}}{B_0}\right)\right]. \tag{8.25}$$

In the lowest order of approximations, the term of $\partial u_r/\partial r$ on the right-hand side of equation (8.10) can be neglected in evaluating h_1, compared with the term of $(\partial/\partial z - z/H^2)u_z$. Hence, by using equation (8.10) we can express $\partial h_1^{(0)}/\partial r$ on the right-hand side of equation (8.25) directly by u_z. Furthermore, the main term on the right-hand side of equation (8.8) is the term with $u_z^{(0)}$. Hence, by using equation (8.8) the term of $\partial(b_\varphi^{(0)}/B)/\partial r$ on the right-hand side of equation (8.25) can be also expressed in terms of $u_z^{(0)}$. Consequently, in the lowest order approximations of nearly vertical oscillations, $u_r^{(0)}$ can be expressed in terms of $u_z^{(0)}$ alone from equation (8.25). After some manipulations we have finally

$$u_r^{(0)} = \mathscr{L}_s\left(\frac{\partial}{\partial\eta} - \eta\right)u_z^{(0)} + \mathscr{L}_A\left(\frac{\partial}{\partial z} - \frac{1}{2}\eta\right)u_z^{(0)}, \tag{8.26}$$

where $\mathscr{L}_{\rm s}$ and $\mathscr{L}_{\rm A}$ are operators defined by

$$\mathscr{L}_{\rm s} = \frac{c_{\rm s}^2/H}{-\tilde{\omega}^2+\kappa^2}\left[\frac{\partial}{\partial r} - \frac{d\ln\tilde{\omega}}{dr}\right] \tag{8.27}$$

and

$$\mathscr{L}_{\rm A} = \frac{c_{\rm A}^2/H}{-\tilde{\omega}^2+\kappa^2}\left[\frac{\partial}{\partial r} - \frac{d\ln\tilde{\omega}}{dr}\right]. \tag{8.28}$$

Now, we return to equation (8.24). The equation is an inhomogeneous equation with respect to $g_z^{(1)}(r,\eta)$:

$$\begin{aligned}\left(\frac{\partial^2}{\partial\eta^2} - \eta\frac{\partial}{\partial\eta} + n - 1\right)g_z^{(1)}(r,\eta) &= -\epsilon(r)g_z^{(0)}(\eta)\\ &-\left(\frac{d}{d\eta} - \frac{c_{\rm A}^2/2}{c_{\rm s}^2+c_{\rm A}^2}\right)\left(\frac{d}{d\eta} - \eta\right)g_z^{(0)}(\eta)\frac{H}{f_z}\frac{d}{dr}\mathscr{L}_{\rm s}(f_z)\\ &-\left(\frac{d}{d\eta} - \frac{c_{\rm A}^2/2}{c_{\rm s}^2+c_{\rm A}^2}\right)\left(\frac{d}{d\eta} - \frac{1}{2}\eta\right)g_z^{(0)}(\eta)\frac{H}{f_z}\frac{d}{dr}\mathscr{L}_{\rm A}(f_z).\end{aligned} \tag{8.29}$$

As is done in a standard perturbation method and also in Chaps. 4 and 5, $g_z^{(1)}(r,\eta)$ is now expressed in a series of orthogonal functions of the zeroth order equation as

$$g_z^{(1)}(r,\eta) = \sum_m c_m(r)\mathscr{H}_m(\eta). \tag{8.30}$$

The solvability condition of equation (8.29) is obtained by using the orthogonality of the Hermite polynomials, and is written as

$$\begin{aligned}&f_z\left\langle\mathscr{H}_{n-1}^2(\eta)\right\rangle\epsilon(r)\\ &+H\frac{d}{dr}\mathscr{L}_{\rm s}(f_z)\left\langle\mathscr{H}_{n-1}\left(\frac{d}{d\eta} - \frac{1}{2}\frac{c_{\rm A}^2}{c_{\rm s}^2+c_{\rm A}^2}\right)\left(\frac{d}{d\eta} - \eta\right)\mathscr{H}_{n-1}\right\rangle\\ &+H\frac{d}{dr}\mathscr{L}_{\rm A}(f_z)\left\langle\mathscr{H}_{n-1}\left(\frac{d}{d\eta} - \frac{1}{2}\frac{c_{\rm A}^2}{c_{\rm s}^2+c_{\rm A}^2}\right)\left(\frac{d}{d\eta} - \frac{1}{2}\eta\right)\mathscr{H}_{n-1}\right\rangle = 0,\end{aligned} \tag{8.31}$$

where $\langle A(\eta)B(\eta)\rangle$ is the integration of $A(\eta)B(\eta)$ with respect to η in the range of $(-\infty,\infty)$ with a weighting function $\exp(-\eta^2/2)$.

This solvability condition leads to an ordinary differential equation of $f_z(r)$, when the integration with respect to η is performed. After some manipulation we can write the results in the form:

$$A\left(c_{\rm s}^2+\frac{1}{2}c_{\rm A}^2\right)\frac{d}{dr}\left[\frac{\tilde{\omega}}{\tilde{\omega}^2-\kappa^2}\frac{d}{dr}\left(\frac{f_z(r)}{\tilde{\omega}}\right)\right]+\epsilon(r)f_z(r)=0, \tag{8.32}$$

where

$$A=n-\frac{1}{2}\frac{c_{\rm A}^2}{c_{\rm s}^2+c_{\rm A}^2}\frac{nc_{\rm s}^2+c_{\rm A}^2/2}{c_{\rm s}^2+c_{\rm A}^2/2}. \tag{8.33}$$

In the previous studies (Chap. 4) on nearly vertical oscillations in non-magnetized disks, we have adopted h_1 (not u_z) as a dependent variable. To compare our present results with those in the previous ones, we introduce here a new variable $\tilde{f}$ defined by $\tilde{f}=f_z/\tilde{\omega}$.[3] Then, equation (8.32) is reduced to (Kato 2011)

$$\frac{1}{\tilde{\omega}}\frac{d}{dr}\left[\frac{\tilde{\omega}}{\tilde{\omega}^2-\kappa^2}\frac{d\tilde{f}}{dr}\right]+\frac{\epsilon}{A\Omega_\perp^2H^2}\tilde{f}=0, \tag{8.34}$$

where $\epsilon(r)$ is written explicitly by using equations (8.16) and (8.20) as

$$\epsilon(r)=\frac{\tilde{\omega}^2-\Omega_\perp^2}{c_{\rm s}^2+c_{\rm A}^2}H^2-(n-1). \tag{8.35}$$

Equation (8.34) is a wave equation describing the radial behavior of c-mode and vertical p-mode oscillations in disks with toroidall magnetic fields. In the limit of $c_{\rm A}^2=0$, equation (8.34) becomes formally the same as that used in Chap. 6.

8.1.2 Radial Eigenvalue Problems

We solve equation (8.34) as an eigenvalue problem to examine where the oscillations are trapped and how much the eigenfrequency of the trapped oscillations depends on strength of toroidal magnetic fields. As we did in Chap. 7, we introduce a new independent variable, $\tau(r)$, defined by

$$\tau(r)=\int_{r_{\rm in}}^{r}\frac{\tilde{\omega}^2(r')-\kappa^2(r')}{-\tilde{\omega}(r')}dr', \quad \tau_{\rm c}\equiv\tau(r_{\rm c}), \tag{8.36}$$

---

[3]In the lowest order of approximations, the equation of continuity gives $i\tilde{\omega}h_1+(c_{\rm s}^2/H)(\partial/\partial\eta-\eta)u_z=0$. If u_z is taken to be proportional to $\mathcal{H}_{n-1}(\eta)$, i.e., $u_z=f_z(r)\mathcal{H}_{n-1}(\eta)$, the above continuity relation shows that h_1 has a component proportional to $\mathcal{H}_n(\eta)$, i.e., $h_1=f_h(r)\mathcal{H}_n(\eta)$ and $_zf(r)$ and $f_h(r)$ is related by $i\tilde{\omega}f_h=(c_{\rm s}^2/H)f_z$.

where $\tau = 0$ at the inner edge of disks and $\tau = \tau_c$ at the capture radius r_c. It is noted that $\tilde{\omega}$ is negative in the inner trapped region of c-mode and vertical p-mode oscillations. Equation (8.34) is written in the form:

$$\frac{d^2\tilde{f}}{d\tau^2} + Q\tilde{f} = 0, \tag{8.37}$$

where

$$Q(\tau) = \frac{\tilde{\omega}^2}{\tilde{\omega}^2 - \kappa^2} \frac{\epsilon}{A\Omega_\perp^2 H^2} \tag{8.38}$$

and $\epsilon(r)$ is given by equation (8.35). It is noted that the dependent variable is changed from f_z to $\tilde{f}$. Equations (8.37) and (8.38) show that the propagation region of oscillations is the region where $Q > 0$, which is the region of $\epsilon > 0$.

We solved equation (8.37) by a standard WKB method with relevant boundary conditions. The outer edge of the propagation region is a turning point where Q vanishes, and as the inner boundary condition, we adopt $\tilde{f} = 0$ (see Sect. 7.1). Then, the trapping condition is (see Chap. 7)

$$\int_0^{\tau_c} Q^{1/2} d\tau = \left(\frac{3}{4} + n_r\right)\pi, \tag{8.39}$$

where $n_r(= 0, 1, 2, \ldots)$ is zero or a positive integer specifying the node number of $\tilde{f}$ in the radial direction. If independent variable is returned to r from τ, the above trapping condition is written as

$$\int_{r_{\rm in}}^{r_c} \frac{(\tilde{\omega}^2 - \kappa^2)^{1/2}}{c_{\rm s}} \left(\frac{\epsilon}{A}\right)^{1/2} dr = \left(\frac{3}{4} + n_r\right)\pi. \tag{8.40}$$

In the limit of $c_{\rm A} = 0$, equation (8.40) is reduced to

$$\int_{r_{\rm in}}^{r_c} \frac{(\tilde{\omega}^2 - \kappa^2)^{1/2}(\tilde{\omega}^2 - n\Omega_\perp^2)^{1/2}}{c_{\rm s}(n\Omega_\perp^2)^{1/2}} dr = \left(\frac{3}{4} + n_r\right)\pi. \tag{8.41}$$

This equation should be compared with equation (7.16). In the limit of $c_{\rm A} = 0$, an equation corresponding to equation (8.41) has been derived without using the approximation of $\epsilon < 1$, which is equation (7.16). Comparison of equations (8.41) and (7.16) shows that the present approximation of $\epsilon(r) < 1$ corresponds to partially regarding $\tilde{\omega}^2$ as $n\Omega_\perp^2$. The validity of the approximation of $\epsilon < 1$ can be checked by calculating the value of ϵ after the final results are obtained. The calculation shows that this approximation is acceptable as the first approximation (see figure 9 of Kato 2012b).

For a given set of parameters, including spin parameter, a_*, and mass of neutron stars, M, any solution of equation (8.40) specifies $r_{\rm c}$, which gives ω of the trapped oscillation through equation (8.21). In other words, ω and r_c are related by equation (8.21), i.e., $\omega = \omega(r_c)$ or $r_c = r_c(\omega)$, and the trapping condition determines r_c or ω as functions of such parameters as c_S^2, c_A^2, a_* and M.

8.1.3 Comparison of c-Mode Oscillations with KHz QPOs

To obtain numerical values of the frequency, ω, and the capture radius, r_c, of trapped oscillations, we need to specify the radial distribution of acoustic speed, $c_s(r)$, and strength of magnetic fields, $c_A^2(r)$. In addition, the mass of the central star, M, and its spin, a_*, are needed. As dimensionless parameters specifying these quantities, we adopt $c_s/c_{s0} \equiv \beta$, c_A^2/c_s^2, $M/M_\odot$, and a_*. Here, $c_{s0}(r)$ is a reference distribution of acoustic speed in disks, which is given by equation (7.24).

We particularly consider two-armed ($m = 2$) c-mode ($n = 1$) oscillations. The inner boundary of oscillations is taken at the radius of $\kappa = 0$, i.e., at $r_{\rm ISCO}$ (the radius of the marginally stable circular orbit), and $u_z = 0$ (i.e., $\tilde{f} = 0$) is adopted there. We had calculated the cases where $du_z/dr = 0$ at the radius. We find that the differences of these boundary conditions bring about small quantitative differences in results, but there is no essential difference in parameter dependences of results.

In Sect. 7.3 we have presented a picture that two-armed ($m = 2$) c-mode ($n = 1$) oscillations of the two fundamental modes ($n_r = 0$ and $n_r = 1$) in the radial direction are the upper and lower kHz QPOs, respectively. Our purpose here is to examine whether this picture is affected if toroidal magnetic fields are present. Kato (2012b) extended this examination to cases where disks with toroidal magnetic fields have finite thickness in the vertical direction. Including this latter case, we summarize here comparison of calculated frequency correlations between two modes ($n_r = 0$ and $n_r = 1$) and observed correlations of twin kHz QPOs.

First, we focus our attention to disks with $M = 1.4\,M_\odot$ and $a_* = 0$, and consider the case where the disk is non-terminated isothermal one ($\eta_s = \infty$). In this case we find that the calculated correlation curve is almost universal in the sense that for a moderate set of $\beta \equiv c_s^2/(c_s^2)_0$ and c_A^2/c_s^2, the calculated correlation curve is almost a part of a unique curve. To demonstrate this, let us first consider the case where β is fixed at $\beta = 3.0$ and c_A^2/c_s^2 is changed from 0 to 64, which is shown in Fig. 8.1. When $c_A^2/c_s^2 = 0$, the point representing the set of frequencies of the $n_r = 0$ and $n_r = 1$ oscillations on the frequency-frequency diagram is the right-end point of the curve in Fig. 8.1. In the another limit of $c_A^2/c_s^2 = 64$, the point is the left-end of the curve. As the value of c_A^2/c_s^2 changes from 0 to 64, the point on the diagram moves along the curve from the right-end to the left-end. As in the cases in Chap. 7 (no magnetic field), observational data of kHz QPOs of some typical sources are plotted, using a part of the figure of Abramowicz (2005). Next, let us consider the case where $\beta = 10.0$ and c_A^2/c_s^2 is changed from 0 to 64, which is shown in Fig. 8.2. Comparison of Figs. 8.1 and 8.2 shows that an increase of β under keeping other parameter values unchanged decreases frequencies of both $n_r = 0$ and $n_r = 1$ oscillations. Hence, in the case of Fig. 8.2, the correlation curve on the frequency-

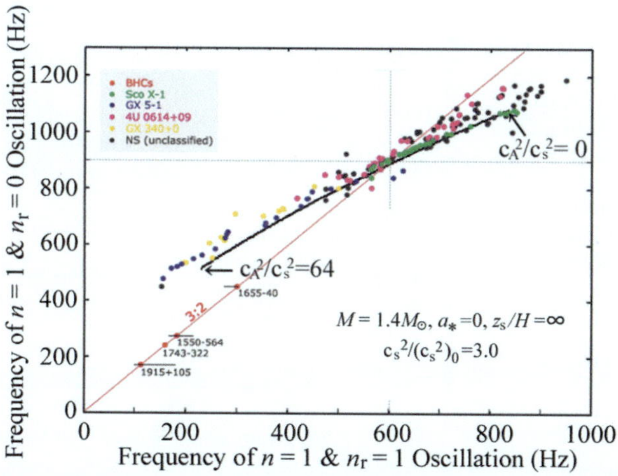

Fig. 8.1 Diagram comparing the calculated frequency-frequency correlation curve with the observed frequency-frequency plots of twin kHz QPOs of some typical NS LMXBs. The plots of observational data are a part of the figure of Abramowicz (2005). Adopted parameters specifying disk structure are $\eta_s = \infty$ and $\beta \equiv c_s^2/(c_s^2)_0 = 3.0$. The value of c_A^2/c_s^2 is changed from $c_A^2/c_s^2 = 0$ to 64.0, and the left and right ends of the correlation curve are for $c_A^2/c_s^2 = 64.0$ and 0, respectively (After Kato 2012a, PASJ ©).

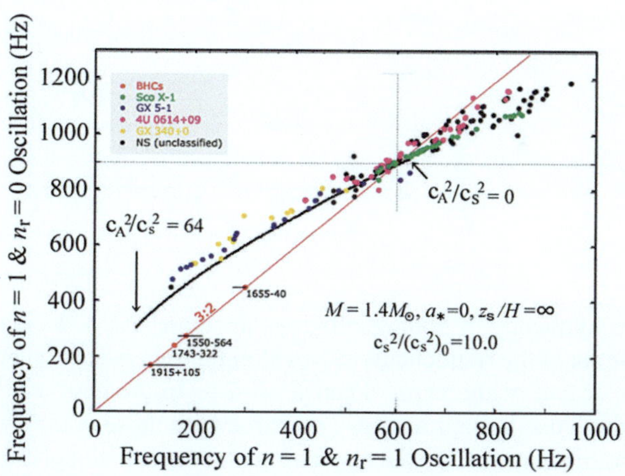

Fig. 8.2 The same as Fig. 8.1, except for $\beta = 10.0$ (After Kato 2012a, PASJ ©).

frequency diagram move leftwards compared with the case of Fig. 8.1. In the part overlapping, however, the curves in Figs. 8.1 and 8.2 are almost the same.

Next, let us consider the case where c_A^2/c_s^2 is fixed and β is changed in the range of $\beta = 1.0 \sim 10.0$, which is shown in Fig. 8.3 for $c_A^2/c_s^2 = 4.0$ and in Fig. 8.4 for $c_A^2/c_s^2 = 16.0$. In both figures, the left-end points of the curves are for $\beta = 10.0$. This is because if the value of β increases under keeping other parameter values

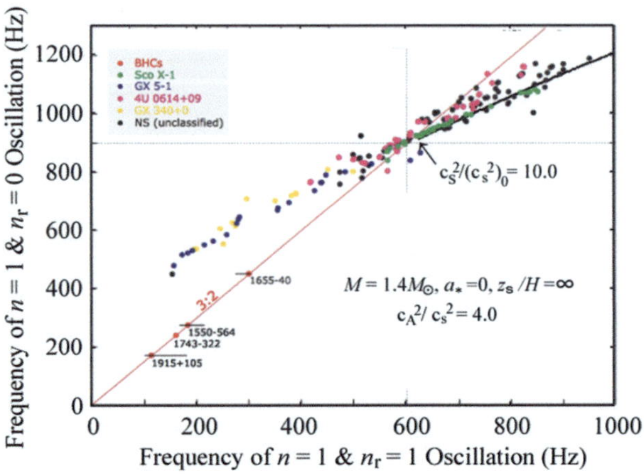

Fig. 8.3 The same as Figs. 8.1 and 8.2, except that c_A^2/c_s^2 is fixed at $c_A^2/c_s^2 = 4.0$ and β is change in the range of $\beta = 1.0$ to $\beta = 10.0$. The right-upper end-point of the correlation curve is for $\beta = 1.0$, and the left-lower end-point of the curve is for $\beta = 10.0$ (After Kato 2012a, PASJ ©).

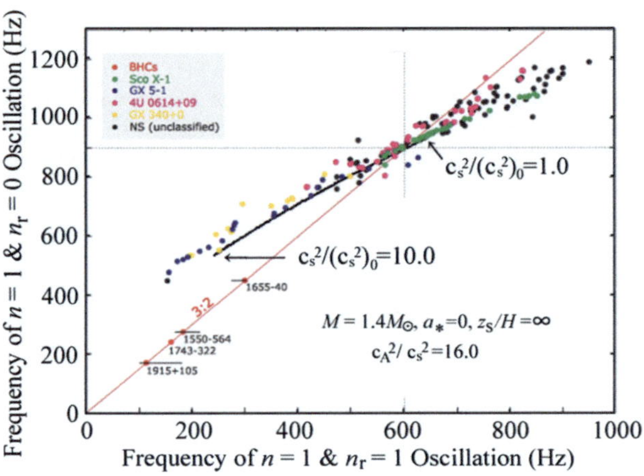

Fig. 8.4 The same as Fig. 8.3, except that c_A^2/c_s^2 is fixed at $c_A^2/c_s^2 = 16.0$ (After Kato 2012a, PASJ ©).

unchanged, the frequency of both oscillations decrease. As β decreases, the point representing the set of the frequencies moves along the correlation curve and reaches at the right-upper end of the curve when $\beta = 1.0$. In the case of $c_A^2/c_s^2 = 16.0$, frequencies of both oscillations are low compared with those in the case of $c_A^2/c_s^2 = 4.0$, if other parameters are fixed. Hence, in Fig. 8.4 the calculated correlation curve shifts to the left-lower direction compared with the case of $c_A^2/c_s^2 = 4.0$. In the region where frequencies are overlapped, however, the correlation curve is almost the same as that in Fig. 8.3. That is, as mentioned before, the correlation curve is almost unique. Of course, if the mass of the central source is different from $M = 1.4\,M_\odot$ the correlation curve on the frequency-frequency diagram is not overlapped (see below).

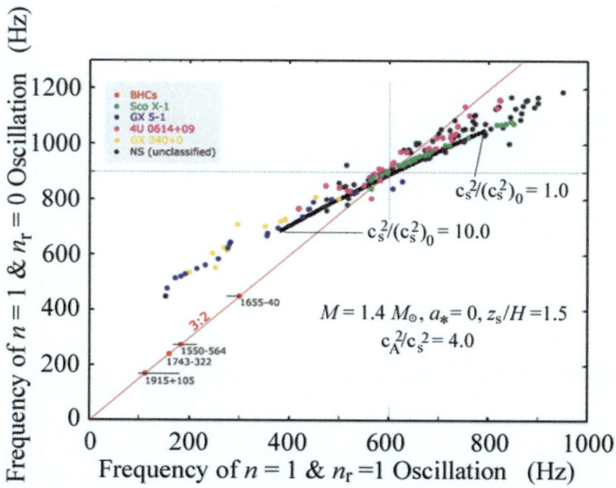

Fig. 8.5 The same as Fig. 8.3, except that the disk is vertically trancated with $\eta_s = 1.5$ (After Kato 2012a, PASJ ©).

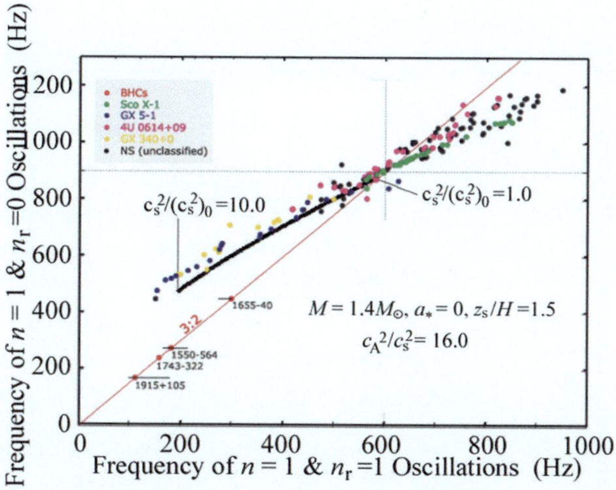

Fig. 8.6 The same as Fig. 8.5 except for $c_A^2/c_s^2 = 16.0$ (After Kato 2012a, PASJ ©).

Here, we consider the case where the disks are terminated in the vertical direction. To demonstrate rather extreme cases, we adopt $\eta_s = 1.5$. The correlation curve is calculated by changing the value of $c_s^2/(c_s^2)_0$ from 0 to 10.0, fixing the value of c_A^2/c_s^2. The case where $c_A^2/c_s^2 = 4.0$ is shown in Fig. 8.5, and the case of $c_A^2/c_s^2 = 16.0$ is in Fig. 8.6. The results show that decrease of the disk thickness brings about no essential characteristic changes on the correlation curves.

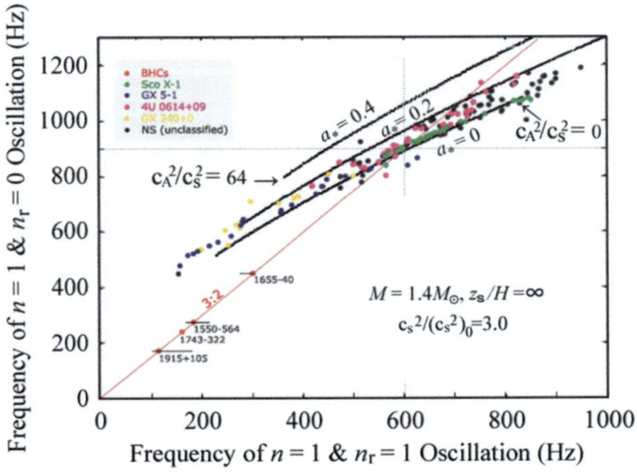

Fig. 8.7 Same as Fig. 8.1, except for cases where the spin parameter, a_*, is non-zero are considered. Among three correlation curves, the upper one is for $a_* = 0.4$, the middle one is for $a_* = 0.2$. The lower one is for $a_* = 0$, and the same as that of Fig. 8.1 (After Kato 2012a, PASJ ©).

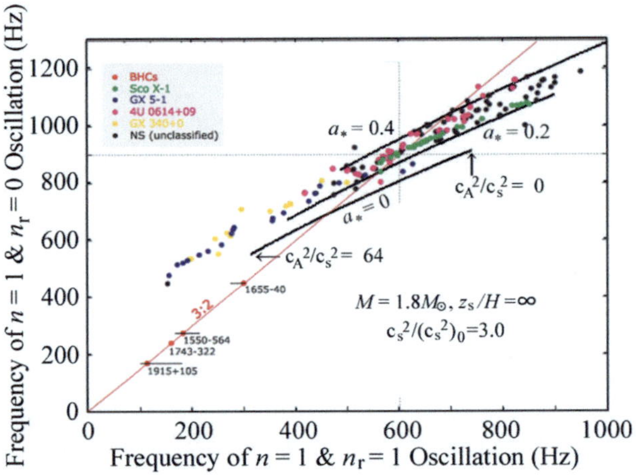

Fig. 8.8 Same as Fig. 8.7, except that the mass of the central source is $1.8\,M_\odot$ (After Kato 2012a, PASJ ©).

Next, we show effects of mass and spin of the central source on the correlation curve. Figure 8.7 is for three cases of $a_* = 0$, 0.2, and 0.4 with $M = 1.4\,M_\odot$. Non-terminated disks with $\beta = 3.0$ are adopted and c_A^2/c_s^2 is changed in the range of $c_A^2/c_s^2 = 0$–64. Figure 8.8 is the same as Fig. 8.7 except that $M = 1.8\,M_\odot$ is adopted. As is easily understood, an increase of spin of the central source increases the frequencies of both the $n_r = 0$ and $n_r = 1$ oscillations, but the increase in the $n_r = 0$ oscillation is larger compared with that of the $n_r = 1$, since the former is more concentrated in the inner region of disks. Hence, the correlation curve shifts in the right-upper direction on the frequency-frequency diagram as a_* increases, as is shown in Figs. 8.7 and 8.8. An increase of mass of the central source shifts the correlation curve in the left-lower direction on the diagram, as is also easily understood from the fact that mass increase decreases the frequencies of oscillations. Comparison of Fig. 8.7 with Fig. 8.8 shows, for example, that if the masses of typical LMXBs presented in figures are $1.4\,M_\odot$, their spin will be around $a_* \sim 0$ and at most $a_* < 0.2$. If their masses are $1.8\,M_\odot$, their spin is larger and $0.2 < a_* < 0.4$.

8.2 Extremely Low-Frequency Global Oscillations and Application to Be-Star Disks

In Chaps. 6 and 7, we have examined disk oscillations, neglecting radial variations of $H(r)$, $\rho_{00}(r)$, and others. This is allowed as the first approximation, because the trapped region of oscillations is narrow in many cases. As was shown in Chap. 6, however, the trapped region of the tilt mode is rather wide in the radial direction. Furthermore, observations of Be stars frequently show the so-called V/R variation in emission line spectra with period of a few to a few ten years, as described in Chap. 1. These V/R variations are considered to be due to global low-frequency oscillations (one-armed p-mode oscillations, i.e., $m = 1$ and $n = 0$, or in other words, one-armed precession modes) in Keplerian disks surrounding Be stars (Kato 1983; Okazaki 1991).

Effects of radial change of disk thickness, H, on disk oscillations were already examined in Chap. 5 by regarding $d\ln H/d\ln r$ as the first-order small perturbed quantities. In these treatments, effects of radial variations of unperturbed disk quantities such as $d\ln\rho_{00}/d\ln r$ on disk oscillations have been taken into account. Examination of disk oscillations by using such wave equations have been made by Okazaki (1991) in the case of one-armed oscillations in order to study the V/R variations of Be stars.

His results show that discrete low-frequency oscillation modes are possible only when the disks have a definite outer boundary, and further that the resulting oscillations are retrograde. These results are, however, not always compatible with observations, since many Be stars are single stars (not binary systems) and thus the disks have no definite outer boundaries (no tidal truncation), and further many observations suggest that one-armed precession waves in many Be stars are prograde, as mentioned in Sect. 1.5.2. This discord between observations and calculations of precession waves will suggest that somethings are missed in theoretical considerations of oscillations. Hence, many attempts have been made to obtain trapped prograde precession modes in non-truncated Be disks. That is, Papaloizou et al. (1992), Papaloizou and Heemskerk (1993), and Savonije and Heemskerk (1993) considered the effects of the quadrupole gravitational potential associated with the rotational deformation of the star. Papaloizou and Savonije (2006) considered the effects of inner boundary condition. Ogilvie (2008) emphasized the importance of 3D behaviors of oscillations. That is, he pointed out the importance of taking into account the quantities of the order of $(d\ln H/d\ln r)^2$. The purpose of this section is, as Ogilvie (2008) did, to proceed to the second order approximation with respect to a small expansion parameter $d\ln H/d\ln r$.

In the following formulations, we use u_r (not h_1) as the dependent variable, since in studies of global disk oscillations in Be stars, u_r are often used instead of h_1 as the dependent variable.[4] In Sect. 5.3 we have derived a wave equation with respect to u_r till the first order of $d\ln H/d\ln r$ for wave modes of arbitrary m and n. What we need here, however, is a wave equation including terms till the second order of $d\ln H/d\ln r$. Derivation of such equations are generally very complicated. Hence, we restrict our attention in this section only to the oscillations of $m = 1$ and $n = 0$, since we are now interested in the V/R variations of Be stars.

8.2.1 Basic Equations and Separation of Variables

We start from the basic equations described in Chap. 3, but here $\tilde{\omega}$ is approximated as $-\Omega$ except when it appears in the form of $(\tilde{\omega}^2 - \kappa^2)$, since in the present case $m = 1$, $\omega \ll \Omega$, and thus $\tilde{\omega} \sim -\Omega$. In the term of $(\tilde{\omega}^2 - \kappa^2)$, however, this approximation cannot be applied, because $(\tilde{\omega}^2 - \kappa^2)$ is a small quantity. The basic equations describing linear perturbations, equations (3.48) and (3.49), are then reduced, respectively, to

$$\frac{1}{r^2}\frac{\partial}{\partial r}(r^2 h_1) = i\frac{\tilde{\omega}^2 - \kappa^2}{\Omega}u_r, \tag{8.42}$$

$$\left(\frac{\partial^2}{\partial z^2} - \frac{z}{H^2}\frac{\partial}{\partial z} + \frac{\Omega^2}{c_s^2}\right)h_1 = -i\Omega\left[\frac{\partial}{\partial r} + \frac{d}{dr}\ln\left(\frac{\rho_{00}}{r\Omega}\right) + \frac{z^2}{H^2}\frac{d\ln H}{dr}\right]u_r. \tag{8.43}$$

In the above equations the independent variables are (r, z), not (r, η).

[4]The wave equation expressed in terms of h_1 is simpler than that expressed in terms of u_r, e.g., compare equations (5.28) and (5.35). See also that the basic wave equation (3.50) has been expressed by h_1. In spite of this, we use here u_r as the dependent variable to represent wave motions.

Our purpose here is to derive a wave equation expressed in terms of u_r from the above two equations. Equation (8.43) is multiplied by r^2H^2 and partial derivative with respect to r is taken. Then, using equation (8.42) we have

$$i\frac{\tilde{\omega}^2-\kappa^2}{\Omega}\left(H^2\frac{\partial^2}{\partial z^2}-z\frac{\partial}{\partial z}+1\right)u_r+\frac{dH^2}{dr}\frac{\partial^2 h_1}{\partial z^2}$$
$$=-i\frac{\partial}{r^2\partial r}\left[r^2\Omega H^2\left\{\frac{\partial}{\partial r}+\frac{d}{dr}\ln\left(\frac{\rho_{00}}{r\Omega}\right)+\frac{z^2}{H^2}\frac{d\ln H}{dr}\right\}u_r\right]. \tag{8.44}$$

The independent variables (r, z) are now changed to (r, η). Then, considering that $\partial/\partial r$ and $\partial/\partial z$ are changed, respectively, to $\partial/\partial r-\eta(d\ln H/dr)\partial/\partial\eta$ and $(1/H)\partial/\partial\eta$, we can arrange equation (8.44) in the form:

$$\left(\frac{\partial^2}{\partial\eta^2}-\eta\frac{\partial}{\partial\eta}+1\right)u_r-\frac{2i}{r}\frac{\Omega}{\tilde{\omega}^2-\kappa^2}\frac{d\ln H}{d\ln r}\frac{\partial^2 h_1}{\partial\eta^2}$$
$$=-\frac{1}{r^2}\frac{c_s^2}{\tilde{\omega}^2-\kappa^2}\left[r^2\left(\frac{\partial}{\partial r}+\frac{d}{dr}\ln(r^2\Omega)\right)\left(\frac{\partial}{\partial r}+\frac{d}{dr}\ln\frac{\rho_{00}}{r\Omega}\right)u_r\right.$$
$$+\frac{d\ln H}{d\ln r}r\left(2-\eta\frac{\partial}{\partial\eta}\right)\left(\frac{\partial}{\partial r}+\frac{d}{dr}\ln\frac{\rho_{00}}{r\Omega}\right)u_r$$
$$+\frac{d\ln H}{d\ln r}\left(\eta^2-\eta\frac{\partial}{\partial\eta}\right)\left(\frac{\partial}{\partial r}(ru_r)+\frac{d\ln\Omega}{d\ln r}u_r\right)$$
$$\left.+\left(\frac{d\ln H}{d\ln r}\right)^2\left(\eta^2\frac{\partial^2}{\partial\eta^2}-\eta\frac{\partial}{\partial\eta}-\eta^3\frac{\partial}{\partial\eta}\right)u_r\right]. \tag{8.45}$$

Here and hereafter, $d\ln H/d\ln r$ is assumed to be constant. Equation (8.45) is practically an equation for u_r, but partially h_1 retains as is shown on the right-hand side. To remove h_1 from this equation we must use again equation (8.42). Then, the order of the derivative with respect to r increases. For the purpose of the present analyses, however, it is unnecessary to proceed to such direction, since the term with h_1 in equation (8.45) is found later to contribute to results only in the order of $(d\ln H/d\ln r)^2$.

Now, u_r and h_1 are separated as

$$u_r(r,\eta)=f(r)g(\eta,r)\quad\text{and}\quad h_1(r,\eta)=f_h(r)g_h(\eta,r), \tag{8.46}$$

where the r-dependences of g and g_h are assumed to be weak. In these separation subscript h is attached to f and g for h_1, but no subscript to f and g for u_r.[5] By

[5] In Chaps. 5, 6, 7, and Sect. 8.1, we have used $f(r)$ in the case where $h_1(r,\eta)$ is separated as $h_1(r,\eta)=f(r)g(\eta,r)$. In this section, however, we use $f(r)$ when $u_r(r,\eta)$ is separated, and f_h is used when $h_1(r,\eta)$ is separated, without confusion.

dividing equation (8.45) by fg, we have

$$
\begin{aligned}
&\frac{1}{g}\left(\frac{\partial^2}{\partial\eta^2}-\eta\frac{\partial}{\partial\eta}+1\right)g-\frac{2i}{r}\frac{\Omega}{\tilde{\omega}^2-\kappa^2}\frac{d\ln H}{d\ln r}\frac{\partial^2 g_h}{\partial\eta^2}\frac{f_h}{fg}\\
&=-\frac{1}{r^2}\frac{c_s^2}{\tilde{\omega}^2-\kappa^2}\left[\frac{r^2}{fg}\left(\frac{\partial}{\partial r}+\frac{d}{dr}\ln(r^2\Omega)\right)\left(\frac{\partial}{\partial r}+\frac{d}{dr}\ln\frac{\rho_{00}}{r\Omega}\right)(fg)\right.\\
&\qquad+\frac{d\ln H}{d\ln r}\frac{r}{fg}\left(2-\eta\frac{\partial}{\partial\eta}\right)\left(\frac{\partial}{\partial r}+\frac{d}{dr}\ln\frac{\rho_{00}}{r\Omega}\right)(fg)\\
&\qquad+\frac{d\ln H}{d\ln r}\frac{1}{fg}\left(\eta^2-\eta\frac{\partial}{\partial\eta}\right)\left(\frac{\partial}{\partial r}(rfg)+\frac{d\ln\Omega}{d\ln r}fg\right)\\
&\qquad\left.+\left(\frac{d\ln H}{d\ln r}\right)^2\frac{1}{g}\left(\eta^2\frac{\partial^2}{\partial\eta^2}-\eta\frac{\partial}{\partial\eta}-\eta^3\frac{\partial}{\partial\eta}\right)g\right]. \qquad (8.47)
\end{aligned}
$$

This equation can be separated as

$$
\begin{aligned}
&\frac{1}{g}\left(\frac{\partial^2}{\partial\eta^2}-\eta\frac{\partial}{\partial\eta}\right)g+\frac{2}{r}\frac{d\ln H}{d\ln r}\frac{\Omega}{i(\tilde{\omega}^2-\kappa^2)}\frac{\partial^2 g_h}{\partial\eta^2}\frac{f_h}{gf}+\frac{1}{r^2}\frac{c_s^2}{\tilde{\omega}^2-\kappa^2}\frac{1}{g}B\\
&+\frac{1}{r^2}\frac{c_s}{\tilde{\omega}^2-\kappa^2}\frac{d\ln H}{d\ln r}\left[2r\frac{1}{g}\frac{\partial g}{\partial r}-\frac{1}{gf}r\left(\frac{\partial}{\partial r}+\frac{d}{dr}\ln\frac{\rho_{00}}{r\Omega}\right)\left(\eta\frac{\partial g}{\partial\eta}f\right)\right.\\
&\qquad\left.+\frac{1}{gf}\frac{1}{\Omega}\frac{\partial}{\partial r}\left\{r\Omega f\left(\eta^2-\eta\frac{\partial}{\partial\eta}\right)g\right\}\right]\\
&+\frac{1}{r^2}\frac{c_s^2}{\tilde{\omega}^2-\kappa^2}\left(\frac{d\ln H}{d\ln r}\right)^2\frac{1}{g}\left(\eta^2\frac{\partial^2}{\partial\eta^2}-(\eta^3+\eta)\frac{\partial}{\partial\eta}\right)g=K(r) \qquad (8.48)
\end{aligned}
$$

and

$$
\begin{aligned}
&1+\frac{1}{r^2}\frac{c_s^2}{\tilde{\omega}^2-\kappa^2}\left[\frac{r^2}{f}\left(\frac{d}{dr}+\frac{d}{dr}\ln(r^2\Omega)\right)\left(\frac{d}{dr}+\frac{d}{dr}\ln\frac{\rho_{00}}{r\Omega}\right)f\right.\\
&\qquad\left.+\frac{d\ln H}{d\ln r}2r\frac{1}{f}\left(\frac{d}{dr}+\frac{d}{dr}\ln\frac{\rho_{00}}{r\Omega}\right)f\right]=-K(r), \qquad (8.49)
\end{aligned}
$$

where $B(r,\eta)$ is defined by

$$
B(r,\eta)\equiv r^2\frac{\partial g}{\partial r}\frac{1}{f}\left[2\frac{d}{dr}+\frac{d}{dr}\ln(r\rho_{00})\right]f+r^2\frac{\partial^2 g}{\partial r^2} \qquad (8.50)
$$

and $K(r)$ is a separation constant. In the limit of $d\ln H/d\ln r=0$, K is obviously a constant. We consider cases where $d\ln H/d\ln r$ is a small expansion parameter in a perturbation method. Then, K depends weakly on r.

8.2.2 Solution Until the Order of $(d\ln H/d\ln r)^1$

The initial part of this subsection is a repetition of Sect. 5.2.2 in the special case of $m = 1$ and $n = 0$.

Equation (8.48) is solved by perturbation method by taking $d\ln H/d\ln r$ to be a small expansion parameter. That is, $g(\eta, r)$ and K are expanded as

$$g(\eta, r) = g^{(0)}(\eta) + g^{(1)}(\eta, r) + g^{(2)}(\eta, r) + \ldots \tag{8.51}$$

and

$$K(r) = K^{(0)} + K^{(1)}(r) + K^{(2)}(r) + \ldots, \tag{8.52}$$

where the superscript (0), (1), and (2) show the order of expansion when g and K are expanded in terms of $d\ln \mathrm{H}/d\ln r$. Then, in the zeroth order with respect to $d\ln H/d\ln r$, equation (8.48) is

$$\frac{d^2 g^{(0)}}{d\eta^2} - \eta \frac{dg^{(0)}}{d\eta} - K^{(0)} = 0, \tag{8.53}$$

where $\partial g^{(0)}/\partial r = \partial^2 g^{(0)}/\partial r^2 = 0$ have been used. Since we are interested here in one-armed p-mode oscillations, we adopt

$$g^{(0)}(\eta) = \mathscr{H}_0(\eta) = 1, \quad \text{and} \quad K^{(0)} = 0. \tag{8.54}$$

Next, we proceed to the order of $d\ln H/d\ln r$. From equation (8.48), as the first order equation, we have

$$\frac{\partial^2 g^{(1)}}{\partial \eta^2} - \eta \frac{\partial g^{(1)}}{\partial \eta}$$
$$= -\frac{1}{r^2}\frac{c_s^2}{\tilde{\omega}^2 - \kappa^2}\left[B^{(1)} + \frac{d\ln H}{d\ln r}\frac{1}{f}\left(r\frac{d}{dr} + \frac{d\ln r\Omega}{d\ln r}\right) f\eta^2 g^{(0)}\right] + K^{(1)} g^{(0)}. \tag{8.55}$$

In deriving equation (8.55), $dg^{(0)}/dr = 0$ and $dg^{(0)}/d\eta = 0$ have been used, since $g^{(0)} = \mathscr{H}_0 = 1$.

The first order expression for g, i.e., $g^{(1)}$ is generally written in the form:

$$g^{(1)}(r, \eta) = \left(\frac{d\ln H}{d\ln r}\right) \sum_{m \neq 0} a_m^{(1)}(r) \mathscr{H}_m(\eta). \tag{8.56}$$

Here, $a_m^{(1)}$ (except for $a_0^{(1)}$) is determined from equation (8.55) as will be shown later. The value of $a_0^{(1)}$ is, however, undetermined, i.e., $a_0^{(1)}$ is arbitrary. As usually done in perturbation method, we adopt $a_0^{(1)} = 0$, since the term has the same η-dependence as the zeroth-order solution and thus we can include it in the zeroth-order solution. Then, the solvability condition of equation (8.55) that the right-hand side of equation (8.55) has no component proportional to $g^{(0)}$ is written as

$$K^{(1)}f = \frac{1}{r^2}\frac{\Omega H^2}{\tilde{\omega}^2 - \kappa^2}\frac{d\ln H}{d\ln r}\frac{d}{dr}(r\Omega f). \tag{8.57}$$

It should be noted that the term $B^{(1)}$ in equation (8.55) has no contribution to the above solvability condition, since $a_0^{(1)} = 0$ has been adopted.

In order to derive a wave equation describing radial behavior of the oscillations, $K(r)$ on the right-hand side of equation (8.49) is written as $K = K^{(0)} + K^{(1)} = K^{(1)}$. Then, we have, from equation (8.49),

$$f + \frac{c_s^2}{\tilde{\omega}^2 - \kappa^2}\Bigg[\left(\frac{d}{dr} + \frac{d}{dr}\ln(r^2\Omega)\right)\left(\frac{d}{dr} + \frac{d}{dr}\ln\frac{\rho_{00}}{r\Omega}\right)f$$
$$+\frac{d\ln H}{d\ln r}\frac{1}{r}\left(3\frac{d}{dr} + \frac{d}{dr}\ln\frac{\rho_{00}^2}{r\Omega}\right)f\Bigg] = 0. \tag{8.58}$$

This equation can be arranged in the form:

$$\frac{d^2f}{dr^2} + \frac{d}{dr}\ln(r\rho_{00}H^3)\frac{df}{dr} + \frac{\tilde{\omega}^2 - \kappa^2}{c_s^2}f$$
$$+\left[\frac{d\ln r^2\Omega}{dr}\frac{d}{dr}\ln\left(\frac{\rho_{00}}{r\Omega}\right) + \frac{d^2}{dr^2}\ln\left(\frac{\rho_{00}}{r\Omega}\right) + \frac{d\ln H}{d\ln r}\frac{d}{rdr}\ln\left(\frac{\rho_{00}^2}{r\Omega}\right)\right]f = 0. \tag{8.59}$$

This is a wave equation describing radial behavior of oscillations in the case where the terms till the order of $d\ln H/d\ln r$ are taken into account.

Equation (8.59) is identical with the wave equation derived by Okazaki (1991) under the approximation that u_r is independent of η and $u_z \propto \eta$. It is further noticed that equation (8.59) is a special case of equation (5.35) in the sense that m and n are restricted here only to $m = 1$, $n = 0$, and $\tilde{\omega} \sim -\Omega$. It is noted that f in the present section is f_u in Chap. 5.[6]

Next, we consider the coefficients $a_m^{(1)}$ ($m \neq 0$) in the expansion of $g^{(1)}$ (see equation (8.56)). Multiplying $\mathscr{H}_2(\eta)$ to equation (8.55) and using the orthogonality

[6]It is noted again that in the present section the subscript u to f is omitted in order to avoid complexity. f in the present section should not be confused with f when $h_1(r,\eta)$ is separated as $h_1(r,\eta) = g(\eta, r)f(r)$.

relation of the Hermite polynomials, we have

$$2a_2^{(1)} = \frac{1}{r^2}\frac{\Omega H^2}{\tilde{\omega}^2-\kappa^2}\frac{1}{f}\frac{d}{dr}(r\Omega f)$$
$$+\frac{\Omega^2 H^2}{\tilde{\omega}^2-\kappa^2}\left[\frac{da_2^{(1)}}{dr}\frac{1}{f}\left(2\frac{d}{dr}+\frac{d}{dr}\ln(r\rho_{00})\right)f+\frac{d^2a_2^{(1)}}{dr^2}\right], \tag{8.60}$$

and $a_m^{(1)}$'s for other than $m = 2$ are zero. In equation (8.60) the terms with $da_2^{(1))}/dr$ and $da_2^{(1)}/dr^2$ comes from $B^{(1)}$. Equation (8.60) can be arranged into the following form:

$$\frac{d^2a_2^{(1)}}{dr^2}+\frac{d}{dr}\ln(f^2r\rho_{00})\frac{da_2^{(1)}}{dr}-\frac{\tilde{\omega}^2-\kappa^2}{c_s^2}2a_2^{(1)} = -\frac{1}{r^2\Omega f}\frac{d}{dr}(r\Omega f). \tag{8.61}$$

This is a second-order differential equation which determines the behavior of $a_2^{(1)}$. This equation is unnecessary in studying the first order wave equation for f (i.e., equation (8.59)), but becomes necessary when we proceed to the wave equation of the next order of approximations, as is shown in Sect. 8.2.3.

In the remaining part of this subsection, we examine characteristics of wave equation (8.59) and show that it is the same as that derived by Okazaki (1991). For simplicity, we consider the case where the disk thickess, H, varies as $H \propto r^{3/2}$ so that $c_s = $ const. in the radial direction,[7] and further $\rho_{00} \propto r^{-\alpha}$ is adopted with a parameter, α. Then, we can express equation (8.59) in the form:

$$\frac{d^2}{dr^2}\tilde{f}+k^2\tilde{f}=0, \tag{8.62}$$

where $\tilde{f}$ and k^2 are defined, respectively, by

$$\tilde{f}=r^{11/4-\alpha/2}f \tag{8.63}$$

and

$$k^2=\frac{(\omega-\Omega)^2-\kappa^2}{c_s^2}-\frac{(\alpha-1/2)(\alpha+3/2)}{4r^2}. \tag{8.64}$$

Okazaki solved equations (8.62), (8.63) and (8.64), assuming that the disk is terminated at a certain radius, say r_D, and imposing the boundary condition that the Lagrangian variation of pressure vanishes there. He obtained retrograde trapped

[7]Exactly speaking, equation (8.59) has been derived under the assumption of $d\ln H/d\ln r < 1$, but we apply here equation (8.59) to more general cases.

oscillations. In the cases where the disk extends infinitely without outer boundary, there is no trapped oscillation.

The above results can be also derived by the WKB method which was often used in Chaps. 6 and 7. To do so, let us introduce the variable defined by

$$\tau(r) = \int_{r_{\rm in}}^{r} \frac{1}{r\rho_{00}H^3}dr. \tag{8.65}$$

Then, equation (8.59) can be written in the form:

$$\frac{d^2f}{d\tau^2} + Q(\tau)f = 0, \tag{8.66}$$

where

$$Q = (r\rho_{00}H^3)^2\biggl[\frac{\tilde{\omega}^2-\kappa^2}{c_{\rm s}^2} + \frac{d{\rm ln}r^2\Omega}{dr}\frac{d}{dr}{\rm ln}\biggl(\frac{\rho_{00}}{r\Omega}\biggr) + \frac{d^2}{dr^2}{\rm ln}\biggl(\frac{\rho_{00}}{r\Omega}\biggr) + \frac{d{\rm ln}H}{d{\rm ln}r}\frac{d}{rdr}{\rm ln}\biggl(\frac{\rho_{00}^2}{r\Omega}\biggr)\biggr]. \tag{8.67}$$

By using the boundary conditions that $f = 0$ at $\tau_{\rm in} = \tau(r_{\rm in})$ and $\tau_{\rm c}[= \tau(r_{\rm c})]$ is a turing point of Q, we have a trapping condition, which is

$$\int_{\tau_{\rm in}}^{\tau_c} Q^{1/2}d\tau = \biggl(\frac{3}{4}+n_r\biggr)\pi. \tag{8.68}$$

Returning the variable from τ to r, we can reduce the condition to

$$\int_{r_{\rm in}}^{r_c} \frac{Q^{1/2}}{r\rho_{00}H^3}dr = \biggl(\frac{3}{4}+n_r\biggr)\pi. \tag{8.69}$$

If we adopt again $H(r) \propto r^{3/2}$ and $\rho_{00}(r) \propto r^{-\alpha}$ as Okazaki (1991) did, equation (8.69) is reduced to

$$\int_{r_{\rm in}}^{r_c}\biggl[\frac{\tilde{\omega}^2-\kappa^2}{c_{\rm s}^2} + \frac{1-5\alpha}{2r^2}\biggr]^{1/2} dr = \biggl(\frac{3}{4}+n_r\biggr)\pi. \tag{8.70}$$

In low-frequency oscillations, $\tilde{\omega}^2-\kappa^2 \sim -2\omega\Omega$, if $\Omega = \kappa$ is adopted. Furthermore, the term of $1/r^2$ in equation (8.70) is negative for $\alpha > 1/5$. Hence, $\omega < 0$ is necessary for the trapping condition is realized. That is, for an oscillation to be trapped, it must be retrograde.

In summary, we cannot have trapped prograde oscillations, although observations suggest that in many cases the V/R variations are due to prograde oscillations.

In Sect. 8.2.3 we examine whether prograde trapped oscillation modes are expected, if we proceed to the next order of approximation of $(d\ln H/d\ln r)^2$.

8.2.3 Solutions Until the Order of $(d\ln H/d\ln r)^2$

We proceed to the terms of the order of $(d\ln H/d\ln r)^2$. From equation (8.48) we have

$$\left(\frac{\partial^2}{\partial\eta^2}-\eta\frac{\partial}{\partial\eta}\right)g^{(2)}+\frac{2}{r}\frac{d\ln H}{d\ln r}\frac{\Omega}{i(\tilde{\omega}^2-\kappa^2)}\frac{\partial^2 g_h^{(1)}}{\partial\eta^2}\frac{f_h}{f}+\frac{1}{r^2}\frac{\Omega H^2}{\tilde{\omega}^2-\kappa^2}B^{(2)}$$
$$+\frac{1}{r^2}\frac{\Omega H^2}{\tilde{\omega}^2-\kappa^2}\frac{d\ln H}{d\ln r}\left[2r\Omega\frac{\partial g^{(1)}}{\partial r}-\frac{r\Omega}{f}\left(\frac{\partial}{\partial r}+\frac{d}{dr}\ln\frac{\rho_{00}}{r}\right)\left(\eta\frac{\partial g^{(1)}}{\partial\eta}f\right)\right.$$
$$\left.-\frac{1}{f}\frac{\partial}{\partial r}\left\{r\Omega f\left(\eta\frac{\partial}{\partial\eta}-\eta^2\right)g^{(1)}\right\}\right]$$
$$=K^{(0)}g^{(2)}+K^{(1)}(r)g^{(1)}+K^{(2)}(r)g^{(0)}. \tag{8.71}$$

In deriving this equation, $\partial g^{(0)}/\partial r=0$ and $\partial g^{(0)}/\partial\eta=0$ have been used.

In calculating the second term on the left-hand side of equation (8.71) careful consideration is necessary, since $g_h^{(1)}$ and f_h are involved there. The pressure perturbation, h_1, and the radial velocity, u_r, are related by (see equation (8.43))

$$\left(\frac{\partial\eta^2}{\partial\eta^2}-\eta\frac{\partial}{\partial\eta}+1\right)h_1$$
$$=-i\Omega H^2\left[\frac{\partial}{\partial r}+\frac{d}{dr}\ln\left(\frac{\rho_{00}}{r\Omega}\right)+\frac{d\ln H}{rd\ln r}\left(-\eta\frac{\partial}{\partial\eta}+\eta^2\right)\right]u_r. \tag{8.72}$$

By substituting $h_1^{(1)}=[a_{h,0}^{(1)}\mathcal{H}_0+(d\ln H/d\ln r)a_{h,2}^{(1)}\mathcal{H}_2+\ldots]f_h$ into equation (8.72), we have[8]

$$f_h\left(a_{h,0}^{(1)}\mathcal{H}_0-\frac{d\ln H}{d\ln r}a_{h,2}^{(1)}\mathcal{H}_2+\ldots\right)=$$
$$-i\Omega H^2\left[\frac{\partial}{\partial r}+\frac{d}{dr}\ln\left(\frac{\rho_{00}}{r\Omega}\right)\right]\left(f\mathcal{H}_0+f\frac{d\ln H}{d\ln r}a_2^{(1)}\mathcal{H}_2\right)$$
$$-i\Omega H^2\frac{d\ln H}{d\ln r}\frac{1}{r}\left(-\eta\frac{\partial}{\partial\eta}+\eta^2\right)\left(f\mathcal{H}_0+f\frac{d\ln H}{d\ln r}a_2^{(1)}\mathcal{H}_2\right)+\ldots. \tag{8.73}$$

[8]The coefficient $a_{h,0}^{(1)}$ is not always zero, although $a_0^{(1)}$ is taken to be zero as mentioned before.

This equation consists of terms proportional to $\mathscr{H}_0$, $\mathscr{H}_2$, and $\mathscr{H}_4$. From the terms proportional to $\mathscr{H}_2$ we have

$$a_{h,2}^{(1)} f_h = i\Omega H^2 \left[\frac{d}{dr} + \frac{d}{dr}\left(\frac{\rho_{00}}{r\Omega} \right) \right] (a_2^{(1)} f) + i\Omega H^2 f, \tag{8.74}$$

neglecting the terms of the order of $d\ln H/d\ln r$ compared with those of unity.

Considering that $\partial^2 g_h^{(1)}/\partial\eta^2 = 2(d\ln/d\ln r)a_{h,2}^{(1)}$ and $a_{h,2}^{(1)}$ is given by equation (8.74), we see that the solvability condition of equation (8.71) (i.e., the condition that the second and subsequent terms of equation (8.71) has no term proportional to $\mathscr{H}_0$) is found to be, after some manipulation,

$$K^{(2)}(r) f = \frac{1}{r^2} \frac{\Omega^2 H^2}{\tilde{\omega}^2 - \kappa^2} \left(\frac{d\ln H}{d\ln r} \right)^2 \times \left[2r \left\{ \frac{d}{dr} + \frac{d}{dr} \ln\left(\frac{\rho_{00}}{r\Omega} \right) \right\} (a_2^{(1)} f) + 4f \right]. \tag{8.75}$$

Consequently, substituting the above expression for $K^{(2)} f$ into equation (8.49) by taking into account the terms till $(d\ln H/d\ln r)^2$, we have, as an extension of equation (8.58),

$$\begin{aligned} & f + \frac{1}{r^2} \frac{\Omega H^2}{\tilde{\omega}^2 - \kappa^2} \frac{d}{dr} \left[r^2 \Omega \left(\frac{d}{dr} + \frac{d}{dr} \ln \frac{\rho_{00}}{r\Omega} \right) f \right] \\ & + \frac{1}{r} \frac{\Omega^2 H^2}{\tilde{\omega}^2 - \kappa^2} \frac{d\ln H}{d\ln r} \left(3\frac{d}{dr} + \frac{d}{dr} \ln \frac{\rho_{00}}{r\Omega} \right) f \\ & + \frac{1}{r^2} \frac{\Omega^2 H^2}{\tilde{\omega}^2 - \kappa^2} \left(\frac{d\ln H}{d\ln r} \right)^2 \left[2r \left\{ \frac{d}{dr} + \frac{d}{dr} \ln\left(\frac{\rho_{00}}{r\Omega} \right) \right\} (a_2^{(1)} f) + 4f \right] = 0. \end{aligned} \tag{8.76}$$

This equation is arranged in the form:

$$\begin{aligned} & \frac{d^2 f}{dr^2} + \frac{d}{dr} \ln\left(r\rho_{00} H^3 \right) \frac{df}{dr} + \frac{\tilde{\omega}^2 - \kappa^2}{c_s^2} f \\ & + \left[\frac{d}{dr} \ln(r^2\Omega) \frac{d}{dr} \ln\left(\frac{\rho_{00}}{r\Omega} \right) + \frac{d^2}{dr^2} \ln\left(\frac{\rho_{00}}{r\Omega} \right) + \left(\frac{d\ln H}{d\ln r} \right) \frac{d}{rdr} \ln\left(\frac{\rho_{00}^2}{r\Omega} \right) \right] f \\ & + \left(\frac{d\ln H}{d\ln r} \right)^2 \frac{1}{r^2} \left[2r \left\{ \frac{d}{dr} + \frac{d}{dr} \ln\left(\frac{\rho_{00}}{r\Omega} \right) \right\} (a_2^{(1)} f) + 4f \right] = 0. \end{aligned} \tag{8.77}$$

This is the wave equation derived by taking into account the terms until the order of $(d\ln H/d\ln r)^2$. If the terms proportional to $(d\ln H/d\ln r)^2$ are neglected, equation (8.77) is reduced to equation (8.59).

One of the additional terms in equation (8.77), compared with equation (8.59), is the term with $a_2^{(1)}f$. The radial behavior of $a_2^{(1)}(r)$ is governed by equation (8.61). Since $a_2^{(1)}$ appears in equation (8.77) in a package of $a_2^{(1)}f$, it is convenience to write down an equation describing $a_2^{(1)}f$, using equation (8.60). In the derivation of equation of $a_2^{(1)}f$, however, the terms of the orders of $d\ln H/d\ln r$ and $(d\ln H/d\ln r)^2$ are unnecessary, since the term with $a_2^{(1)}f$ in equation (8.77) is already on the order of $(d\ln H/d\ln r)^2$. The results show that

$$\begin{aligned}
&\frac{d^2}{dr^2}(a_2 f) + \frac{d}{dr}\ln(r\rho_{00})\frac{d}{dr}(a_2 f) \\
&+\left[\frac{d\ln r^2\Omega}{dr}\frac{d}{dr}\ln\left(\frac{\rho_{00}}{r\Omega}\right) + \frac{d^2}{dr^2}\ln\left(\frac{\rho_{00}}{r\Omega}\right) - \frac{\tilde{\omega}^2 - \kappa^2}{c_s^2}\right](a_2 f) \\
&+\frac{1}{r^2\Omega}\frac{d}{dr}(r\Omega f) = 0.
\end{aligned} \tag{8.78}$$

The set of equations (8.77) and (8.78) describes wave motions of the one-armed low-frequency p-mode oscillations in the case where the terms till the order of $(d\ln H/d\ln r)^2$ are considered. By solving simultaneously equations (8.77) and (8.78), we can study whether one-armed, low-frequency p-mode oscillations can be trapped in a finite region of an infinitely extended Keplerian disks and whether the oscillations are prograde.

To examine the above problem, numerical calculations will be necessary. Here, however, we are satisfied by suggesting a possibility of presence of prograde oscillations, if the effects of the second term with respect to $(d\ln H/d\ln r)^2$ in equation (8.77) are considered. This term is $(4/r^2)(d\ln H/d\ln r)^2 f$. It is obvious from the following considerations that this term acts in the direction so as to make the oscillations prograde and trapped in a finite region of disks. That is, if the terms with $a_2^{(1)}f$ are neglected, the trapping condition becomes roughly as (compare with equation (8.70))

$$\int_{r_{\rm in}}^{r_{\rm c}}\left[\frac{\tilde{\omega}^2 - \kappa^2}{c_s^2} + \frac{1-5\alpha}{2r^2} + \frac{4}{r^2}\left(\frac{d\ln H}{d\ln r}\right)^2\right]^{1/2} dr \sim \frac{3}{4}\pi, \tag{8.79}$$

where $H(r) \propto r^{3/2}$ and $\rho_{00} \propto r^{-\alpha}$ are adopted. Compared with equation (8.70), equation (8.79) has an additional term, $(4/r^2)(d\ln H/d\ln r)^2$. This additional term obviously acts in the direction to make ω positive. This is because the additional term is positive, and thus equation (8.79) has a tendency to be satisfied with $\tilde{\omega}^2 - \kappa^2 < 0$. This means $\omega > 0$, because $\tilde{\omega}^2 - \kappa^2 \sim -2\omega\Omega$.

8.2.4 Derivation of Wave Equation by Galerkin's Method

In Sect. 8.2.3 we have examined one-armed p-mode oscillations by a perturbation method till the order of $(d\ln H/d\ln r)^2$, where $d\ln H/d\ln r$ is a small expansion parameter. The results obtained are applicable in principle only when $d\ln H/d\ln r$ is smaller than unity. Hence, it will be worthwhile to consider an alternative method to approximately solve the basic partial differential equation of wave motions. To do so, we had considered a kind of Galerkin's method in Chap. 5. Here, we extend the method to cases where u_r is approximated by two terms. That is, the velocity perturbations associated with one-armed p-mode oscillations, u_r, are now approximated by two Hermite polynomials as

$$u_r = f_0(r)\mathscr{H}_0(\eta) + f_2(r)\mathscr{H}_2(\eta). \tag{8.80}$$

The above expression for u_r is substituted into the partial differential equation governing $u_r(r, \eta)$, i.e., equation (8.45). Then, the resulting equation is integrated over the vertical direction after being multiplied by $\mathscr{H}_0(\eta)$ with the weighting function $\exp(-\eta^2/2)$. The Hermite polynomials, $\mathscr{H}_i(\eta)$, with different order are orthogonal each other with the weighting function $\exp(-\eta^2/2)$, when their products are integrated over $\eta = -\infty$ to $\eta = \infty$. Using this fact, after substituting equation (8.80) into equation (8.45), we equate the sum of the terms proportional to $\mathscr{H}_0$ in the resulting equation to zero, and also those proportional to $\mathscr{H}_2$ to zero. From the terms proportional to $\mathscr{H}_0$, we have

$$\begin{aligned}
&-\frac{\tilde{\omega}^2 - \kappa^2}{c_s^2} f_0 + i\frac{4}{r\Omega H^2}\frac{d\ln H}{d\ln r} f_{h2} \\
&= \left(\frac{d}{dr} + \frac{d\ln r^2\Omega}{dr}\right)\left(\frac{d}{dr} + \frac{d}{dr}\ln\frac{\rho_{00}}{r\Omega}\right) f_0 \\
&+ \frac{1}{r^2}\frac{d\ln H}{d\ln r}\left[2r\left(\frac{d}{dr} + \frac{d}{dr}\ln\frac{\rho_{00}}{r\Omega}\right)(f_0 - f_2) + \frac{d}{dr}(rf_0) + \frac{d\ln H}{d\ln r} f_0\right] \\
&- \frac{6}{r^2}\left(\frac{d\ln H}{d\ln r}\right)^2 f_2.
\end{aligned} \tag{8.81}$$

Similarly, from the terms proportional to $\mathscr{H}_2$, we have

$$\frac{\tilde{\omega}^2-\kappa^2}{c_s^2}f_2+i\frac{24}{r\Omega H^2}f_{h4}$$
$$=\left(\frac{d}{dr}+\frac{d\ln r^2\Omega}{dr}\right)\left(\frac{d}{dr}+\frac{d}{dr}\ln\frac{\rho_{00}}{r\Omega}\right)f_2$$
$$+\frac{1}{r^2}\frac{d\ln H}{d\ln r}\left[\frac{1}{\Omega}\frac{d}{dr}(r\Omega f_0)+3\frac{d\ln\Omega}{d\ln r}f_2\right]$$
$$-\frac{12}{r^2}\left(\frac{d\ln H}{d\ln r}\right)^2 f_2. \tag{8.82}$$

where f_{h2} and f_{h4} are the coefficients when $h_1(r,\eta)$ are expanded by the Hermite polynomials as

$$h_1=f_{h0}(r)\mathscr{H}_0(\eta)+f_{h2}(r)\mathscr{H}_2(\eta)+f_{h4}(r)\mathscr{H}_4(\eta)+\ldots \tag{8.83}$$

In order to express f_{h2} and f_{h4} in terms of f_0 and f_2, we consider the following equation (equation (8.43))

$$\left(\frac{\partial^2}{\partial\eta^2}-\eta\frac{\partial}{\partial\eta}+1\right)h_1$$
$$=-i\Omega H^2\left[\left(\frac{d}{dr}+\frac{d}{dr}\ln\frac{\rho_{00}}{r\Omega}\right)u_r+\frac{d\ln H}{d\ln r}\frac{1}{r}\left(\eta^2-\eta\frac{\partial}{\partial\eta}\right)u_r\right]. \tag{8.84}$$

The $\mathscr{H}_0$-dependent terms of this equation give

$$f_{h0}=-i\Omega H^2\left[\left(\frac{d}{dr}+\frac{d}{dr}\ln\frac{\rho_{00}}{r\Omega}\right)f_0+\frac{1}{r}\frac{d\ln H}{d\ln r}f_0\right]. \tag{8.85}$$

The $\mathscr{H}_2$-dependent terms of equation (8.84) lead to

$$f_{h2}=i\Omega H^2\left[\left(\frac{d}{dr}+\frac{d}{dr}\ln\frac{\rho_{00}}{r\Omega}\right)f_2+\frac{1}{r}\frac{d\ln H}{d\ln r}(f_0-3f_2)\right]. \tag{8.86}$$

Furthermore, the $\mathscr{H}_4$-dependent terms give

$$3f_{h4}=i\Omega H^2\frac{1}{r}\frac{d\ln H}{d\ln r}f_2. \tag{8.87}$$

Substitution of f_{h2} given by equation (8.86) into equation (8.81) gives an equation governing the radial variation of $f_0(r)$, which is

$$\left(\frac{d}{dr}+\frac{d\ln r^2\Omega}{dr}\right)\left(\frac{d}{dr}+\frac{d}{dr}\ln\frac{\rho_{00}}{r\Omega}\right)f_0+\frac{\tilde{\omega}^2-\kappa^2}{c_s^2}f_0$$
$$+\frac{1}{r^2}\frac{d\ln H}{d\ln r}\left[3r\frac{d}{dr}+\frac{d}{d\ln r}\left(\frac{\rho_{00}^2}{r\Omega}\right)\right]f_0+4\left(\frac{d\ln H}{d\ln r}\right)^2 f_0$$
$$=-2\frac{d\ln H}{d\ln r}\frac{1}{r}\left(\frac{d}{dr}+\frac{d}{dr}\ln\frac{\rho_{00}}{r\Omega}\right)f_2-\frac{6}{r^2}\left(\frac{d\ln H}{d\ln r}\right)^2 f_2. \tag{8.88}$$

The right-hand side of this equation shows that f_2 has effects on the radial variation of f_0. To solve this equation, we need another relation between f_0 and f_2, which is equation (8.82). That is, substituting equation (8.87) into equation (8.82), we have

$$\left(\frac{d}{dr}+\frac{d\ln r^2\Omega}{dr}\right)\left(\frac{d}{dr}+\frac{d}{dr}\ln\frac{\rho_{00}}{r\Omega}\right)f_2-\frac{\tilde{\omega}^2-\kappa^2}{c_s^2}f_2$$
$$+\frac{1}{r^2}\frac{d\ln H}{d\ln r}\left[3\frac{d\ln\Omega}{d\ln r}f_2-12\frac{1}{r^2}\left(\frac{d\ln H}{d\ln r}\right)^2 f_2\right]$$
$$=-\frac{1}{\Omega}\frac{d}{dr}(r\Omega f_0). \tag{8.89}$$

Equations (8.88) and (8.89) are simultaneous differential equation describing radial variations of f_0 and f_2. By imposing relevant boundary conditions, the eigenvalue ω is obtained.

The next issue is to compare the set of equations (8.88) and (8.89) with the set of equations (8.77) and (8.78) derived by the perturbation method.

In the perturbation method described in Sect. 8.2.3, f_0 and f_2 are written as f and $(d\ln H/d\ln r)(\alpha_2^{(1)}f)$, respectively.[9] If these expressions are used here,

[9] In Sect. 8.2.3, $u_r(r,\eta)$ was expanded as

$$u_r(r,\eta)=(g^{(0)}+g^{(2)}+\ldots)f(r)=\left[\mathscr{H}_0+\frac{d\ln H}{d\ln r}(a_2^{(1)}\mathscr{H}_2)+\left(\frac{d\ln H}{d\ln r}\right)^2(a_2^{(2)}\mathscr{H}_2+a_4^{(2)}\mathscr{H}_4)+\ldots\right]f.$$

In the present subsection, $u_r(r,\eta)$ is expressed as (see equation (8.80))

$$u_r(r,\eta)=\mathscr{H}_0(\eta)f_0(r)+\mathscr{H}_2(\eta)f_2(r).$$

equation (8.88) is written in the form:

$$
\begin{aligned}
&\frac{d^2f}{dr^2}+\frac{d}{dr}\ln\left(r\rho_{00}H^3\right)\frac{df}{dr}+\frac{\tilde{\omega}^2-\kappa^2}{c_s^2}f \\
&+\left[\frac{d}{dr}\ln(r^2\Omega)\frac{d}{dr}\ln\left(\frac{\rho_{00}}{r\Omega}\right)+\frac{d^2}{dr^2}\ln\left(\frac{\rho_{00}}{r\Omega}\right)+\left(\frac{d\ln H}{d\ln r}\right)\frac{d}{rdr}\ln\left(\frac{\rho_{00}^2}{r\Omega}\right)\right]f \\
&+\left(\frac{d\ln H}{d\ln r}\right)^2\frac{1}{r^2}\left[2r\left\{\frac{d}{dr}+\frac{d}{dr}\ln\left(\frac{\rho_{00}}{r\Omega}\right)\right\}(\alpha_2^{(1)}f)+4f\right] \\
&+\frac{6}{r^2}\left(\frac{d\ln H}{d\ln r}\right)^3(\alpha_2^{(1)}f)=0. \qquad (8.90)
\end{aligned}
$$

Next, rewriting f_0 and f_2 again by f and $(d\ln H/d\ln r)\alpha_2^{(1)}f$, we have from equation (8.89)

$$
\begin{aligned}
&\left[\frac{d^2}{dr^2}+\frac{d}{dr}\ln(r\rho_{00})\frac{d}{dr}+\frac{\tilde{\omega}^2-\kappa^2}{c_s^2}\right](\alpha_2^{(1)}f) \\
&+\left[\frac{d}{dr}\ln(r^2\Omega)\frac{d}{dr}\ln\left(\frac{\rho_{00}}{r\Omega}\right)+\frac{d^2}{dr^2}\ln\left(\frac{\rho_{00}}{r\Omega}\right)+\left(\frac{d\ln H}{d\ln r}\right)\frac{d}{rdr}\ln\left(\frac{\rho_{00}^2}{r\Omega}\right)\right](\alpha_2^{(1)}f) \\
&+3\left(\frac{d\ln H}{d\ln r}\right)(\alpha_2^{(1)}f)-4\left(\frac{d\ln H}{d\ln r}\right)^2(\alpha_2^{(1)}f)+\frac{1}{r^2\Omega}\frac{d}{dr}(r\Omega f)=0. \qquad (8.91)
\end{aligned}
$$

The set of equations (8.90) and (8.91) is the final results obtained by using a Galerkin method. These equations should be compared with the set of equations (8.77) and (8.78). Equation (8.90) is the same as equation (8.77) except the last term of the order of $(d\ln H/d\ln r)^3$. In perturbation method all terms of the order of $(d\ln H/d\ln r)^3$ have been neglected, since $(d\ln H/d\ln r)$ is assumed to be less than unity and only the terms till the order of $(d\ln H/d\ln r)^2$ had been taken into account. Next, If the two terms with $d\ln H/d\ln r$ and the one term with $(d\ln H/d\ln r)^2$ in equation (8.91) are neglected, the resulting equation becomes the same as equation (8.78) derived by the perturbation method. In the second order perturbation method considered in Sect. 8.2.3, the equation which we need for $\alpha_2^{(1)}f$ was that till the zeroth-order terms with respect to $(d\ln H/d\ln r)$. Hence, the terms with $(d\ln H/d\ln r)$ and $(d\ln H/d\ln r)^2$ did not considered from the beginning in deriving wave equation for $\alpha_2^{(1)}f$.

It is not clear which approximation (perturbation method and Galerkin's method) better represents real situations when $d\ln H/d\ln r$ is not small. We suppose that the perturbation method can represent real situations even when $d\ln H/d\ln r$ is close to unity or slightly larger than unity, but the Galerkin's method will be better when $d\ln H/d\ln r$ is larger than unity.

References

Abramowicz, M. A. 2005, Astro. Nachr., 326, 782
Kato, S. 1983, Publ. Astron. Soc. Jpn., 35, 249
Kato, S. 2011, Publ. Astron. Soc. Jpn., 63, 125
Kato, S. 2012a, Publ. Astron. Soc. Jpn., 64, 62
Kato, S. 2012b, Publ. Astron. Soc. Jpn., 64, 78
Ogilvie, G. I., 2008, Mon. Not. R. Astron. Soc., 388, 1372
Okazaki, A. T. 1991, Publ. Astron. Soc. Jpn., 43, 75
Papaloizou, J. C. B. & Heemskerk, M. H. M. 1993, Astron. Astrophys. 276, 409
Papaloizou, J. C. B. & Savonije, G. I. 2006, Astron. Astrophys., 456, 1097
Papaloizou, J. C. B., Savonije, G. J., & Henrichs, H. F. 1992, Astron. Astrophys., 265, L45
Savonije, G. J. & Heemskerk, M. H. M. 1993, Astron. Astrophys. 276, 409

Part II
Excitation Processes of Disk Oscillations

Chapter 9
Overstability of Oscillations by Viscosity

Abstract For disk oscillations to be observed, excitation processes are necessary. Otherwise, they will be damped with time by dissipative processes which usually act so as to dampen oscillations. It is well-known, however, that in thermally non-equilibrium open systems dissipative processes can often excite oscillations. The excitation of stellar pulsation by non-adiabatic processes is one of these typical examples in classical astrophysics. Three typical excitation processes are known in stellar pulsation, which are κ-, ϵ-, and δ-mechanisms (see Unno et al. (Nonradial oscillations of stars. University of Tokyo Press, Tokyo, 1989) for details of stellar pulsation theory).

The above excitation mechanisms of oscillations can also operate in accretion disks, if favorable situations are realized. In case of accretion disks, however, another important mechanism exists, which does not work in stellar pulsation. This is a process due to angular momentum transport in disks. In accretion disks angular momentum is transported outward by viscous processes. If this outward angular momentum flow is modulated by oscillations, it can work so as to excite the oscillations when the modulation occurs in a phase relevant to excitation of oscillations.

Keywords Angular momentum flow • p-mode oscillations • Viscous overstability

9.1 A Mathematical Derivation of Criterion of Viscous Oscillatory Instability (Overstability)

Before deriving a mathematical expression for stability criterion, the physical cause of presence of overstability is briefly mentioned. Accretion disks are differentially rotating systems. Hence, azimuthal forces work on fluid elements through viscosity. These azimuthal forces transport angular momentum outwards, while fluid elements fall inwards (this is accretion). Let us now impose an oscillation on such systems. The oscillation modulates the azimuthal force by changing viscosity by the oscillation itself. If this modulation of the azimuthal force occurs in a proper phase with the change of azimuthal velocity associated with the oscillation, a positive (or negative) work is done on the oscillation, and the oscillation grows. This is the essence of the viscous overstability of oscillations (Kato 1978).

S. Kato, *Oscillations of Disks*, Astrophysics and Space Science Library 437,
DOI 10.1007/978-4-431-56208-5_9

This amplification process is similar with the κ-mechanism in stellar pulsation. In κ-mechanism the radiative energy flow through stellar interior is modulated by change of opacity due to oscillations and a part of the radiative energy is converted into energy of oscillations. In accretion disks angular momentum is transported outwards in the unperturbed state. This angular momentum flow through the disks is modulated by oscillations. If a part of this modulation is converted to angular momentum of oscillations, the oscillations can grow. It is noted that such excitation process is absent in stars, since there is no angular momentum flow in the unperturbed stars.

This viscous oscillatory instability was first derived by an Eulerian description (Kato 1978), using a perturbation method. A Lagrangian description, however, seems to be more instructive and perspective. Hence, we adopt here the latter one.

A Lagrangian description of small amplitude perturbation on equilibrium states has been presented in Chap. 3 under conditions of adiabatic and inviscid perturbations. If viscous processes are present, the equation of motion describing perturbations, equation (3.1), is modified to

$$\frac{D_0\boldsymbol{\xi}}{Dt^2}+\delta\left(\frac{1}{\rho}\nabla p+\nabla\psi\right)=\delta\boldsymbol{N}, \tag{9.1}$$

where $\boldsymbol{N}$ is the viscous stress force per unit mass. The symbol δ denotes Lagrangian variation.

As an energy equation describing non-adiabatic processes, we adopt here

$$\frac{D_0}{Dt}\delta p-c_{\mathrm{s}}^2\frac{D_0}{Dt}\delta\rho=\delta[(\Gamma_3-1)(-\mathrm{div}\boldsymbol{F}+\Phi)], \tag{9.2}$$

where $\boldsymbol{F}$ and Φ are thermal energy flux and thermal energy generation rate (per unit volume), respectively. Furthermore, Γ_3 in equation (9.2) is an exponent specifying the ratio of the specific heats in the case where the effects of radiation pressure are taken into account, and c_{s} is the speed of sound, defined by $c_{\mathrm{s}}^2=\Gamma_1 p_0/\rho_0$, where Γ_1 is another exponent specifying the ratio of the specific heats (see, e.g., Chandrasekhar (1938) for adiabatic exponents). If the Lagrangian pressure variation is decomposed into the adiabatic part, $\Gamma_1 p_0/\rho_0$, and non-adiabatic part, $(\delta p)_{\mathrm{na}}$, the latter is

$$(\delta p)_{\mathrm{na}}\equiv\delta p-\Gamma_1\frac{p_0}{\rho_0}\delta\rho \tag{9.3}$$

and the energy equation (9.2) is written as

$$\frac{D_0}{Dt}(\delta p)_{\mathrm{na}}=\delta[(\Gamma_3-1)(-\mathrm{div}\boldsymbol{F}+\Phi)]. \tag{9.4}$$

Finally, the equation of continuity is

$$\delta\rho + \rho_0 \mathrm{div}\boldsymbol{\xi} = 0, \tag{9.5}$$

which is unchanged from the case of adiabatic and inviscid perturbations.

By performing the same procedure as in Chap. 3, we can reduce equation (9.1) to[1]

$$\rho_0 \frac{\partial^2 \boldsymbol{\xi}}{\partial t^2} + 2\rho_0(\boldsymbol{u}_0 \cdot \nabla)\frac{\partial}{\partial t}\boldsymbol{\xi} + \mathscr{L}(\boldsymbol{\xi}) = \rho_0 \delta \boldsymbol{N} - \nabla(\delta p)_{\mathrm{na}}. \tag{9.6}$$

This equation is an extension of equation (3.6) in Chap. 3 to the cases of non-adiabatic and viscous perturbations. In Chap. 3, the unperturbed flow, $\boldsymbol{u}_0$, was considered to be rotation alone, but this Lagrangian expression can be applied in more general cases (Lynden-Bell and Ostriker 1867) and here accretion flows can be included in $\boldsymbol{u}_0$.

All perturbed quantities are assumed to be proportional to $\exp[i(\omega t - m\varphi)]$, and the displacement vector, $\boldsymbol{\xi}$, for example, is written as

$$\boldsymbol{\xi} = \Re[\hat{\boldsymbol{\xi}}\exp(i\omega t)] = \Re\Big[\breve{\boldsymbol{\xi}}\exp[i(\omega t - m\varphi)]\Big], \tag{9.7}$$

where $\Re$ represents the real part. Then, equation (9.6) is written as

$$-\omega^2\rho_0\hat{\boldsymbol{\xi}} + 2i\omega\rho_0(\boldsymbol{u}_0 \cdot \nabla)\hat{\boldsymbol{\xi}} + \mathscr{L}(\hat{\boldsymbol{\xi}}) = \rho_0\widehat{\delta \boldsymbol{N}} - \nabla\widehat{(\delta p)}_{\mathrm{na}}. \tag{9.8}$$

Multiplying equation (9.8) by the complex conjugate of $\hat{\boldsymbol{\xi}}$, i.e., $\hat{\boldsymbol{\xi}}^*$, and integrating the resulting equation over the whole volume, we have

$$\Im\Big[-\omega^2\int \rho_0\hat{\boldsymbol{\xi}}^*\hat{\boldsymbol{\xi}}d^3r + 2i\omega\int \rho_0\hat{\boldsymbol{\xi}}^*(\boldsymbol{u}_0 \cdot \nabla)\hat{\boldsymbol{\xi}}d^3r\Big]$$
$$= \Im\int \hat{\boldsymbol{\xi}}^* \cdot \Big[\rho_0\widehat{\delta \boldsymbol{N}} - \nabla\widehat{(\delta p)}_{\mathrm{na}}\Big]d^3r, \tag{9.9}$$

where $\Im$ represents the imaginary part and we have used the fact that $\mathscr{L}$ is a Hermite operator under the condition that ρ_0 vanishes on disk surfaces.

---

[1]It is noted that

$$\delta\Big(\frac{1}{\rho}\nabla p\Big) = -\frac{\delta\rho}{\rho_0^2}\nabla p_0 + \frac{1}{\rho_0}[\nabla p_1 + (\boldsymbol{\xi}\cdot\nabla)\nabla p_0]$$
$$= -\frac{\delta\rho}{\rho_0^2}\nabla p_0 + \frac{1}{\rho_0}[\nabla(\delta p) - \nabla(\boldsymbol{\xi}\cdot\nabla)p_0 + (\boldsymbol{\xi}\cdot\nabla)\nabla p_0].$$

If δp in the above equation is written as $\delta p = \Gamma_1(p_0/\rho_0)\delta + (\delta p)_{\mathrm{na}}$, we have $\rho_0\delta[(1/\rho_0)\nabla p] = \nabla(\delta p)_{\mathrm{na}} +$ parts of $\mathscr{L}(\boldsymbol{\xi})$.

Due to non-adiabatic and viscous processes, the frequency of oscillation, ω, becomes complex with imaginary part $\omega_{\rm i}$. If we write ω as[2]

$$\omega = \omega_0 + i\omega_{\rm i}, \tag{9.10}$$

equation (9.9) is written as

$$-2\omega_{\rm i}\left[\omega_0 \int \rho_0\hat{\boldsymbol{\xi}}^* \cdot \hat{\boldsymbol{\xi}} d^3r - i\int \rho_0\hat{\boldsymbol{\xi}}^*(\boldsymbol{u}_0\cdot\nabla)\hat{\boldsymbol{\xi}} d^3r\right] = \Im\int \hat{\boldsymbol{\xi}}^*\cdot\left[\rho_0\widehat{\delta\boldsymbol{N}} - \nabla\widehat{(\delta p)}_{\rm na}\right]d^3r, \tag{9.11}$$

where the fact that $i\rho_0(\boldsymbol{u}_0\cdot\nabla)$ is a Hermite operator has been used. In Chap. 3 we have defined wave energy, E,[3] which is

$$E = \frac{1}{2}\int\left[\omega_0^2\rho_0\hat{\boldsymbol{\xi}}^*\cdot\hat{\boldsymbol{\xi}} - i\omega_0\rho_0\hat{\boldsymbol{\xi}}^*(\boldsymbol{u}_0\cdot\nabla)\hat{\boldsymbol{\xi}}\right]d^3r. \tag{9.12}$$

By using this expression for wave energy, we find that the growth rate of oscillations, $-\omega_{\rm i}$, is given by

$$-\omega_{\rm i} = \frac{\omega_0}{4E}\Im\int \hat{\boldsymbol{\xi}}^*\cdot\left[\rho_0\widehat{\delta\boldsymbol{N}} - \nabla\widehat{(\delta p)}_{\rm na}\right]d^3r. \tag{9.13}$$

Equation (9.13) has a clear physical meaning. Roughly speaking, the work done on oscillations is proportional to the volume integration of $\boldsymbol{u}\cdot\rho_0\boldsymbol{f}$, where $\boldsymbol{u}$ is the velocity associated with the oscillations and $\boldsymbol{f}$ is the force acting on the oscillations. In the present problem, $\boldsymbol{u} \sim \partial\boldsymbol{\xi}/\partial t$ and $\rho_0\boldsymbol{f}$ is $\rho_0\delta\boldsymbol{N} - \nabla(\delta p)_{\rm na}$. The product $\boldsymbol{u}\cdot\rho_0\boldsymbol{f}$ is thus $\Re[i\omega_0\hat{\boldsymbol{\xi}}\exp(i\omega_0 t)]\cdot\Re[\rho_0\hat{\boldsymbol{f}}\exp(i\omega_0 t)]$, whose the time-independent part is $(1/2)\Re[-i\omega_0\hat{\boldsymbol{\xi}}^*\cdot\hat{\boldsymbol{f}}]$.[4] Hence, the growth time of wave energy (i.e., $-2\omega_{\rm i}$) is found to be given by equation (9.13).

The second term of the right-hand side of equation (9.13) comes from non-adiabatic processes. The stability criterion coming from this term is compared with that used in stellar pulsation theory. Performing integration by part and using

[2]Non-adiabatic and viscous processes introduce not only an imaginary part of frequencies but also a slight change of the real part. By neglecting the latter change, however, we regard ω_0 as the frequency of non-adiabatic and inviscid oscillations.

[3]The expression for wave energy is now slightly modified by the presence of accretion flows, but it is negligible.

[4]The formula

$$\Re(A)\Re(B) = \frac{1}{2}\Re[AB + AB^*] = \frac{1}{2}\Re[AB + A^*B]$$

is used, where A and B are complex variables and B^* is the complex conjugate of B.

equation (9.4) and $(\Gamma_3 - 1)\delta\rho/\rho_0 = \delta T/T_0$, we have

$$-\Im\int \hat{\boldsymbol{\xi}}^* \cdot \left[\nabla(\widehat{\delta p})_{\rm na}\right] d^3r = \Re\int (\omega_0 - m\Omega)^{-1}\left(\frac{\delta \breve{T}}{T_0}\right)^* \delta(-{\rm div}\breve{\boldsymbol{F}} + \breve{\Phi}) d^3r. \tag{9.14}$$

Equation (9.14) shows that the growth rate by non-adiabatic processes, $(-\omega_{\rm i})_{\rm nonad}$, is given by

$$(-\omega_{\rm i})_{\rm nonad} = \frac{1}{4E}\Re\int \frac{\omega_0}{\omega_0 - m\Omega}\left(\frac{\delta \breve{T}}{T_0}\right)^* \delta(-{\rm div}\breve{\boldsymbol{F}} + \breve{\Phi}) d^3r. \tag{9.15}$$

In the case of non-rotating stars, $\omega_0/(\omega_0 - m\Omega)$ is unity, and the real part of the integral in equation (9.15) is two times the mechanical work done on stars by thermal processes (see, for example, Eddington 1926). Hence, dividing it by $4E$ we obtain the growth rate of oscillations. This pulsation instability is well-known since Eddington by thermodynamical considerations (see Unno et al. 1989).

9.2 Overstability by Viscous Stress Force

Because the overstability by thermal processes (see equation (9.15)) is well known, we restrict our attention hereafter only to overstability by viscous force. The growth rate by viscous force, $-\omega_{\rm i}$, is given by[5]

$$-\omega_{\rm i} = \frac{\omega_0}{4E}\Im\int \rho_0 \hat{\boldsymbol{\xi}}^* \cdot \widehat{\delta \boldsymbol{N}} d^3r. \tag{9.16}$$

Lagrangian variation of viscous force per unit mass, $\delta\hat{\boldsymbol{N}}$, consists of two parts as

$$\delta\hat{\boldsymbol{N}} = \hat{\boldsymbol{N}}_1 + (\hat{\boldsymbol{\xi}} \cdot \nabla)\boldsymbol{N}_0, \tag{9.17}$$

where $\hat{\boldsymbol{N}}_1$ is the Eulerian variation of $\hat{\boldsymbol{N}}$, and $\boldsymbol{N}_0$ is the imbalance of viscous force in the unperturbed state. It is noted that the presence of $\boldsymbol{N}_0$ is the cause of accretion flow (advection flow) in the unperturbed disks.

The components of $\delta\hat{\boldsymbol{N}}$ in the cylindrical coordinates are

$$(\delta\hat{\boldsymbol{N}})_r = \hat{N}_{r1} + \left(\hat{\xi}_r\frac{\partial}{\partial r} + \hat{\xi}_z\frac{\partial}{\partial z}\right)N_{0r} - \frac{\hat{\xi}_\varphi}{r}N_{0\varphi}, \tag{9.18}$$

[5]The growth rate of oscillations comes from both non-adiabatic and viscous processes. Hereafter, however, $-\omega_{\rm i}$ is used to represent the growth rate due to viscous process, in order to avoid many subscripts.

$$(\delta\hat{\boldsymbol{N}})_\varphi = \hat{N}_{\varphi 1} + \left(\hat{\xi}_r\frac{\partial}{\partial r} + \hat{\xi}_z\frac{\partial}{\partial z}\right)N_{0\varphi} + \frac{\hat{\xi}_\varphi}{r}N_{0r}, \tag{9.19}$$

$$(\delta\hat{\boldsymbol{N}})_z = \hat{N}_{z1} + \left(\hat{\xi}_r\frac{\partial}{\partial r} + \hat{\xi}_z\frac{\partial}{\partial z}\right)N_{0z}. \tag{9.20}$$

Among the various terms including components of $\boldsymbol{N}_0$ in equations (9.18), (9.19) and (9.20), the main ones are $-(\hat{\xi}_\varphi/r)N_{0\varphi}$ in the r-component and $\hat{\xi}_r\partial N_{0\varphi}/\partial r$ in the φ-component. Further, $\hat{\xi}_r$ and $\hat{\xi}_\varphi$ are related by[6]

$$(\omega_0 - m\Omega)\breve{\xi}_\varphi = i2\Omega\breve{\xi}_r. \tag{9.21}$$

Based on the above considerations, we decompose $-\omega_\mathrm{i}$ given by equation (9.16) into two terms:

$$-\omega_\mathrm{i} = (-\omega_\mathrm{i})_\mathrm{viscosity} + (-\omega_\mathrm{i})_\mathrm{imbalance}, \tag{9.22}$$

where

$$(-\omega_\mathrm{i})_\mathrm{viscosity} = \frac{\omega_0}{4E}\Im\int\rho_0(\breve{\xi}_r^*\breve{N}_{1r} + \breve{\xi}_\varphi^*\breve{N}_{1\varphi} + \breve{\xi}_z^*\breve{N}_{1z})d^3r, \tag{9.23}$$

and

$$(-\omega_\mathrm{i})_\mathrm{imbalance} = -\frac{\omega_0}{4E}\Re\int\rho_0\frac{2\Omega}{\omega_0 - m\Omega}\breve{\xi}_r^*\breve{\xi}_r\frac{\partial}{r\partial r}(rN_{o\varphi})d^3r. \tag{9.24}$$

Equation (9.23) is the main term of $-\omega_\mathrm{i}$, and equation (9.24) is supplementary. First we consider the former.

[6] It is noted that in the case where the force acting on perturbations in the azimuthal direction is neglected, $\breve{u}_\varphi$ and $\breve{u}_r$ are related by

$$i(\omega - m\Omega)\breve{u}_\varphi + \frac{\kappa^2}{2\Omega}\breve{u}_r = 0,$$

but $\breve{\xi}_\varphi$ and $\breve{\xi}_r$ are related by (see equation (3.13))

$$i(\omega - m\Omega)\breve{\xi}_\varphi + 2\Omega\breve{\xi}_r = 0.$$

9.2.1 Viscous Overstability by Viscous Shear

It is convenient to change variables from displacement vector to velocity perturbation. The relations between $\breve{\boldsymbol{\xi}}$ and $\breve{\boldsymbol{u}}$ are already given in Chap. 3, which are

$$\begin{aligned} i(\omega - m\Omega)\breve{\xi}_r &= \breve{u}_r, \\ i(\omega - m\Omega)\breve{\xi}_\varphi &= \breve{u}_\varphi + \breve{\xi}_r r\frac{d\Omega}{dr} = \frac{4\Omega^2}{\kappa^2}\breve{u}_\varphi, \\ i(\omega - m\Omega)\breve{\xi}_z &= \breve{u}_z. \end{aligned} \tag{9.25}$$

Using these relations, we have

$$(-\omega_{\rm i})_{\rm viscosity} = \frac{\omega_0}{4E}\Re\int\rho_0\frac{1}{\omega_0 - m\Omega}\biggl(\breve{u}_r^*\breve{N}_{1r} + \frac{4\Omega^2}{\kappa^2}\breve{u}_\varphi^*\breve{N}_{1\varphi} + \breve{u}_z^*\breve{N}_{1z}\biggr)d^3r. \tag{9.26}$$

9.2.1.1 Expressions for $\rho_0 N_1$

To calculate $\rho_0 N_{1r}$, $\rho_0 N_{1\varphi}$, and $\rho_0 N_{1z}$, it is convenient to use the relation:

$$\rho_0\boldsymbol{N}_1 = (\rho\boldsymbol{N})_1 - \rho_1\boldsymbol{N}_0, \tag{9.27}$$

since $(\rho\boldsymbol{N})_1$ is directly related to stress tensor t_{ij} as shown in Appendix A, where the subscript 1 denotes the Eulerian perturbation. Here, to demonstrate the presence of overstability, we consider oscillations whose radial wavelength is sufficiently short compared with the characteristic radial length of unperturbed disk quantities. We have, then,

$$(\rho N_r)_1 = ik(t_{rr})_1 + \frac{\partial}{\partial z}(t_{rz})_1, \tag{9.28}$$

where k is the radial wavelength of the perturbation. For example, $\partial(t_{rr})_1/\partial r$ is written as $ik(t_{rr})_1$. From the expressions for stress tensor given in Appendix A, we have

$$\begin{aligned} (t_{rr})_1 &= \frac{4}{3}ik\eta_0 u_r - \frac{2}{3}\eta_0\frac{\partial u_z}{\partial z}, \\ (t_{rz})_1 &= \eta_0\biggl(iku_z + \frac{\partial u_r}{\partial z}\biggr), \end{aligned} \tag{9.29}$$

where η_0 is viscosity in the unperturbed disks. Then, substituting equations (9.29) into equation (9.28), from equation (9.27) we have

$$\rho_0 N_{1r} = -\frac{4}{3}k^2\eta_0 u_r - \frac{2}{3}ik\eta_0\frac{\partial u_z}{\partial z} + \frac{\partial}{\partial z}(t_{rz})_1, \tag{9.30}$$

where the second term on the right-hand side of equation (9.27) has been neglected since N_{0r} is small.

Under the same approximations, we have

$$(\rho N_\varphi)_1 = ik(t_{r\varphi})_1 + \frac{\partial}{\partial z}(t_{z\varphi})_1, \tag{9.31}$$

where

$$(t_{r\varphi})_1 = ik\eta_0 u_\varphi + \eta_1 r\frac{d\Omega}{dr}, \quad \text{and} \quad (t_{z\varphi})_1 = \eta_0\frac{\partial u_\varphi}{\partial z}. \tag{9.32}$$

It is noted that in $(t_{r\varphi})_1$ given above, we have a term resulting from viscosity variation. This is because in the unperturbed state $(t_{r\varphi})_0$ (which is proportional to η) is present, and thus a term of variation of viscosity is present in $(t_{r\varphi})_1$. This is of importance in relation to the cause of overstability. Using the above expressions for $(t_{r\varphi})_1$ and $(t_{z\varphi})_1$, from equation (9.27) we have

$$\rho_0 N_{1\varphi} = -k^2\eta_0 u_\varphi + \frac{\partial}{\partial z}(t_{z\varphi})_1 + ik\eta_1 r\frac{d\Omega}{dr} - \rho_1 N_{0\varphi}. \tag{9.33}$$

Furthermore, the Eulerian variation of ρN_z is given by (Appendix A)

$$(\rho N_z)_1 = ik(t_{rz})_1 + \frac{\partial}{\partial z}(t_{zz})_1, \tag{9.34}$$

where

$$(t_{rz})_1 = ik\eta_0 u_z + \eta_0\frac{\partial u_\varphi}{\partial z},$$
$$(t_{zz})_1 = 2\eta_0\frac{\partial u_z}{\partial z} - \frac{2}{3}\eta_0\left(iku_r + \frac{\partial u_z}{\partial z}\right). \tag{9.35}$$

Substituting the above expressions in equation (9.34), from equation (9.27) we have

$$\rho_0 N_{1z} = -k^2\eta_0 u_z + ik\eta_0\frac{\partial u_\varphi}{\partial z} + \frac{\partial}{\partial z}(t_{zz})_1, \tag{9.36}$$

where the last term on the right-hand side of equation (9.27) has been neglected.

9.2.1.2 Expressions for Growth Rate

In the framework of the local approximations in the radial direction, the z-dependence of perturbed quantities are all expressed in terms of a single Hermite polynomial (see Chap. 5). That is, in the oscillation mode with n node(s) in the vertical direction, u_r, u_φ, and ρ_1/ρ_0 are all proportional to $\mathcal{H}_n(z/H)$, while u_z is proportional to $\mathcal{H}_{n-1}(z/H)$, where H is the scale height representing disk half thickness, and $n(=0,1,2,\ldots)$ represents the node number of oscillations in the vertical direction. Here, we write $\eta=\rho\nu$, where ν is the kinematic viscosity and assumed to be independent of z. Furthermore, we remember that $d\mathcal{H}_n(z/H)/dz=(n/H)\mathcal{H}_{n-1}(z/H)$ and the Hermite polynomials with different index are orthogonal with a weighting function $\exp(-z^2/2H^2)$. This weighting function has the same z-dependence as that of the unperturbed density in the vertically isothermal disks. In addition, in the framework of the local approximations, u_r and h_1 (or ρ_1) are the same phase but they are anti-phase with u_φ and u_z (see Sect. 3.2). For example, we have $\Re\breve{u}_\varphi^*\breve{u}_r=0$.

Considering these properties, we have, from equation (9.30),

$$\Re\int\rho_0\breve{u}_r^*\breve{N}_{1r}dz=-\int\eta_0\left(\frac{4}{3}k^2\breve{u}_r^*\breve{u}_r+\frac{\partial\breve{u}_r^*}{\partial z}\frac{\partial\breve{u}_r}{\partial z}\right)dz. \tag{9.37}$$

In deriving this relation, we have performed the integration of $\breve{u}_r^*\partial(\breve{t}_{rz})_1/\partial z$ with respect to z by part under the condition that η_0 vanishes on the disk surface.

By similar procedures we have, from equation (9.33),

$$\Re\int\rho_0\breve{u}_\varphi^*\breve{N}_{1\varphi}dz=-\int\eta_0\left(k^2\breve{u}_\varphi^*\breve{u}_\varphi+\frac{\partial\breve{u}_\varphi^*}{\partial z}\frac{\partial\breve{u}_\varphi}{\partial z}\right)dz+\Re\int ik\breve{\eta}_1\breve{u}_\varphi^*r\frac{d\Omega}{dr}dz. \tag{9.38}$$

It is noted that the term of $\breve{\rho}_1N_{0\varphi}$ in equation (9.33) does not contribute to the integration of equation (9.38), because in the local approximations in the radial direction, $\breve{u}_\varphi$ and $\breve{\rho}_1/\rho_0$ are anti-phase as

$$c_{\rm s}^2\frac{\breve{\rho}_1}{\rho_0}=-\frac{(\omega-m\Omega)^2-\kappa^2}{k(\omega-m\Omega)}\breve{u}_r=i\frac{(\omega-m\Omega)^2-\kappa^2}{k\kappa^2}\frac{2\Omega}{\kappa^2}\breve{u}_\varphi. \tag{9.39}$$

Different from the term of $\breve{\rho}_1N_{0\varphi}$, the term with $\breve{\eta}_1$ in equation (9.33) does not disappear in equation (9.38). Here, for simplicity, we assume $\eta_1/\eta_0=\rho_1/\rho_0$. Then, using equation (9.39), we can reduce equation (9.38) to

$$\begin{aligned}\Re\int\rho_0\breve{u}_\varphi^*\breve{N}_{1\varphi}dz=-\int\eta_0\biggl[&k^2\breve{u}_\varphi^*\breve{u}_\varphi+\frac{\partial\breve{u}_\varphi^*}{\partial z}\frac{\partial\breve{u}_\varphi}{\partial z}\\&+\frac{(\omega-m\Omega)^2-\kappa^2}{c_{\rm s}^2}\frac{rd\Omega^2/dr}{\kappa^2}\breve{u}_\varphi^*\breve{u}_\varphi\biggr]dz,\end{aligned} \tag{9.40}$$

or using the local dispersion relation given by $[(\omega - m\Omega)^2 - \kappa^2][(\omega - m\Omega)^2 - n\Omega_\perp^2] = k^2 c_s^2 (\omega - m\Omega)^2$, we have

$$\Re \int \rho_0 \breve{u}_\varphi^* \breve{N}_{1\varphi} dz = -\int \eta_0 \biggl[k^2 \breve{u}_\varphi^* \breve{u}_\varphi + \frac{\partial \breve{u}_\varphi^*}{\partial z} \frac{\partial \breve{u}_\varphi}{\partial z} + \frac{k^2(\omega - m\Omega)^2}{(\omega - m\Omega)^2 - n\Omega_\perp^2} \frac{r d\Omega^2/dr}{\kappa^2} \breve{u}_\varphi^* \breve{u}_\varphi \biggr] dz. \tag{9.41}$$

Finally, we have, from equation (9.36),

$$\Re \int \rho_0 \breve{u}_z^* \breve{N}_{1z} d^3 r = -\int \eta_0 \left(k^2 \breve{u}_z^* \breve{u}_z + \frac{4}{3} \frac{\partial \breve{u}_z^*}{\partial z} \frac{\partial \breve{u}_z}{\partial z} \right) dz, \tag{9.42}$$

where integration by part has been performed in the vertical direction and the surface integral has been neglected.

Summing the above results (equations (9.37), (9.41), and (9.42)), we can express the growth rate by viscous processes, equation (9.26), in the form

$$(-\omega_{\rm i})_{\rm viscosity} = (-\omega_{\rm i})_{\rm damping} + (-\omega_{\rm i})_{\rm shear}, \tag{9.43}$$

where

$$(-\omega_{\rm i})_{\rm damping} = -\frac{\omega_0}{4E} \int \frac{\eta_0}{\omega_0 - m\Omega} \times \left[k^2 \left(\frac{4}{3} \breve{u}_r^* \breve{u}_r + \frac{4\Omega^2}{\kappa^2} \breve{u}_\varphi^* \breve{u}_\varphi + \breve{u}_z^* \breve{u}_z \right) + \frac{\partial \breve{u}_r^*}{\partial z} \frac{\partial \breve{u}_r}{\partial z} + \frac{\partial \breve{u}_\varphi^*}{\partial z} \frac{\partial \breve{u}_\varphi}{\partial z} + \frac{4}{3} \frac{\partial \breve{u}_z^*}{\partial z} \frac{\partial \breve{u}_z}{\partial z} \right] d^3 r, \tag{9.44}$$

and

$$(-\omega_{\rm i})_{\rm shear} = -\frac{\omega_0}{4E} \int \frac{\eta_0}{\omega_0 - m\Omega} \left[\frac{k^2(\omega - m\Omega)^2}{(\omega - m\Omega)^2 - n\Omega_\perp^2} \frac{4\Omega^4}{\kappa^4} \frac{d\ln\Omega^2}{d\ln r} \breve{u}_\varphi^* \breve{u}_\varphi \right] d^3 r. \tag{9.45}$$

Equation (9.44) shows the usual viscous damping of oscillations. The sign of the integral in equation (9.44) is the same as the sign of $\omega_0 - m\Omega$ in the region where the oscillations predominantly exist. That is, if the oscillations exist inside the corotation point of $\omega_0 - m\Omega = 0$, the sign of the integral is negative. In these cases, the sign of E/ω_0, is also negative, and we have $(-\omega_{\rm i})_{\rm damping} < 0$. If the oscillations exist in the region of $\omega_0 - m\Omega > 0$, the integral is positive and E/ω_0 is also positive. This leads again to $(-\omega_{\rm i})_{\rm damping} < 0$.

Different from $(-\omega_{\rm i})_{\rm damping}$, the sign of $(-\omega_{\rm i})_{\rm shear}$ can become positive. The term $(-\omega_{\rm i})_{\rm shear}$ comes from a combined action of time change of viscosity (i.e.,

η_1) and the shear flow (see equation (9.38)).[7] As shown in equation (9.45), in the disk systems where $d\ln\Omega^2/d\ln r < 0$, $(-\omega_{\rm i})_{\rm shear}$ becomes positive (overstable) for oscillation modes with $(\omega_0 - m\Omega)^2 - n\Omega_\perp^2 > 0$.

9.2.2 Application to Various Oscillation Modes

The results obtained by equations (9.43), (9.44) and (9.45) are now applied to typical oscillation modes.

(i) p-Mode Oscillations ($n = 0$)

In the case of p-mode oscillations ($n = 0$), there is no vertical motion in the lowest order of approximations, i.e., $u_z = 0$, and furthermore $\partial u_r/\partial z = \partial u_\varphi \partial z = 0$. Then, using the relation between $\breve{u}_r$ and $\breve{u}_\varphi$ given by $\breve{u}_r = -i(2\Omega/\kappa^2)(\omega - m\Omega)\breve{u}_\varphi$, we have, from equations (9.43), (9.44) and (9.45),

$$(-\omega_{\rm i})_{\rm viscosity} = -\frac{\omega}{4E}\int \frac{\eta_0}{\omega - m\Omega}k^2\left[\frac{4}{3} + \frac{\kappa^2 + d\Omega^2/d\ln r}{(\omega - m\Omega)^2}\right]\breve{u}_r\breve{u}_r^* d^3r. \tag{9.46}$$

The corotation region, $\omega - m\Omega \sim 0$, is an evanescent region of oscillations (see Chap. 6). We are thus interested in oscillations which are either in the region of $\omega - m\Omega < 0$ or in the region of $\omega - m\Omega > 0$. In each region the sign of E/ω has the same sign as the sign of $\omega - m\Omega$, since

$$E = \frac{1}{2}\int \frac{\omega}{\omega - m\Omega}\rho_0(\breve{u}_r\breve{u}_r^* + \breve{u}_z\breve{u}_z^*)d^3r, \tag{9.47}$$

(see equation (3.58)).

Considering these situations, we see from equation (9.46) that growth of the oscillations is determined by the sign of the terms in the large brackets of equation (9.46). That is, the condition of growth, $(-\omega_{\rm i})_{\rm viscosity} > 0$, is

$$\frac{4}{3}\frac{(\omega - m\Omega)^2}{\kappa^2} + 1 < -\frac{\Omega^2}{\kappa^2}\frac{d\ln\Omega^2}{d\ln r}. \tag{9.48}$$

In the p-mode oscillations with $n = 0$, the dispersion relation for local perturbations is $(\omega - m\Omega)^2 = \kappa^2 + k^2c_{\rm s}^2$. Equation (9.48) is thus reduced to

$$\frac{7}{3} < \frac{\Omega^2}{\kappa^2}\left[-\frac{7}{3}(kH)^2 - \frac{d\ln\Omega^2}{d\ln r}\right] \sim -\frac{\Omega^2}{\kappa^2}\frac{d\ln\Omega^2}{d\ln r}. \tag{9.49}$$

In deriving the last relation, $H < 1/k(\sim \lambda) < r$ is adopted.

---

[7] In the expression for $(-\omega_{\rm i})_{\rm shear}$ given by equation (9.45), η_1 does not appear explicitly, because η_1 is written as $\eta_1 = \eta_0\rho_1/\rho_0$.

In Newtonian Keplerian disks, $\kappa^2 = \Omega^2$ and $d\ln\Omega^2/d\ln r = -3$. Hence, the above condition of overstability, equation (9.49), is satisfied, and the p-mode oscillations are excited by the viscous process. One of possible applications of this overstability is one-armed low-frequency global oscillations in Be-star disks. In Be-stars long term V/R variations in double-peaked emission lines have been observed (see Chap. 1). This variation is usually considered to be due to one-armed low frequency p-mode oscillations (Kato 1983). The oscillations may be excited by the present viscous processes, although the present results cannot be quantitatively applied, because the local approximation is involved in our present derivation of inequality (9.48) or (9.49). Recently, viscous excitation of one-armed global p-mode oscillations with long wavelength in Be-star disks has been shown by Ogilvie (2008).

In relativistic disks, p-mode oscillations with $(\omega - m\Omega) < -\kappa$ (i.e., $(\omega - m\Omega)^2 > \kappa^2$) are trapped in the innermost region of the disks if the inner edge of disks is a reflection boundary (see Chap. 6). In the trapped region, κ is smaller than Ω (especially, much smaller in the region near the inner edge). In addition, $|d\ln\Omega^2/d\ln r|$ is much larger than 3. Hence, the condition of overstability (9.48) is safely satisfied (see Kato 2001).

In neutron-star and black-hole X-ray binaries, quasi-periodic high-frequency oscillations (HFQPOs) whose frequencies are comparable with the Keplerian frequencies of the innermost region of disks are often observed (see Chap. 1). Their origin is still a question and is under debate. One of the possibilities is p-mode oscillations trapped in the innermost region of relativistic disks. The present excitation mechanism supports this idea from the viewpoint of excitation of the oscillations. It is noted that the trapped p-mode oscillations can be also excited by corotation resonance (see Chap. 10), but the p-mode oscillations which can be excited by corotation resonance are non-axisymmetric ones ($m \neq 0$) alone, while the present viscous process can excite axisymmetric ones ($m = 0$), not only non-axisymmetric ones.

The order of the growth rate of oscillations is obtained from equation (9.46) as

$$(-\omega_{\rm i})_{\rm viscosity} = \emptyset\left(\frac{\nu_0}{\lambda^2}\right), \tag{9.50}$$

where ν_0 is the kinematic viscosity defined by $\nu_0 = \eta_0/\rho_0$, and $\lambda(\sim 1/k)$ is the wavelength of oscillations, which is roughly the width of trapped region of oscillations.

In studies of accretion disks, we introduce various timescales in order to clarify behaviors of various time-dependent phenomena in disks. One of them is the viscous timescale, $\tau_{\rm vis}$. This is the timescale by which a global disk structure evolves by the effects of viscosity. As the viscous term of the Navier-Stokes equation shows, $\tau_{\rm vis}$ can be defined by $\tau_{\rm vis}^{-1} = \nu_0/r^2$. Equation (9.50) shows that by use of $\tau_{\rm vis}$ we can express $(-\omega_{\rm i})_{\rm viscous}$ as

$$(-\omega_{\rm i})_{\rm viscous} \sim \tau_{\rm vis}^{-1}\left(\frac{r^2}{\lambda^2}\right). \tag{9.51}$$

Since we are now considering local perturbations ($\lambda < r$), the growth time is shorter than the viscous timescale by the factor $(\lambda/r)^2$. This growth time given by equation (9.51) is comparable with that of the so-called secular instability in accretion disks (for the secular instability, see, e.g., Kato et al. 2008).

An important issue to be remembered is whether viscosity behaves like $\eta_1/\eta_0 = \rho_1/\rho_0$ when the equilibrium state is perturbed. In the region of relativistic disks the frequencies of oscillations are high. Hence, it is uncertain whether turbulent viscosity responds quickly to rapid time variations and whether the response is described by $\eta_1/\eta_0 = \rho_1/\rho_0$.

As will be discussed in Sect. 13.1, in the innermost region of relativistic disks transonic accretion flows are unstable against axisymmetric p-mode oscillations, if viscosity is larger than a critical value. This instability of transonic flows seems to be related to the present viscous instability.

(ii) c-Mode Oscillations ($n = 1$)

In c-mode ($n = 1$) and vertical p-mode ($n \geq 2$) oscillations, we have $(-\omega_{\rm i})_{\rm shear} > 0$, because $(\omega - m\Omega)^2 - n\Omega_\perp^2 > 0$ for these oscillations (see equation (9.45)). Thus, we examine whether this positive value of $(-\omega_{\rm i})_{\rm shear} > 0$ can overcome the negative value of $(-\omega_{\rm i})_{\rm damping} < 0$. If so, the oscillations will be amplified. First, we consider c-mode oscillations.

In this case the oscillations have z-components as well as z-dependences, i.e., u_r, $u_\varphi \propto \mathscr{H}_1(z/H)$ and $u_z \propto \mathscr{H}_0(z/H)$ (see Chap. 6), where $\mathscr{H}_0$ and $\mathscr{H}_1$ are the Hermite polynomials of index 0 and 1, respectively. Furthermore, the magnitude of u_z is larger than that of u_r (and u_φ) by $1/kH(\sim \lambda/H)$, where λ is the radial wavelength of oscillations.

Let us estimate the magnitudes of the terms in the large brackets in equation (9.44). We can neglect the term of $(4/3)(\partial\breve{u}_z^*/\partial z)(\partial\breve{u}_z/\partial z)$. Hence, we see that the magnitude of the terms in the large brackets in equation (9.44) is on the order of $(u_r^* u_r)/H^2$, i.e.,

$$\text{order of terms in [ ] in equation (9.44)} \sim \frac{1}{H^2}u_r^* u_r, \tag{9.52}$$

which comes from the terms of $k^2 u_z^* u_z$, $(\partial u_r^*/\partial z)(\partial u_r/\partial z)$, and $(\partial u_\varphi^*/\partial z)(\partial u_\varphi/\partial z)$. The magnitude of the corresponding terms in equation (9.45) is $k^2(u_r^* u_r)$. That is,

$$\text{order of terms in [ ] in equation (9.45)} \sim k^2 u_r^* u_r, \tag{9.53}$$

Comparison of the above two equations show that

$$(-\omega_{\rm i})_{\rm shear}\ :\ (\omega_{\rm i})_{\rm damping}\simeq k^2\ :\ 1/H^2. \tag{9.54}$$

That is, the magnitude of the damping term, $(-\omega_{\rm i})_{\rm damping}$, is larger than that of the amplification term, $(-\omega_{\rm i})_{\rm shear}$, by $1/(kH)^2$, which is $\sim(\lambda/H)^2$.

In the case of one-armed ($m = 1$) oscillations (i.e., tilts), situations are somewhat different. That is, in the above estimate of $(-\omega_{\rm i})_{\rm shear}$, the magnitude of $(\omega-m\Omega)^2/[(\omega-m\Omega)^2-n\Omega_\perp^2]$ in equation (9.45) has been taken to be unity. In the case of one-armed oscillations, however, we have $(\omega-m\Omega)^2-\Omega_\perp^2\sim-2\omega\Omega$, because the oscillations have low negative frequencies (see Sect. 6.3), which leads to $(\omega-m\Omega)^2/[(\omega-m\Omega)^2-n\Omega_\perp^2]\sim\Omega/(-2\omega)$. The frequency of the tilt mode is negative and its absolute value is smaller than the Keplerian frequency of the outer edge of disks (see Sect. 7.4.2 for the order of ω in the case of binaries). This means that in typical propagation region of tilt mode oscillations, $\Omega/(-\omega)$ is much larger than unity. This implies that in the case of tilt modes, $(-\omega_{\rm i})_{\rm shear}$ is larger than the estimate in the above paragraph, and the tilt mode might be excited. More careful arguments will be worthwhile.

Finally, possible excitation of c-mode oscillations in tori is mentioned. In this book, we concern only with geometrically thin disks. In geometrically thick disks (tori), however, H is comparable with λ, i.e., $H\sim\lambda$. In such cases, the viscous damping becomes week and we might expect excitation of c-mode oscillations as well as p-mode oscillations.

(iii) Vertical p-Mode Oscillations ($n\geq 1$)

In the oscillations with $n\geq 2$, the vertical component of velocity, u_z, has node(s) in the vertical direction, i.e., $u_z\propto\mathcal{H}_{n-1}$. Hence, the main term in the large brackets of equation (9.44) is $(4/3)(\partial\breve{u}_z^*/\partial z)(\partial\breve{u}_z/\partial z)$ and it is on the order of $(\breve{u}_z^*\breve{u}_z)/H^2$. The corresponding term in equation (9.45) is on the order of $k^2(\breve{u}_\varphi^*\breve{u}_\varphi)$. Comparison of these two terms gives $|(-\omega_{\rm i})_{\rm damping}|\ :\ (-\omega_{\rm i})_{\rm shear}\simeq(\breve{u}_z^*\breve{u}_z)\!:\ (kH)^2(\breve{u}_\varphi^*\breve{u}_\varphi)\sim 1\ :\ (kH)^4\gg 1$. That is, the oscillations are strongly damped by viscosity.

Finally, it is noted that the above arguments are based on the assumption of isotropic viscosity. In real disks turbulence will be strongly anisotropic. It is known that viscous processes in disks are due to turbulent magnetic field resulting from the magneto-rotational instability (magneto-rotational instability MRI (Balbus and Hawley 1991)). In this case, momentum transport in the vertical direction might be less efficient compared with those in the horizontal directions, because the eddies of magnetic turbulence are elongated in the horizontal direction. If so, the above arguments concerning the viscous damping of c-mode and vertical p-mode oscillations might be much overestimated, compared with the viscous amplification (see also Nowak and Wagoner 1992). More careful discussions will be required.

(iii) g-Mode Oscillations

The propagation region of g-mode oscillations is characterized by

$$(\omega - m\Omega)^2 - \kappa^2 < 0, \tag{9.55}$$

which means $(\omega - m\Omega)^2 - n\Omega_\perp^2 < 0$, and thus we have $(-\omega_{\rm i})_{\rm shear} < 0$. That is, the viscosity always acts so as to dampen the g-mode oscillations.

9.2.3 *Effects of Viscous Imbalance*

As a supplement, we briefly consider the effects of angular momentum imbalance in the unperturbed disks on overstability by examining the sign of $(-\omega_{\rm i})_{\rm imbalance}$ given by equation (9.24). In the unperturbed state disks rotate differentially in such a way as the inner part rotates faster than the outer part does. If we consider the disk gas at radius r, the gas gets angular momentum from the gas of the inner region and transport it toward the gas of the outer region by viscous processes. The net gain of angular momentum is negative and the gas falls inward. The angular momentum balance in the unperturbed state is then described by (e.g., Kato et al. 2008)

$$U_0 \frac{d}{dr}(r^2\Omega) = rN_{0\varphi}, \tag{9.56}$$

where $U_0(r)$, which is negative, is the accretion flow in the unperturbed disks.

The accretion flow, $U_0(r)$, is related to mass accretion rate, $\dot{M}$, by

$$\dot{M} = -(2\pi)^{3/2} rH(r)\rho_{00}(r)U_0(r). \tag{9.57}$$

It should be noted that in the unperturbed disks the radial distributions of $U_0(r)$, $H(r)$, and $\rho_{00}(r)$ are adjusted so that $\dot{M}$ becomes independent of r. Using the fact that $\dot{M} = \text{const}$, from equations (9.56) and (9.57), we have

$$\frac{1}{r}\frac{d}{dr}(rN_{\varphi 0}) = -U_0 \frac{d}{dr}(r^2\Omega)\frac{d}{rdr}\left[\ln\left(\frac{d(r^2\Omega)/dr}{rH\rho_{00}}\right)\right]. \tag{9.58}$$

In standard disks the term of $d[\ln\ldots]/dr$ in equation (9.58) is negative. For example, in the innermost region of the standard disks where radiation pressure dominates over gas pressure ($p \sim p_{\rm rad}$) and opacity comes from electron scattering ($\bar{\kappa} \sim \kappa_{\rm es}$), we have $\rho_{00}H \propto r^{3/2}$ and thus $rH\rho_{00} \propto r^{5/2}$. In the middle region where gas pressure dominates over radiation pressure ($p \sim p_{\rm gas}$) with $\bar{\kappa} \sim \kappa_{\rm es}$, we have $rH\rho_{00} \propto r^{2/5}$, and in the outer region where $p \sim p_{\rm gas}$ and opacity comes from free-free processes, $\bar{\kappa} \sim \kappa_{\rm ff}$, we have $rH\rho_{00} \propto r^{1/4}$ (see, e.g., by Kato et al. 2008, p. 111). If we consider that $d(r^2\Omega)/dr \propto r^{1/2}$ in Keplerian disks, the term of $d[\ln\ldots]/dr$ in equation (9.58) is negative and we have $d(rN_{\varphi 0})/dr < 0$, except in the innermost region. This shows that $(-\omega_{\rm i})_{\rm imbalance} > 0$, except in the innermost region. That is, the imbalance of the viscous force in the unperturbed flow acts so as to amplify the oscillations except in the innermost region. This growth is, however, weak compared with the growth or damping resulting from $(-\omega_{\rm i})_{\rm viscosity}$. That is, roughly speaking, from equations (9.24) and (9.12), using equation (9.58), we have

$$(-\omega_{\rm i})_{\rm imbalance} \sim \alpha\frac{|U_0|}{r} \sim \emptyset\left(\frac{\nu_0}{r^2}\right), \tag{9.59}$$

where α is a numerical factor of order of unity and its sigh depends on disk structure and on the part of disks. In deriving the last relation of equation (9.59), we have used $\emptyset(U_0) = \emptyset(\nu_0/t)$. The value of $(-\omega_{\rm i})_{\rm imbalance}$ is smaller than $(-\omega_{\rm i})_{\rm viscosity}$ by the factor of L^2/r^2.

Up to the present, we focused our attention on geometrically thin disks. It is of importance to extend our attention to tori. It is noted here that Horák et al. (2012) studied viscous overstability of slender tori. They show that the "inertial mode" in terms of terminology of Blaes et al. (2006) is overstable. The mode will be related to c-mode.

References

Balbus, S. A., & Hawley, J. F. 1991, Astrophys. J., 376, 214

Blaes, O., Arras, P., & Fragile. P. C. 2006, Mon. Not. R. Astron. Soc., 369, 1235

Chandrasekhar, S., 1938, *Stellar Structure* (University of Chicago Press, Chicago)

Eddington, A. S. 1926, *The Internal Constitution of the Stars* (Cambridge University Press, Cambridge)

Kato, S. 1978, Mon. Not. R. Astron. Soc., 185, 629

Kato, S. 1983, Publ. Astron. Soc. Jpn., 35, 249

Horák, J., Abramowicz, M. A., Levin, L., Slapak, R., & Staub, O. 2012, Publ. Astron. Soc. Jpn., 64, 76

Kato, S., 2001, Publ. Astron. Soc. Jpn., 53, 1

Kato, S., Fukue, J., & Mineshige, S., 2008, *Black-Hole Accretion Disks – Towards a new Paradigm –* (Kyoto University Press, Kyoto)

Lynden-Bell, D. & Ostriker, J. P. 1867, Monthly Notices Roy. Astron. Soc, 136, 293

Nowak, M.A. & Wagoner, R.V., 1992, Astrophys. J., 393, 697.

Ogilvie, G. I., 2008, Mon. Not. R. Astron. Soc., 388, 1372

Unno, W., Osaki, Y., Ando, H., Saio, H., & Shibahashi, H., 1989, *Nonradial Oscillations of Stars* (University of Tokyo Press, Tokyo)

Chapter 10
Corotation Instability

Abstract Corotation resonance is one of important processes which excite or dampen disk oscillations. Especially, in the case of non-axisymmetric p-mode oscillations, this is one of important excitation processes. The corotation instability had been recognized in fields outside the accretion disk dynamics (i.e., fluid dynamics, oceanography, meteorology, and galactic dynamics), before its importance in tori was found by Papaloizou and Pringle (Mon Not R Astron Soc 208:721, 1984) (see Chap. 2). This work by Paparoizou and Pringle stimulated many subsequent studies on corotation instability in tori and disks. In this chapter, we will describe the essence of corotation instability in geometrically thin disks by presenting Drury's work (Mon Not R Astron Soc 217:821, 1985).

Keywords Corotation resonance • Overreflection • P-mode oscillations • Specific vorticity

10.1 A Brief Historical Review of Corotational Instability in Disk Dynamics

In astrophysics, perhaps, the first recognition of corotation amplification was in density wave theory of spiral galaxies. As mentioned in Chap. 2, the density wave theory made great success in describing the spiral structure of disk galaxies (started by Lin and Shu 1964). Soon after its success, however, a problem of maintenance of the spiral wave was pointed out by Toomre (1969). This problem was partially solved by the finding of excitation process of density waves at the corotation resonance by C.C. Lin's group (especially by energetic work by Mark 1974, 1976a,b). A series of Mark's work were done for collisionless stellar systems, and thus it is very hard for persons in other fields to understand the essence of the instability.

In the middle of 1970s, structure and stability of tori took much attention of many researchers, because Papaloizou and Pringle (1984, 1985, 1987) found that geometrically thick tori are dynamically unstable and may be destructed. This instability was the corotation instability, and is related to the corotation amplification known in galactic dynamics. After Papaloizou and Pringle, stimulated by their work, many researchers extensively studied the corotation instability (e.g., Blaes 1985,

S. Kato, *Oscillations of Disks*, Astrophysics and Space Science Library 437,
DOI 10.1007/978-4-431-56208-5_10

1987, Goldreich and Narayan 1985, Goldreich et al. 1986, Kojima 1986, Hanawa 1986, Kato 1987,[1] Narayan et al. 1987, and Glatzel 1887a,b). Especially, the instability was extensively examined by Goldreich and Narayan (1985), Goldreich et al. (1986), and Narayan et al. (1987) under simplification of shearing sheet approximation introduced by Goldreich and Lynden-Bell (1965).

After quasi-periodic oscillations were observed in X-ray binaries in the end of 1990s, the corotation instability was taken again much attention of many researchers as one of possible processes of excitation of disk oscillations. Lai's group are extensively studying the corotation instability in disks in order to apply to QPOs (Lai and Tsang 2009; Tsang and Lai 2008, 2009a,b). Studies in general relativistic disks are made by Horák and Lai (2013).

It should be noted here that the corotation instability may lead to destruction of tori in the nonlinear stage of its evolution. In geometrically thin disks, however, this is not the case, because the growth rate is much smaller in the case of thin disks. This is related to the fact that the corotation region is located in the evanescent region of the oscillations, and the penetration of oscillations to the corotation region is weak in geometrically thin disks. Because of this, the growth rate of oscillations is small, while in the case of tori, it is not the case.

Effects of coronation resonance on oscillations can be briefly summarized as follows. We need to separate two cases where oscillations have node(s) in the vertical direction or not (i.e., $n \geq 1$ or $n = 0$). In the case of $n = 0$ (i.e., p-mode), wave amplification at the coronation resonance occurs. This amplification can be divided into two cases: The coronation point is really singular or not singular. If specific vorticity, $d(\kappa^2/2\Omega\rho_{00})/dr$, just vanishes at the corotation point (radius), the singularity at the coronation point disappears (see Sect. 10.2). In this case a wave incident to the corotation point from each side of the corotation point can be overreflected at the corotation point (see Sect. 10.3). The case of shearing sheet approximation is a case where the specific vorticity vanishes at the corotation point. In the case where the corotation point is a singular (i.e., $d(\kappa^2/2\Omega\rho_{00})/dr \neq 0$), the overreflection occurs only in one side of the corotation point, depending on the sign of the specific vorticity at the corotation point (see Sect. 10.3).

In the case where the oscillations have node(s) in the vertical direction (i.e., $n \geq 1$), the corotation point is always singular, and the waves are underreflected there (Drury 1985; Kato 2003; Li et al. 2003).

Soon after the work of Papaloizou and Pringle (1984), Drury (1985) published an important paper, which seem to imply that Drury had recognized the corotational instability before Papaloizou and Pringle, and present mathematical bases of the instability. His analyses include the cases where oscillations have node(s) in the vertical direction (i.e., $n \neq 0$).

Examination of the corotational instability in realistic cases is mathematically complicated. Hence, we shall be satisfied here with presenting general arguments made by Drury (1985).

[1]This paper seems to have errors.

10.2 Preliminary Remarks on Wave Equation in Studying Corotation Resonance

The corotation instability occurs for p-mode oscillations ($n = 0$). Hence, many studies on this instability have been made by using vertically integrated quantities. In this chapter, however, in addition to the corotation instability of p-mode oscillations, we show that the corotation resonance make oscillations with node(s) in the vertical direction ($n \neq 0$) dampen. To demonstrate the latter, it is convenient to start from wave equations approximately separating radial and vertical behaviors of oscillations. To do so, we consider vertically isothermal disks and neglect the radial change of disk thickness for simplicity, i.e., $d\ln H/d\ln r = 0$.

Under the above approximations, we decompose variables describing oscillations into r- and z-dependent parts. That is, variables, $h_1(r, z)$ and $u_r(r, z)$, are written as

$$h_1(r,z) = f_h(r)\mathscr{H}_n(\eta), \quad u_r(r,z) = f_u(r)\mathscr{H}_n(\eta), \tag{10.1}$$

where $\eta = z/H$, $\mathscr{H}_n(\eta)$ is the Hermite polynomial of index n with argument η, and $n(= 0, 1, 2, \ldots)$ is the node number in the vertical direction. In cases of p-mode oscillations, we have $n = 0$.

From equation (5.32) (or equation (3.48)) we have

$$\frac{df_h}{dr} = \frac{2m\Omega}{r\tilde{\omega}}f_h - \frac{i}{\tilde{\omega}}(\tilde{\omega}^2 - \kappa^2)f_u. \tag{10.2}$$

Furthermore, equation (5.33) leads to

$$\frac{df_u}{dr} = -\left(\frac{d\ln r\rho_{00}}{dr} + \frac{m\kappa^2}{2r\tilde{\omega}\Omega}\right)f_u + \frac{i}{\tilde{\omega}}\left[\frac{\tilde{\omega}^2 - n\Omega_\perp^2}{c_s^2} - \frac{m^2}{r^2}\right]f_h. \tag{10.3}$$

Elimination of f_u from the above two equations leads to an equation with respect to f_h, which is

$$\frac{d^2f_h}{dr^2} + \frac{d}{dr}\ln\left(\frac{r\rho_{00}}{\tilde{\omega}^2 - \kappa^2}\right)\frac{df_h}{dr}$$
$$\left[\frac{2m\Omega}{r\tilde{\omega}}\frac{d}{dr}\ln\left(\frac{\tilde{\omega}^2 - \kappa^2}{\rho_{00}\Omega}\right) + \frac{\tilde{\omega}^2 - \kappa^2}{c_s^2}\left(1 - n\frac{\Omega_\perp^2}{\tilde{\omega}^2}\right) - \frac{m^2}{r^2}\right]f_h = 0. \tag{10.4}$$

This equation is the limit of $d\ln H/d\ln r = 0$ of equation (5.28). By introducing a new variable defined by

$$\bar{f}_h = \left(\frac{r\rho_{00}}{\tilde{\omega}^2 - \kappa^2}\right)^{1/2} f_h, \tag{10.5}$$

we can reduce equation (10.4) to

$$\frac{d^2}{dr^2}\bar{f}_h = K_h^2\bar{f}_h, \tag{10.6}$$

where

$$K_h^2 = -\frac{1}{4}\left[\frac{d}{dr}\ln\left(\frac{r\rho_{00}}{\tilde{\omega}^2-\kappa^2}\right)\right]^2 - \frac{1}{2}\frac{d^2}{dr^2}\ln\left(\frac{r\rho_{00}}{\tilde{\omega}^2-\kappa^2}\right)$$
$$+\left[\frac{2m\Omega}{r\tilde{\omega}}\frac{d}{dr}\ln\left(\frac{\tilde{\omega}^2-\kappa^2}{\rho_{00}\Omega}\right) + \frac{\tilde{\omega}^2-\kappa^2}{c_s^2}\left(1-n\frac{\Omega_\perp^2}{\tilde{\omega}^2}\right) - \frac{m^2}{r^2}\right]. \tag{10.7}$$

Hereafter, p-mode oscillations and other type oscillations are considered separately.

10.2.1 p-Mode Oscillations

Let us consider p-mode oscillations ($n = 0$). Equation (10.7) then shows that the corotation radius, $\tilde{\omega} = 0$, is a regular singular point of equation (10.6), if $(d/dr)\ln[(\tilde{\omega}^2-\kappa^2)/\rho_{00}\Omega]_c \neq 0$, where and hereafter the subscript c denotes the value at the corotation radius. Using $\tilde{\omega} = 0$, we can reduce this condition of the presence of singularity to

$$\left[\frac{d}{dr}\left(\frac{\kappa^2/2\Omega}{\rho_{00}}\right)\right]_c \neq 0. \tag{10.8}$$

This is non-zero of specific vorticity at the coratation resonance.[2]

A glance of equation (10.7) might suggest that the radii of the Lindblad resonances, i.e., the radii of $\tilde{\omega}^2 - \kappa^2 = 0$, are also singular points. This is, however, not the case. This can be shown by examining the solution of equation (10.6) around the radii by a power series of $r - r_L$, where r_L is the radius of one of the Lindblad resonances. By this procedure we can find that two independent solutions of power series have no singularity, i.e., the radii of the Lindblad resonances are not real singularities (see Sect. 2.3). The apparent singularity at the Lindblad resonances can be removed from wave equations, if we use wave equations expressed in terms of f_u, instead of f_h. Such a wave equation can be derived by eliminating f_h from the

[2]If vertically integrated disks are considered, instead of H-constant disks, the condition of presence of corotation singularity is

$$\left[\frac{d}{dr}\left(\frac{\kappa^2/2\Omega}{\Sigma_0}\right)\right]_c \neq 0,$$

where Σ_0 is the surface density in the unperturbed disks.

set of equations (10.2) and (10.3). The resulting wave equation is the special case of equation (5.35) of $d\ln H/d\ln r = 0$. It is noted that even in wave equation with respect to f_u, the corotation radius is singular point if inequality (10.8) is realized.

Now we discuss approximate forms of wave equations in studying p-mode oscillations. The disk is geometrically thin (i.e., $r^2\Omega^2/c_s^2 \gg 1$). Hence, if the terms related to the apparent singularities are neglected, the main terms of K_h^2 in equation (10.7) are two. One is $(\tilde{\omega}^2 - \kappa^2)/c_s^2$, and the other is the term related to corotation singularity. The latter term is $-(2m\Omega/r\tilde{\omega})d\ln[\rho_{00}\Omega/(\tilde{\omega}^2 - \kappa^2)]/dr$, which is important only in a narrow region around $\tilde{\omega} = 0$, and thus written approximately in the form

$$\frac{2m\Omega}{r}\frac{d}{dr}\ln\left(\frac{\kappa^2/2\Omega}{\rho_{00}}\right)\frac{1}{\tilde{\omega}}. \tag{10.9}$$

As mentioned before, the gradient of specific vorticity determines the sign of the coefficient of $1/\tilde{\omega}$.

In order to make the wave equation a dimensionless form and to change the origin of the radial coordinate to the corotation radius, r_c, we introduce a dimensionless radial coordinate defined by

$$x = \frac{r - r_c}{r_c}. \tag{10.10}$$

Then, an approximate dimensionless equation describing the behavior of $\bar{h}_1$ is

$$\frac{d\bar{h}_1}{dx^2} + k^2\bar{h}_1 = 0, \tag{10.11}$$

with

$$k^2 = \frac{\tilde{\omega}^2 - \kappa^2}{c_s^2}r_c^2 + \frac{1}{\tilde{\omega}}\left[\frac{mr\rho_{00}}{(\kappa/2\Omega)^2}\frac{d}{dr}\left(\frac{\kappa^2/2\Omega}{\rho_{00}}\right)\right]_c. \tag{10.12}$$

A brief comment is presented here. Terms related to apparent singularities at Lindblad resonances ($\tilde{\omega}^2 - \kappa^2 = 0$) are neglected in the above equation of k^2.

Figure 10.1 is a schematic diagram showing the radial distribution of $k^2(x)$ in the case of $[d(\kappa^2/2\rho_{00}\Omega)/dr]_c < 0$. Roughly speaking, the region between two Lindblad resonances, i.e., $x_{IL}[\equiv (r_{IL} - r_c)/r_c] < x < x_{OL}[\equiv (r_{OL} - r_c)/r_c]$, is an evanescent region of oscillations, and the corotation point is inside this evanescent region. Near the corotation point, however, a point of $k^2 = 0$ appears, which is denoted x_0. In the case of $[d(\kappa^2/2\rho_{00}\Omega)/dr]_c < 0$, x_0 is negative. The region between x_0 and $x = 0$ is a narrow propagation region of oscillation. The waves trapped in this narrow region are called *Rossby-type waves* by analogy with the Rossby waves in meteorology. If $[d(\kappa^2/2\rho_{00}\Omega)/dr]_c > 0$, the point x_0 is positive, and a narrow propagation region appears outside the corotation radius, i.e., $x_0 > 0$.

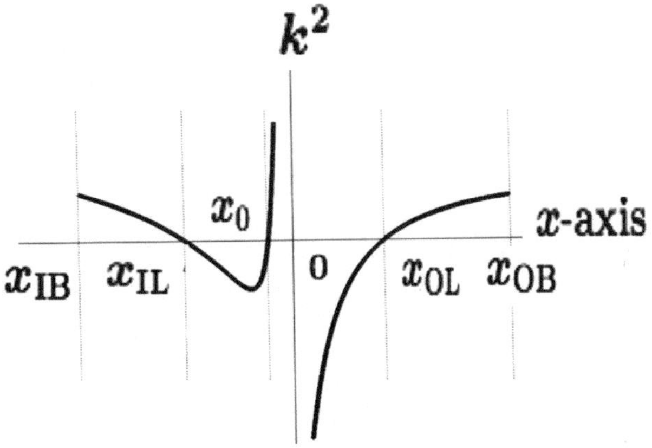

Fig. 10.1 A schematic diagram showing the radial distribution of $k^2(x)$. The case of $[d(\kappa^2/2\rho_{00}\Omega)/dr]_c < 0$ is shown. The symbols x_{IL} and x_{OL} show the radii of the Lindblad resonances, i.e., x_{IL} represents the radius where $\omega - m\Omega = -\kappa$, and x_{OL} does the radius of $\omega - m\Omega = \kappa$. The radius x_0 is the position where $k^2 = 0$, which is negative ($x_0 < 0$), in the case of $[d(\kappa^2/2\rho_{00}\Omega)/dr]_c < 0$. Inner and outer reflecting boundaries are sometimes imposed, which are shown by x_{IB} and x_{OB}, respectively.

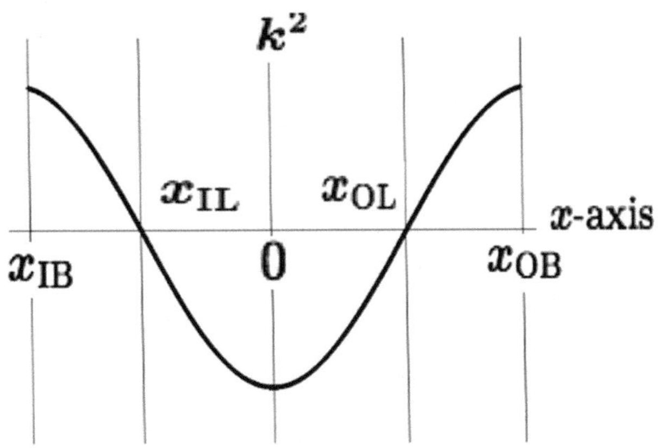

Fig. 10.2 A schematic diagram showing the radial distribution of $k^2(x)$ in the case of $[d(\kappa^2/2\rho_{00}\Omega)/dr]_c = 0$. Different from the cases of $[d(\kappa^2/2\rho_{00}\Omega)/dr]_c \neq 0$, the whole region between x_{IL} and x_{OL} is an evanescent region of oscillations.

In the special case of $[d(\kappa^2/2\rho_{00}\Omega)/dr]_c = 0$, the narrow propagation region near $x = 0$ disappears and the whole region between x_{IL} and x_{OL} is an evanescent region of oscillations, which is schematically shown in Fig. 10.2.

As will be shown later in Sect. 10.3 (see also Fig. 10.3), in the case of Fig. 10.1, an incident wave from the inner propagation region of $x < x_{IL}$ towards the corotation point is overreflected, while an incident wave from the outer propagation region of $x > x_{OL}$ towards the corotation point is underreflected. In the case of $[d(\kappa^2/2\rho_{00}\Omega)/dr]_c > 0$, the situations are opposite. That is, an incident wave from the propagation region of $x > x_{OL}$ is overreflected. In the case of Fig. 10.2 of $[d(\kappa^2/2\rho_{00}\Omega)/dr]_c = 0$, waves towards the corotation point are overreflected back whether they come from inside or outside (see Sect. 10.3). It is noted that in the case of $[d(\kappa^2/2\rho_{00}\Omega)/dr]_c = 0$, the corotation point is not a singular point of k^2, but overreflection of waves is present at the corotation point.

10.2.2 Oscillations of Other Than p-Modes (i.e., $n \neq 0$)

In oscillations other than p-mode ones (i.e., $n \neq 0$), k^2 has a pole of the second order, as equation (10.7) shows, i.e., k^2 has a term proportional to $1/\tilde{\omega}^2$. In this case, if the wave equation is written in the form of equation (10.11), k^2 has an another singular term which is proportional to $1/\tilde{\omega}^2$. Then, as shown in Sect. 10.3, the oscillations coming to the corotation point are underreflected.

10.3 Drury's Argument on Overreflection

Following Drury (1985), we introduced a new variable $\zeta(r)$ defined by $\zeta \equiv irf_u/f_h$. This variable has an important meaning. If we multiply the complex conjugate of f_u, i.e., f_u^*, both to the denominator and numerator of the above expression for ζ, then we have $\zeta = irf_h^* f_u/|f_h|^2$. That is, the sign of the imaginary part of ζ shows the direction of the wave energy flux.[3] By using equations (10.2) and (10.3), we have (Drury 1985)

$$\frac{d\zeta}{dr} = -\frac{d\ln\rho_{00}}{dr}\zeta - \frac{1}{r\tilde{\omega}}\left[r^2\frac{n\Omega_\perp^2 - \tilde{\omega}^2}{c_s^2} + \left(m + 2\Omega\zeta\right)\left(m + \frac{\kappa^2}{2\Omega}\zeta\right) - \tilde{\omega}^2\zeta^2\right]. \tag{10.13}$$

This is the basic equation to study the overreflection at corotation resonance (Drury 1985).

If we integrate equation (10.13) along a segment of the real axis and $\Im(\zeta)$ is zero at any point, then because all coefficients in equation (10.13) are real, $\Im(\zeta)$ is zero at all points of the segment. This means that the direction of the energy flux is not changed along the segment, since if it is changed at a point, the energy flux must be zero throughout the segment and is inconsistent with the above assumptions. This argument is, however, violated at a corotation point since at the point the path of the integration is forced to leave the real axis since the point is a singular point. This brings about an important conclusion (Drury 1985) that if there are places where wave amplification or wave dampening occurs they are corotation points alone. Furthermore, it enables us to relate local results valid in the neibourhood of the corotation singularity to the global behavior of the system.

---

[3] Equation (3.62), for example, shows that the energy flux is given by $\rho_0 h_1 \boldsymbol{u}$.

Based on the above considerations, Drury derived important conclusions from equation (10.13). (i) A *necessary* condition for the corotation region to act as a wave amplifier is the existence of a mode where the energy flux is directed away from the corotation region on both sides. (ii) A *sufficient* condition for waves incident from interior to be amplified is that all modes with a positive energy flux outside corotation have a negative energy flux inside corotation. Conversely, for a wave incident from the exterior, a sufficient condition for overreflection is that all modes with a negative energy flux inside corotation exhibit a positive energy flux outside corotation.

The next subject here is to explicitly show what cases the above amplification condition is satisfied. To do so, we change the independent variable in equation (10.13) from r to $\tilde{\omega}$. Then, we have

$$\tilde{\omega}\frac{d\zeta}{d\tilde{\omega}} = a\zeta^2 + b\zeta + c, \tag{10.14}$$

where

$$\begin{aligned}
a(\tilde{\omega}) &= \frac{1}{mrd\Omega/dr}(\kappa^2 - \tilde{\omega}^2), \\
b(\tilde{\omega}) &= \frac{1}{rd\Omega/dr}\left(2\Omega + \frac{\kappa^2}{2\Omega} + \frac{1}{m}\frac{d\ln\rho_{00}}{d\ln r}\tilde{\omega}\right), \\
c(\tilde{\omega}) &= \frac{1}{mrd\Omega/dr}\left[m^2 + \frac{r^2}{c_s^2}(n\Omega_\perp^2 - \tilde{\omega}^2)\right].
\end{aligned} \tag{10.15}$$

The purpose here is to examine how the sign of $\Im(\zeta)$ changes at corotation point $\tilde{\omega} = 0$.

First, we examine a solution of equation (10.14) which can be obtained by expanding ζ by a power series of $\tilde{\omega}$ as

$$\zeta(\tilde{\omega}) = \sum_{n=0} \zeta_n \tilde{\omega}^n. \tag{10.16}$$

The coefficients a, b, and c are also expanded by power series of $\tilde{\omega}$, for example, as $a(\tilde{\omega}) = a_0 + a_1\tilde{\omega} + a_2\tilde{\omega}^2 + \ldots$. Then, in the present problem we have

$$\begin{aligned}
a_0 &= \frac{\kappa^2}{mrd\Omega/dr}, \qquad a_1 = 0, \ldots \\
b_0 &= \frac{1}{rd\Omega/dr}\left(2\Omega + \frac{\kappa^2}{2\Omega}\right), \qquad b_1 = \frac{1}{mrd\Omega/dr}\frac{d\ln\rho_{00}}{d\ln r}, \ldots \\
c_0 &= \frac{1}{mrd\Omega/dr}\left(m^2 + n\frac{r^2\Omega_\perp^2}{c_s^2}\right), \qquad c_1 = 0, \ldots
\end{aligned} \tag{10.17}$$

Equating the terms of $\tilde{\omega}^0$ in equation (10.14), we have

$$a_0\zeta_0^2 + b_0\zeta_0 + c_0 = 0, \tag{10.18}$$

which gives ζ_0. Next, the terms of $\tilde{\omega}$ in equation (10.14) leads to

$$\zeta_1 = (2a_0\zeta_0 + b_0)\zeta_1 + (a_1\zeta_0^2 + b_1\zeta_0 + c_1), \tag{10.19}$$

which gives ζ_1, unless $2a_0\zeta_0 + b_0 = 1$. Proceeding to the terms of $\tilde{\omega}^2$ in equation (10.14), we have

$$2\zeta_2 = (2a_0\zeta_0 + b_0)\zeta_2 + (a_0\zeta_1^2 + 2a_1\zeta_0\zeta_1 + b_1\zeta_1) + (a_2\zeta_0^2 + b_2\zeta_0 + c_2), \tag{10.20}$$

which gives ζ_2, unless $2a_0\zeta_0 + b_0 = 2$. In this way, if $2a_0\zeta_0 + b_0$ is not equal to a positive integer, we can calculate ζ_n sucessively from ζ_{n-1}, and obtain a formal power series of $\tilde{\omega}$ as a solution of equation (10.14).

Next, let us consider what case $2a_0\zeta_0 + b_0 = 1$ is. Taking the square of $2a_0\zeta_0 + b_0 = 1$ and substituting equation (10.18), we see that $(2a_0\zeta_0+b_0)^2 = 1$ is equivalent with $b_0^2 - 4a_0c_0 = 1$, where $b_0^2 - 4a_0c_0$ is written by using equations (10.17) as

$$b_0^2 - 4a_0c_0 = 1 - 4\mathscr{R}_i, \tag{10.21}$$

$\mathscr{R}_i$ being the Richardson number given, in the present case, by

$$\mathscr{R}_i = \frac{1}{(md\Omega/dr)^2}\frac{n\kappa^2\Omega_\perp^2}{c_s^2} = \frac{n}{(md\ln\Omega/d\ln r)^2}\frac{\kappa^2}{\Omega^2}\frac{r^2}{H^2}. \tag{10.22}$$

In deriving the second relation we have adopted $c_s^2 = \Omega_\perp^2 H^2$. The most interesting case is horizontal p-mode oscillations, i.e., $n = 0$. In this case, $\mathscr{R}_i = 0$ and $(2a_0\zeta_0 + b_0)^2 = 1$. Even in this case, we can have a power series solution to equation (10.14) by adopting ζ_0 which satisfies $2a_0\zeta_0 + b_0 = -1$ (not $2a_0\zeta_0 + b_0 = 1$). It is noted that in the case of $\mathscr{R}_{\rm i} = 0$, $b_0^2 - 4a_0c_0 = 1$ and thus the solution of equation (10.18) is $2a_0\zeta_0 + b_0 = \pm 1$. It is also noted that there is no case where $2a_0\zeta_0 + b_0 = 2, 3, 4, \ldots$, since $b_0^2 - 4a_0c_0 = 1$ in the case of $n = 0$.

In the cases of $n \geq 1$ oscillations (g-mode, c-mode, and vertical p-modes), $\mathscr{R}_i$ is usually much larger than unity, since $\kappa/\Omega \sim 1$ and $(r/H)^2 \gg 1$ except for region very close to the inner edge of disks where $\kappa^2 \ll \Omega^2$. That is, we find that $b_0^2 - 4a_0c_0 < 1$, and ζ_0 satisfying (10.18) does not make $a_0\zeta_0 + b_0$ a positive integer.

The above argument shows that in all cases one of the solutions of equation (10.14) can be expressed by a power series of $\tilde{\omega}$, which is denoted hereafter $\tilde{\zeta}$. This solution is regular at corotation point. This means that this solution does not contribute to overrefletion.

Once one particular solution of ζ, i.e., $\tilde{\zeta}$, is known, the general solution of equation (10.14) , i.e., ζ, is obtained by writing it as $\zeta = \tilde{\zeta} + 1/\xi$. Substitution of $\zeta = \tilde{\zeta} + 1/\xi$ into equation (10.14) shows that ξ is the solution of the linear equation:

$$\tilde{\omega}\frac{d\xi}{d\tilde{\omega}} + (2a\tilde{\zeta} + b)\xi + a = 0. \tag{10.23}$$

By solving this equation we obtain

$$\xi = -\frac{1}{B}\int_{\tilde{\omega}_*}^{\tilde{\omega}} aB\frac{d\tilde{\omega}}{\tilde{\omega}}, \tag{10.24}$$

where

$$B(\tilde{\omega}) = \exp\int_{\tilde{\omega}_*}^{\tilde{\omega}}(2a\tilde{\zeta} + b)\frac{d\tilde{\omega}}{\tilde{\omega}}, \tag{10.25}$$

and $\tilde{\omega}_*$ is the integration constant. Equation (10.24) is easily shown to be the solution of equation (10.23) by substituting it into equation (10.23). It should be noted here that we are interested in the behavior of ξ near to corotation point, and thus $\tilde{\omega}_*$ is taken to be small as well as $\tilde{\omega}$ so that the integrands in equations (10.24) and (10.25) can be expanded by power series of $\tilde{\omega}$.

In summary, the general solution of equation (10.14) is (Drury 1985)

$$\zeta = \tilde{\zeta} - \frac{B}{\int aB(d\tilde{\omega}/\tilde{\omega})}. \tag{10.26}$$

In the followings, we first present results on $\Im(\zeta)$ in the case of $\mathscr{R}_{\rm i} > 1/4$. Next, the case of $\mathscr{R}_{\rm i} = 0$ is considered. The cases between the above two ones are only briefly noted, because these cases will be less interesting.

10.3.1 The Case of $\mathscr{R}_{\rm i} > 1/4$

Waves other than p-mode ones belong to this class.[4] In this case, ν^2 defined by $\nu^2 = b_0^2 - 4a_0c_0$ is negative, and $\nu \equiv i\mu$ is a purely imaginary quantity. As a regular solution in neighborhood of corotation we adopt $\tilde{\zeta} = \tilde{\zeta}_0 + \ldots$, where

$$\tilde{\zeta}_0 = \frac{-b_0 + i\mu}{2a_0}. \tag{10.27}$$

[4]Equation (10.22) shows that $\mathscr{R}_{\rm i} \gg 1/4$ in the case where $n \geq 1$, because $r^2/H^2 \gg 1$.

Then, $B(\tilde{\omega})$ defined by equation (10.25) is found to be, in the lowest order approximations,

$$B(\tilde{\omega}) = \left(\frac{\tilde{\omega}}{\tilde{\omega}_*}\right)^\nu. \tag{10.28}$$

After some manipulation, we find that ξ defined by equation (10.24) becomes, in the lowest order approximations,

$$\xi = -\frac{a_0}{\nu}\left(\frac{\tilde{\omega}_*}{\tilde{\omega}}\right)^\nu\left[1-\left(\frac{\tilde{\omega}}{\tilde{\omega}_*}\right)^\nu\right]. \tag{10.29}$$

Finally, we have

$$\zeta = \tilde{\xi} + 1/\xi = -\frac{b_0}{2a_0} - \frac{\nu}{2a_0}\frac{(\tilde{\omega}/\tilde{\omega}_*)^\nu + 1}{(\tilde{\omega}/\tilde{\omega}_*)^\nu - 1} + \dots. \tag{10.30}$$

We are interested here in whether $\Im(\zeta)$ can change its sign at corotation point. Neglecting terms which do not contribute to the change, we have (Drury 1985)

$$\begin{aligned}
\mathrm{Sgn}\Im(\zeta) &= -\mathrm{Sgn}\,\frac{\mu}{2a_0}\Re\left[\frac{k\tilde{\omega}^\nu + 1}{k\tilde{\omega}^\nu - 1}\right] \\
&= -\mathrm{Sgn}\,\frac{\mu}{2a_0}\left(|k\tilde{\omega}^\nu|^2 - 1\right) \\
&= -\mathrm{Sgn}\,\frac{\mu}{2a_0}\left[|\tilde{k}\exp(-\mu\arg\tilde{\omega})|^2 - 1\right],
\end{aligned} \tag{10.31}$$

where

$$k = 1/\tilde{\omega}_*^\nu \quad \text{and} \quad \tilde{k} = k\exp[i\mu\ \ln|\tilde{\omega}|]. \tag{10.32}$$

In deriving the last equality in equation (10.31), we have used

$$\tilde{\omega}^\nu = \exp(i\mu\ln\tilde{\omega}) = \exp\left[i\mu(\ln|\tilde{\omega}| + i\arg\tilde{\omega})\right]. \tag{10.33}$$

Our purpose here is to examine whether Sgn $\Im(\zeta)$ changes around $\tilde{\omega} = 0$. In the case where angular velocity of rotation decreases outwards ($d\Omega/dr < 0$), the value of $\tilde{\omega}[= \omega - m\Omega]$ on the real axis of r is negative when r is smaller than the corotation radius, say r_c, and becomes positive for $r > r_c$. The corotation point $\tilde{\omega} = 0$ is a branch point of ln $\tilde{\omega}$. When considering the value of ln $\tilde{\omega}$, we must consider the complex r-plane. As shown by Lin (1945a,b), the correct procedure to treat branch points (as well as singular points) is to consider tentatively a growing mode (in the present problem it means that the imaginary part of $\tilde{\omega}$ is negative), and then to examine the limiting case where the imaginary part tends to zero. In the present case of $d\Omega/dr < 0$, the point of $\tilde{\omega} = 0$ is just above the point of $r = r_c$ on the complex r-plane.[5] Hence, the imaginary part of $\ln \tilde{\omega}$ increases by π when r passes the corotation point, $r = r_c$, along the real axis from inside to outside.

Let us consider the case of $\mu > 0$. If Sgn $\Im(\zeta) < 0$ inside corotation, it means that Sgn $[|\tilde{k}\exp(-\mu\arg\tilde{\omega})|^2 - 1] < 0$ (notice that $a_0 < 0$). Then, outside corotation we have Sgn $\Im(\zeta) < 0$, because Sgn $[|\tilde{k}\exp(-\mu\arg\tilde{\omega})|^2 - 1]$ is changed to Sgn $[|\tilde{k}\exp(-\mu\arg\tilde{\omega} - \mu\pi)|^2 - 1]$ = Sgn $[\exp(-2\mu\pi)|\tilde{k}\exp(-\mu\arg\tilde{\omega})|^2 - 1]$, which is negative. If Sgn $\Im(\zeta) > 0$ outside corotation, it means that Sgn $[|\tilde{k}\exp(-\mu\arg\tilde{\omega})|^2 - 1] > 0$. Then, inside corotation Sgn $\Im(\zeta) > 0$, because Sgn $[|\tilde{k}\exp(-\mu\arg\tilde{\omega} + \mu\pi)|^2 - 1] > 0$. Situations are the same when $\mu < 0$, as will be shown by similar considerations. That is, if the energy flux interior to corotation is zero or negative, it must be negative outside corotation. If it is positive or zero outside corotation, it must be positive or zero inside corotation. This means that the corotation region cannot be a wave amplifier if $\mathscr{R}_i > 1/4$. That is, waves with node(s) in the vertical direction ($n \neq 0$) are not amplified at corotation. Really, Kato (2003) and Li et al. (2003) showed that g-mode oscillations are damped at corotation. The damping of c-mode oscillations is shown by Tsang and Lai (2009a).

10.3.2 The Case of $\mathscr{R}_i = 0$

This is the most important case. In this case it is convenience to rewrite equation (10.14) into the following form (Drury 1985):

$$\tilde{\omega}\frac{d\zeta}{d\tilde{\omega}} = a(\zeta - f)(\zeta - g), \tag{10.34}$$

Of course, f and g are related to a, b, and c by

$$b = -a(f + g), \quad \text{and} \quad c = afg. \tag{10.35}$$

Now, f and g are expanded in terms of $\tilde{\omega}$ as

$$f = \sum_{n=0} f_n\tilde{\omega}^n, \quad g = \sum_{n=0} g_n\tilde{\omega}^n. \tag{10.36}$$

Equation (10.34) then shows that the solution regular at corotation point, i.e., $\tilde{\zeta}$, is f_0 or g_0 in the lowest order approximation with respect to $\tilde{\omega}$. Here we take it to be

[5] We consider the point of $\tilde{\omega} = 0$ on the complex r-plane. The point is written as (r_c, ϵ) on the plane. Then, because $\tilde{\omega} = \omega - m\Omega = \omega_c - m\Omega_c + i\omega_i - m(d\Omega/dr)i\epsilon = 0$, we have $\epsilon \sim \omega_i(md\Omega/dr)^{-1} > 0$ for $\omega_i < 0$.

f_0, i.e., $\tilde{\zeta}_0 = f_0$. Then, in the next order of approximation, equation (10.34) leads to

$$\tilde{\zeta}_1 = a_0(\tilde{\zeta}_0 - g_0)(\tilde{\zeta}_1 - f_1). \tag{10.37}$$

In previous discussions we showed that in the case of $\mathscr{R}_i = 0$, we adopted $2a_0\tilde{\zeta}_0 + b_0 = -1$. From this and equations (10.35) we see that $\tilde{\zeta}_1 = (1/2)f_1$. That is, the solution regular at corotation point is written as

$$\tilde{\zeta} = f_0 + \frac{1}{2}f_1\tilde{\omega} + \emptyset(\tilde{\omega}^2). \tag{10.38}$$

Once $\tilde{\zeta}$ is obtained, we can calculate the whole solution of equation (10.34) by the same procedure as that in the beginning of Sect. 10.3. To do so, we first perform the integration in equation (10.25) by using $b = -a(f+g)$, the power series of $\tilde{\zeta}$, f and g, and $a_0(f_0 - g_0) = -1$. The result is

$$B = \frac{\tilde{\omega}_*}{\tilde{\omega}}\left[1 - a_0\left(g_1 + \frac{a_1}{a_0^2}\right)(\tilde{\omega} - \tilde{\omega}_*) + \emptyset(\tilde{\omega}^2, \tilde{\omega}_*^2)\right]. \tag{10.39}$$

Then, ξ given by equation (10.24) becomes, after some manipulations,

$$\xi = a_0\left[1 + a_0 g_1\tilde{\omega}\ \ln\left(\frac{\tilde{\omega}}{\tilde{\omega}_*}\right) - \frac{\tilde{\omega}}{\tilde{\omega}_*} + \ldots\right], \tag{10.40}$$

where $\tilde{\omega}_*$ has been taken to be sufficiently small and, for example, $\ln\tilde{\omega}_*$ is neglected compared with $1/\tilde{\omega}_*$. Although $\tilde{\omega}_*$ is small, $\tilde{\omega}$ has been taken to be much smaller than $\tilde{\omega}_*$,[6] and finally we have

$$\zeta = \left(f_0 + \frac{1}{a_0} + \frac{1}{2}f_1\tilde{\omega} + \ldots\right) - g_1\tilde{\omega}\ln\tilde{\omega} + \frac{1}{a_0}\frac{\tilde{\omega}}{\tilde{\omega}_*} + \ldots \tag{10.41}$$

Equation (10.41) shows that

$$\begin{aligned} \text{Sgn}\ \Im(\zeta) &= -\text{Sgn}\left[g_1\Im(\tilde{\omega}\ln\tilde{\omega})\right] + \text{Sgn}\,\tilde{\omega}\ \Im(k) \\ &= -\text{Sgn}\ \tilde{\omega}[g_1 \arg\tilde{\omega} - \Im(k)]. \end{aligned} \tag{10.42}$$

where

$$k = \frac{1}{a_0\tilde{\omega}_*}. \tag{10.43}$$

[6] As mentioned before, we can study overreflection at the corotation point by examining local behavior at corotation point.

In deriving equation (10.42), terms in $\Im(\zeta)$ which do not change obviously their sign across corotation point have been neglected, because we are interested in the change of sign of $\Im(\zeta)$ across corotation point.

The imaginary part of ln $\tilde{\omega}$ increases by π when r passes corotation point along the real axis from inside to outside, as mentioned before. This means that when Sgn $[\tilde{\omega}\arg\tilde{\omega}] < 0$ outside corotation radius (Sgn $\tilde{\omega} > 0$ and $\arg\tilde{\omega} < 0$), we have Sgn $[\tilde{\omega}\arg\tilde{\omega}] > 0$ inside corotation radius, because Sgn $\tilde{\omega} < 0$ there and $\arg\tilde{\omega}$ is smaller than that in the outside by $-\pi$ and thus $\arg\tilde{\omega} < 0$.

The above arguments give an important sufficient condition for overreflection. That is, if $g_1 > 0$, a positive energy flux outside corotation implies a negative energy flux inside corotation. That is, corotation point is a source of wave energy for waves transmitted from interior to exterior. This means that a transmitted wave in the exterior from interior will be overreflected at the corotation point and it will be amplified if there is a reflecting boundary inside (see panel labeled "$g_1 > 0$" in Fig. 10.3).

We can also say that when Sgn $[\tilde{\omega}\arg\tilde{\omega}] < 0$ inside corotation (Sgn $\tilde{\omega} < 0$ and Sgn $[\arg\tilde{\omega}] > 0$) we have then Sgn$[\tilde{\omega}\arg\tilde{\omega}] > 0$ outside (Sgn $\tilde{\omega} > 0$ and Sgn $[\arg\tilde{\omega}] > 0$). Hence, if $g_1 < 0$, a negative energy flux inside implies a positive energy flux outside, i.e., corotation point is a source of energy. This means that a transmitted wave in interior from exterior will be overreflected at corotation and will be amplified if there is a reflecting boundary outside (see panel labeled "$g_1 < 0$" in Fig. 10.3).

In the exceptional case of $g_1 = 0$, the corotation singularity is removed, but the energy flux changes sign at corotation point in both cases where waves approach to corotation from inside or outside, when $\Im(k) \neq 0$. If Sgn $\zeta < 0$ inside coronation, it means that $\Im(k) > 0$, because Sgn $\tilde{\omega} < 0$. Then, outside coronation Sgn $\zeta > 0$, because Sgn $\tilde{\omega} > 0$. That is, if the energy flux interior to corotation is negative, it must be positive outside corotation. Similarly, if Sgn $\zeta > 0$ outside coronation, Sgn $\zeta < 0$ inside coronation, because the sgn of $\tilde{\omega}$ changes at corotation. That is, overreflection occurs on both sides of corotation (see panel labeled "$g_1 = 0$" in Fig. 10.3).

The next problem is to derive an explicit expression for g_1. To do so Drury (1985) introduced an elegant way. Let us introduce an variable χ defined by $\chi = \rho_{00}\zeta$ and change equation (10.13) into an equation of χ as, in the case of $n = 0$,

$$\tilde{\omega}\frac{d\chi}{d\tilde{\omega}} = \frac{\rho_{00}}{mrd\Omega/dr}\left(m + \frac{2\Omega}{\rho_{00}}\chi\right)\left(m + \frac{\kappa^2}{2\Omega\rho_{00}}\chi\right) + \emptyset(\tilde{\omega}^2). \tag{10.44}$$

Since the whole theory depends only on the sign of the imaginary part of ζ, the same argument can be made for χ. Then, we see that the quantity corresponding to g is $-m\rho_{00}/(\kappa^2/2\Omega)$ in the present case of χ (compare equation (10.44) with equation (10.34)). This means that

$$\begin{aligned}\text{Sgn}\,g_1 = \text{Sgn}\left(\frac{dg}{d\tilde{\omega}}\right)_{\tilde{\omega}=0} &= \text{Sgn}\left[-\frac{1}{md\Omega/dr}\frac{dg}{dr}\right]_{r_c}\\ &= \text{Sgn}\left[-\frac{1}{d\Omega/dr}\frac{d}{dr}\left(\frac{\kappa^2/2\Omega}{\rho_{00}}\right)\right]_{r_c}.\end{aligned} \tag{10.45}$$

Since we are interested in disks with $d\Omega/dr < 0$, we have

$$\mathrm{Sgn}(g_1) = \mathrm{Sgn}\left[\frac{d}{dr}\left(\frac{\kappa^2/2\Omega}{\rho_{00}}\right)\right]_{r_c}. \tag{10.46}$$

The quantity, $\kappa^2/2\Omega$, is called vorticity of the flow, and $(\kappa^2/2\Omega)/\rho_{00}$ is the specific vorticity.

In summary, for $n = 0$ oscillations (the p-mode oscillations), corotation point is a singular point, unless the sign of the gradient of specific vorticity at the point is zero. That is, waves ingoing to corotation point from inside are overreflected there, if the gradient of specific vorticity at the corotation point is positive ($g_1 > 0$) (see the upper-left corner of Fig. 10.3). When the gradient of specific vorticity is negative ($g_1 < 0$) at the corotation point, the situations are changed, and waves ingoing there from outside are overreflected (see the upper-right corner of Fig. 10.3). If the gradient of the specific vorticity vanishes at the corotation point, the singularity at

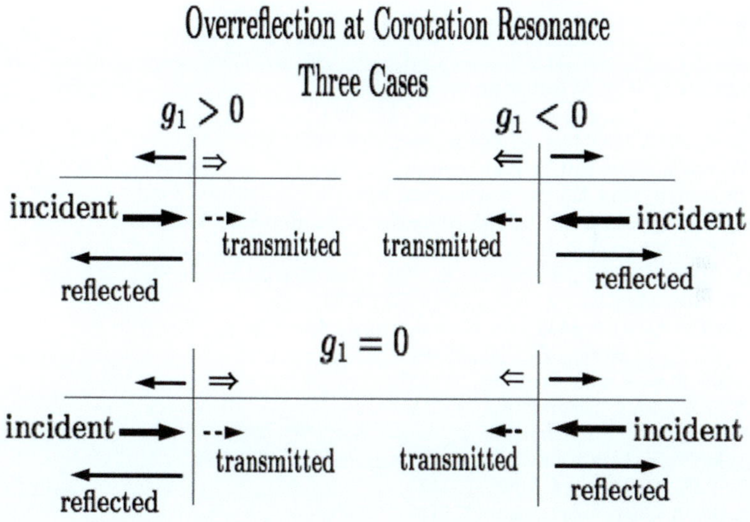

Fig. 10.3 Schematic picture showing overreflection in three cases of $g_1 > 0$, $g_1 < 0$, and $g_1 = 0$. In the case of $g_1 > 0$ (the *upper-left part* of this figure), if an outgoing wave from the corotation point (shown by $\Rightarrow$) exists, there is an ingoing wave (shown by $\leftarrow$) from the corotation point. That is, the corotation point works as a source of energy. This means that a wave from inside to the corotation point is partially transmitted outwards and the remaining part is overreflected inwards. If there is a reflecting boundary inside the corotation point, the wave trapped between the boundary and the corotation point is amplified. In the case of $g_1 < 0$, the situation is changed. If an ingoing wave from the coronation point (shown by $\Leftarrow$) exists, there is an outgoing wave (shown by $\rightarrow$) from the coronation point. That is, a wave coming to the corotation point from outside is overreflected at the coronation point. In the case of $g_1 = 0$, waves from both sides of the corotation point are overreflected as shown in the *lower part* of this figure.

the pint is removed, but waves approaching to corotation point are overreflected, no matter what the waves come to the corotation point from inside nor outside (see the lower part of Fig. 10.3).

10.3.3 Cases of $0 < \mathscr{R}_i \leq 1/4$

Since these cases are less interesting, Drury's results are briefly summaries below.

- $0 < \mathscr{R}_i < 1/4$
  The corotation region can be a source or sink of energy for perturbations, depending on the perturbations.
- $\mathscr{R}_i = 1/4$
  Amplification of perturbations at corotation point is impossible when $\mathscr{R}_i = 1/4$.

References

Blaes, O. M. 1985, Mon. Not. R. Astron. Soc., 216, 533
Blaes, O.M. 1987, Mon. Not. R. Astron. Soc., 227, 975
Drury, L.O'C. 1985, Mon. Not. R. Astron. Soc., 217, 821
Glatzel, W. 1887a, Mon. Not. R. Astron. Soc., 225, 227
Glatzel, W., 1887b, Mon. Not. R. Astron. Soc., 228, 77
Goldreich, P., & Lynden-Bell, D. 1965, Mon. Not. R. Astron. Soc., 130, 125
Goldreich, P., & Narayan, R. 1985, Mon. Not. R. Astron. Soc., 1985, 213, 7
Goldreich, P., Goodman, J., & Narayan, R. 1986, Mon. Not. R. Astron. Soc., 221, 339
Hanawa, T. 1986, Mon. Not. R. Astron. Soc., 223, 859
Horák, J., & Lai, D. 2013, Mon. Not. R. Astron. Soc., 434, 276
Kato, S. 1987, Publ. Astron. Soc. Jpn., 39, 645
Kato, S. 2003, Publ. Astron. Soc. Jpn., 55, 257
Kojima, Y., 1986, Prog. Theor. Phys., 75, 251
Lai, D., & Tsang, D., 2009, Mon. Not. R. Astron. Soc., 393, 979
Li, L. X., Goodman, J., & Narayan, R., 2003, Astrophys. J., 593, 980
Lin, C. C. 1945a, Quart. Appl. Math., 3, 117
Lin, C. C. 1945b, Quart. Appl. Math., 3, 218
Lin, C. C., & Shu, F. H., 1964, Astrophys. J., 140, 646
Mark, J. W.-K., 1974, Astrophys. J., 193, 539
Mark, J. W.-K., 1976a, Astrophys. J., 203, 81
Mark, J. W.-K., 1976b, Astrophys. J., 205, 363
Narayan, R., Goldreich, P., & Goodman, J., 1987, Mon. Not. R. Astron. Soc., 228, 1
Papaloizou, J. C. B., & Pringle, J. E. 1984, Mon. Not. R. Astron. Soc., 208, 721
Papaloizou, J. C. B., & Pringle, J. E. 1985, Mon. Not. R. Astron. Soc., 213, 799
Papaloizou, J. C. B., & Pringle, J. E. 1987, Mon. Not. R. Astron. Soc., 225, 267
Toomre, A. Astrophys. J.,1969, 158, 899
Tsang, D., & Lai, D. 2008, Mon. Not. R. Astron. Soc., 387, 446
Tsang, D., & Lai, D. 2009a, Mon. Not. R. Astron. Soc., 393, 992
Tsang, D., & Lai, D. 2009b, Mon. Not. R. Astron. Soc., 400, 470

Chapter 11
Wave-Wave Resonant Instability in Deformed Disks

Abstract In addition to excitation processes considered in Chaps. 9 and 10, we have another important excitation process of disk oscillations. This is a wave-wave resonant excitation in deformed disks. Disk deformation from axisymmetric state is widely expected in tidally deformed disks. Even in a single star, long-living (time-periodic) deformations may be present on disks (e.g., warps).

Basics on a wave-wave resonant instability in deformed disks was studied by Kato (Publ Astron Soc Jpn 56:905, 2004; 60:111, 2008), and later examined from different ways by Ferreria and Ogilvie (Mon Not R Astron Soc 386:2297, 2008) and Oktariani et al. (Publ Astron Soc Jpn 62:709, 2010). Subsequently, Kato (Publ Astron Soc Jpn 65:75, 2013; 66:24, 2014, see also Kato et al. (Publ Astron Soc Jpn 63:363, 2011)) formulated the instability in a perspective way. In this chapter, we outline the formulation by Kato (Publ Astron Soc Jpn 65:75, 2013). Applications to time variations in dwarf novae and X-ray binaries are presented in Chap. 12.

Keywords Disk deformation • Tidal wave • Warp • Wave energy • Wave-wave coupling

11.1 Brief Outline of Wave-Wave Resonant Instability

In Chaps. 9 and 10 we have considered two important excitation mechanisms of disk oscillations. They are (i) viscous overstability and (ii) excitation by corotation resonance. Types of oscillations excited by these mechanisms, however, are rather limited. They are p-mode oscillations alone.

Observations, however, suggest the presence of other types of oscillations in disks. For example, tilt mode (one-armed vertical p-mode oscillation) seems to be excited on dwarf novae disks, in addition to one-armed eccentric precession mode (one-armed p-mode). The origins of quasi-periodic oscillations observed in low-mass X-ray binaries are still in debate, but some of them might be due to disk oscillations other than p-mode oscillations (see Chap. 7).

Before examining the resonant instability condition in detail, we present here the situations which we consider and briefly summarize the resulting condition of the resonant instability.

S. Kato, *Oscillations of Disks*, Astrophysics and Space Science Library 437,
DOI 10.1007/978-4-431-56208-5_11

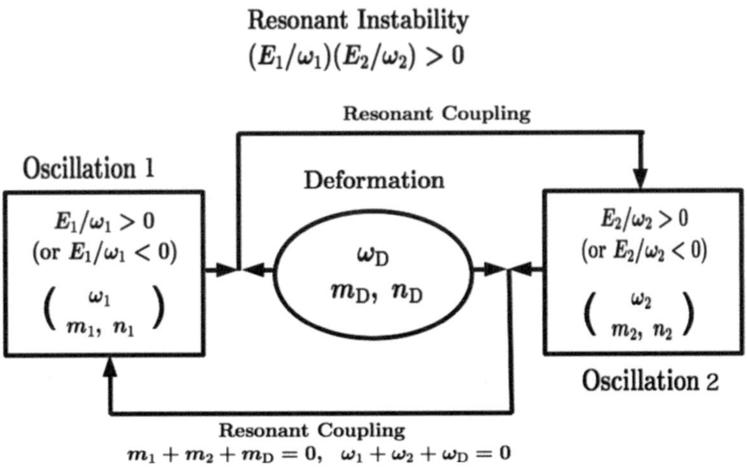

Fig. 11.1 Schematic diagram showing simultaneous amplification of two oscillations which are resonantly coupled through disk deformation. The set of frequency, azimuthal wave number, and vertical node number are represented by (ω, m, n), and the subscripts 1 and 2 represent the two oscillations, and the subscript D does the disk deformation. The resonant conditions are $m_1 + m_2 + m_D = 0$, and $\omega_1 + \omega_2 + \omega_D = 0$. The condition of resonant amplification is $(E/\omega)_1(E/\omega)_2 > 0$, where E is wave energy of oscillations. That is, for amplification to occur two oscillations must have the same signs of $E/\omega > 0$. This amplification condition is derived in Sect. 11.4.

Let us consider a disk deformed from an axisymmetric state. The deformation is assumed to be time-periodic with frequency, ω_D, and azimuthal wavenumber m_D. On such deformed disks, two small-amplitude normal mode oscillations are superposed. The set of frequency and azimuthal wavenumber, (ω, m), of these two normal mode oscillations are (ω_1, m_1) and (ω_2, m_2). If there are the following relations:

$$\omega_1 + \omega_2 + \omega_D = 0, \quad m_1 + m_2 + m_D = 0, \tag{11.1}$$

the two oscillations with (ω_1, m_1) and (ω_2, m_2) can resonantly couple each other through the disk deformation with (ω_D, m_D).[1] This coupling is schematically shown in Fig. 11.1.

In addition to the above, the vertical motions also need to couple for the resonant to occur. For example, if the variable $h_1(= p_1/\rho_0)$ associated with disk deformation, say $h_{1,D}$, has no node in the vertical direction, i.e., $h_{1,D}(r, z) \propto \mathcal{H}_0(z/H)$, h_1's in the ω_1- and ω_2-oscillations must have the same node number in the vertical direction, i.e., $n_1 = n_2$, since the product of $\mathcal{H}_n$ and $\mathcal{H}_0$ is proportional to $\mathcal{H}_n$. In the case where the disk deformation has one node in the vertical direction, i.e., $n_D = 1$, the

[1] Adoption of equations (11.1) as resonant conditions means that we do not restrict ω's and m's to positive values.

difference of vertical node numbers of ω_1- and ω_2-oscillations needs to be unity, i.e., $n_1 \pm 1 = n_2$, since $\mathscr{H}_n$ times $\mathscr{H}_1$ consists of terms proportional to $\mathscr{H}_{n-1}$ and $\mathscr{H}_{n+1}$.

Furthermore, for couplings among two modes really to occur, the radial domains where waves and disk deformations exist need to be overlapped. Otherwise, these modes cannot have nonlinear interactions.

The above conditions on frequency (ω), azimuthal wavenumber (m), vertical node number (n), and radial propagation regions are all necessary conditions for resonant couplings. If these conditions are satisfied, the ω_1- and ω_2-oscillations resonantly couple through the disk deformation. As the results of the coupling the oscillations grow or damp. The problem to be examined is what is the condition of the amplification. This is the subject of this chapter.

The final results of analyses show that the condition of amplification is rather simple. That is, if

$$\left(\frac{E_1}{\omega_1}\right)\left(\frac{E_2}{\omega_2}\right) > 0 \tag{11.2}$$

is realized, both of the two oscillations grow with time, where E_1 and E_2 are wave energies of the ω_1- and ω_2-oscillations, respectively (Kato 2013, see also Kato 2004, 2008; Ferreria and Ogilvie 2008; Oktariani et al. 2010; Kato et al. 2011).

One of basic assumptions involved in our analyses is that the disk deformation is a small perturbation on the unperturbed disks, but its amplitude does not change by resonant coupling with the oscillations imposed on disks. This means that the amplitude of disk deformation is larger than those of oscillations imposed. Cases where this assumption is relaxed are discussed in Sect. 11.4 in relation to the mechanism of amplifications.

11.2 Derivation of Quasi-nonlinear Wave Equation

To examine wave-wave resonant couplings, we must extend the linear wave equation derived in Chap. 3:

$$\rho_0\frac{\partial^2\boldsymbol{\xi}}{\partial t^2} + 2\rho_0(\boldsymbol{u}_0\cdot\nabla)\frac{\partial\boldsymbol{\xi}}{\partial t} + \mathscr{L}(\boldsymbol{\xi}) = 0, \tag{11.3}$$

to quasi-nonlinear cases. Here, $\boldsymbol{\xi}$ is Lagrangian displacement associated with perturbations, $\mathscr{L}(\boldsymbol{\xi})$ is a linear Hermitian operator with respect to $\boldsymbol{\xi}$, and $\boldsymbol{u}_0$ is the velocity in unperturbed disks and is taken to be rotation alone in the present treatments, i.e., $\boldsymbol{u}_0 = (0, r\Omega, 0)$.

Returning to a Lagrangian formulation of full nonlinear wave equation, we start here from (Lynden-Bell and Ostriker 1967, see also chapter 3)

$$\frac{D_0^2\boldsymbol{\xi}}{Dt^2} = \delta\Big(-\nabla\psi - \frac{1}{\rho}\nabla p\Big), \tag{11.4}$$

where D_0/Dt is the time-derivative along an unperturbed flow, $\boldsymbol{u}_0$, and related to the Eulerian time derivative, $\partial/\partial t$, by

$$\frac{D_0}{Dt} = \frac{\partial}{\partial t} + \boldsymbol{u}_0 \cdot \nabla. \tag{11.5}$$

The Lagrangian change of a quantity, $Q(\boldsymbol{r},t)$, i.e., δQ, is defined by

$$\delta Q = Q[\boldsymbol{r} + \boldsymbol{\xi}(\boldsymbol{r},t), t] - Q_0(\boldsymbol{r},t), \tag{11.6}$$

where the subscript 0 to Q represents unperturbed quantity. The Eulerian change of Q, i.e., Q_1, on the other hand, is defined by

$$Q_1 = Q(\boldsymbol{r},t) - Q_0(\boldsymbol{r},t). \tag{11.7}$$

If $Q[\boldsymbol{r} + \boldsymbol{\xi}(\boldsymbol{r},t), t]$ is Taylor-expanded around $Q(\boldsymbol{\xi},t)$ up to the second-order terms with respect to perturbations, we have from equations (11.6) and (11.7),

$$\delta Q = Q_1 + \xi_j \frac{\partial Q_0}{\partial r_j} + \xi_j \frac{\partial Q_1}{\partial r_j} + \frac{1}{2}\xi_i \xi_j \frac{\partial^2 Q_0}{\partial r_i \partial r_j}. \tag{11.8}$$

This is a relation between δQ and Q_1, up to the second-order small quantities with respect to perturbations. Here and hereafter, the summation abbreviation is adopted, using Cartesian coordinates.

Our purpose here is to explicitly write down equation (11.4) up to the second-order small quantities with respect to perturbations. Since the self-gravity of disks is neglected here, i.e., $\psi_1 = 0$, we have easily

$$\delta(\nabla\psi) = \xi_j \frac{\partial}{\partial r_j}(\nabla\psi_0) + \frac{1}{2}\xi_i \xi_j \frac{\partial^2}{\partial r_i \partial_j}(\nabla\psi_0). \tag{11.9}$$

The second term on the right-hand side of equation (11.9) represents nonlinear terms.

Expressing $\delta(\nabla p/\rho)$ in terms of Lagrangian displacements is somewhat complicated. Using equation (11.8) and the definition of Lagrangian change, we have

$$\delta\left(\frac{1}{\rho}\nabla p\right) = \frac{1}{\rho_0 + \delta\rho}\biggl[\nabla p_0 + \nabla p_1 + \xi_j \frac{\partial}{\partial r_j}\nabla(p_0 + p_1) + \frac{1}{2}\xi_i \xi_j \frac{\partial^2}{\partial r_i \partial r_j}\nabla p_0 + \dots\biggr] - \frac{1}{\rho_0}\nabla p_0. \tag{11.10}$$

The Eulerian pressure variation, p_1, which appears on the right-hand side of equation (11.10), is expressed in terms of δp and $\boldsymbol{\xi}$ by using

$$\delta p = p_1 + \xi_j \frac{\partial}{\partial r_j}(p_0 + p_1) + \frac{1}{2}\xi_i\xi_j \frac{\partial^2 p_0}{\partial r_i \partial r_j}. \tag{11.11}$$

We have then (Kato 2004)

$$\begin{aligned}
\rho_0 \delta\left(\frac{1}{\rho}\nabla p\right) = &\nabla\left[\delta p - \xi_j \frac{\partial p_0}{\partial r_j} - \xi_j \frac{\partial}{\partial r_j}\left(\delta p - \xi_i \frac{\partial p_0}{\partial r_i}\right) - \frac{1}{2}\xi_i\xi_j \frac{\partial^2 p_0}{\partial r_i \partial r_j}\right] \\
&+\xi_j \frac{\partial}{\partial r_j}\nabla\left[p_0 + \delta p - \xi_i \frac{\partial p_0}{\partial r_i}\right] + \frac{1}{2}\xi_i\xi_j \frac{\partial^2}{\partial r_i \partial r_j}\nabla p_0 \\
&-\frac{\delta\rho}{\rho_0}\left[\nabla p_0 + \nabla(\delta p) - (\nabla \xi_j)\frac{\partial p_0}{\partial r_j}\right] + \left(\frac{\delta\rho}{\rho_0}\right)^2 \nabla p_0.
\end{aligned} \tag{11.12}$$

This explicitly expresses $\rho_0\delta(\nabla p/\rho)$ up to the second-order quantities in terms of Lagrangian quantities, δp, $\delta\rho$, and $\boldsymbol{\xi}$.

Next, δp and $\delta\rho$ on the right-hand side of equation (11.12) are explicitly expressed in terms of $\boldsymbol{\xi}$. To do so we use the equation of continuity and adiabatic relation. The equation of continuity is expressed as (e.g., Kato and Unno 1967)

$$\delta\rho + \rho_0 \frac{\partial \xi_i}{\partial r_i} = \rho_0 N_c, \tag{11.13}$$

where

$$N_c = \frac{1}{2}\left[\left(\frac{\partial \xi_j}{\partial \xi_j}\right)^2 + \frac{\partial \xi_i}{\partial \xi_j}\frac{\partial \xi_j}{\partial r_i}\right]. \tag{11.14}$$

The adiabatic relation is written as (e.g., Kato and Unno 1967)

$$\delta p - \Gamma_1 \frac{p_0}{\rho_0}\delta\rho = p_0 N_p, \tag{11.15}$$

or

$$\delta p + \Gamma_1 p_0 \frac{\partial \xi_j}{\partial r_j} = p_0(\Gamma_1 N_c + N_p), \tag{11.16}$$

where

$$N_p = \frac{1}{2}\Gamma_1(\Gamma_1 - 1)\left(\frac{\delta\rho}{\rho_0}\right)^2. \tag{11.17}$$

Substitution of equations (11.13) and (11.15) into equation (11.12) finally gives an expression for $\rho_0\delta(\nabla p/\rho)$, expressed in terms of $\boldsymbol{\xi}$ alone. This is summarized as

$$\rho_0\delta\left(\frac{1}{\rho}\nabla p\right) = \boldsymbol{P}_{\rm linear} + \boldsymbol{P}_{\rm nonlinear}, \tag{11.18}$$

where $\boldsymbol{P}_{\rm linear}$ and $\boldsymbol{P}_{\rm nonlinear}$ are the linear and quasi-nonlinear parts of $\rho_0\delta(\rho^{-1}\nabla p)$, respectively. An explicit form of the linear part is given by (Lynden-Bell and Ostriker 1967)

$$\boldsymbol{P}_{\rm linear} = \nabla\left[(1-\Gamma_1)p_0\frac{\partial\xi_i}{\partial r_i}\right] - p_0\nabla\left(\frac{\partial\xi_i}{\partial r_i}\right) - \nabla\left(\xi_j\frac{\partial p_0}{\partial r_j}\right) + \xi_j\frac{\partial}{\partial r_j}\nabla p_0. \tag{11.19}$$

At the stage where equations (11.13) and (11.15) are substituted into equation (11.12) the quasi-nonlinear part of $\rho_0\delta[(1/\rho)\nabla p]$ has various terms, but after some manipulation the part can be summarized into the following form (Kato 2008)[2]:

$$\begin{aligned} -(P_{\rm nonlinear})_k = &-\frac{\partial}{\partial r_j}\left(p_0\frac{\partial\xi_i}{\partial r_k}\frac{\partial\xi_j}{\partial r_i}\right) + \frac{\partial}{\partial r_j}\left[(\Gamma_1-1)p_0\frac{\partial\xi_j}{\partial r_k}\frac{\partial\xi_i}{\partial r_i}\right] \\ &+\frac{1}{2}\frac{\partial}{\partial r_k}\left[(\Gamma_1-1)p_0\frac{\partial\xi_i}{\partial r_j}\frac{\partial\xi_j}{\partial r_i}\right] + \frac{1}{2}\frac{\partial}{\partial r_k}\left[(\Gamma_1-1)^2p_0\frac{\partial\xi_i}{\partial r_i}\frac{\partial\xi_j}{\partial r_j}\right], \end{aligned} \tag{11.20}$$

where $(P_{\rm nonlinear})_k$ represents the k-component of $\boldsymbol{P}_{\rm nonlinear}$.

Substituting equations (11.9), (11.18), (11.19), and (11.20) into equation (11.4), we find that the linear wave equation (11.3) is now extended into a quasi-nonlinear equation as

$$\rho_0\frac{\partial^2\boldsymbol{\xi}}{\partial t^2} + 2\rho_0(\boldsymbol{u}_0\cdot\nabla)\frac{\partial\boldsymbol{\xi}}{\partial t} + \mathscr{L}(\boldsymbol{\xi}) = \boldsymbol{C}(\boldsymbol{\xi}), \tag{11.21}$$

where $\boldsymbol{C}(\boldsymbol{\xi})$ is the quasi-nonlinear term and its Cartesian component is written as

$$C_k(\boldsymbol{\xi}) = -\frac{1}{2}\rho_0\xi_i\xi_j\frac{\partial^2\psi_0}{\partial r_i\partial r_j\partial r_k} - (P_{\rm nonlinear})_k. \tag{11.22}$$

[2]In expression for $\boldsymbol{P}_{\rm nonlinear}$ given by equation (82) by Kato (2008) there are typographical errors.

An explicit expression for $\mathscr{L}(\boldsymbol{\xi})$ is presented in Chap. 3, but here it is presented again for completeness:

$$\mathscr{L}(\boldsymbol{\xi}) = \rho_0(\boldsymbol{u}_0 \cdot \nabla)(\boldsymbol{u}_0 \cdot \nabla)\boldsymbol{\xi} + \rho_0(\boldsymbol{\xi} \cdot \nabla)(\nabla\psi_0) + \boldsymbol{P}_{\rm linear}(\boldsymbol{\xi}). \tag{11.23}$$

11.3 Quasi-nonlinear Coupling Among Oscillations and Disk Deformation

After the above general arguments on quasi-nonlinear equations, we proceed to quasi-nonlinear couplings among disk oscillations and deformation. Before that, however, we briefly mention on normal mode oscillations of linear perturbations and on disk deformation.

11.3.1 Linear Oscillations and Disk Deformation

On an axisymmetric steady disk a deformation is superposed. The origin of the deformation is not discussed here, because it is outside of our present interest. However, if the deformation is due to tidal force, for example, the disk deformation, $\boldsymbol{\xi}_{\rm D}$, is related to the tidal potential, say $\psi_{\rm D}$, by

$$\rho_0\frac{\partial^2\boldsymbol{\xi}_{\rm D}}{\partial t^2} + 2\rho_0(\boldsymbol{u}_0 \cdot \nabla)\frac{\partial\boldsymbol{\xi}_{\rm D}}{\partial t} + \mathscr{L}(\boldsymbol{\xi}_{\rm D}) = -\rho_0\nabla\psi_{\rm D}. \tag{11.24}$$

The deformation is assumed to be time-periodic with frequency $\omega_{\rm D}$ and with azimuthal wavenumber $m_{\rm D}$. Our basic assumption here is that such a time-periodic deformation is maintained on disks over a timescale longer than the growing or damping timescale of oscillatory perturbations on disks. The displacement vector, $\boldsymbol{\xi}_{\rm D}(\boldsymbol{r}, t)$, associated with the deformation over the steady state is denoted by

$$\boldsymbol{\xi}_{\rm D}(\boldsymbol{r}, t) = \Re\left[\hat{\boldsymbol{\xi}}_{\rm D}(\boldsymbol{r})\exp(i\omega_{\rm D}t)\right] = \Re\left[\breve{\boldsymbol{\xi}}_{\rm D}\exp[i(\omega_{\rm D}t - m_{\rm D}\varphi)]\right], \tag{11.25}$$

where $\Re$ denotes the real part, and φ is the azimuthal coordinate of the cylindrical coordinates (r, φ, z) whose center is at the disk center and the z-axis is the rotating axis of the disk. In the case of tidal deformation, for example, the deformation is the forced oscillations resulting from the $\exp[i(\omega_{\rm D}t - m_{\rm D}\varphi)]$-component of the tidal force.

In addition to such a deformation, a set of two normal modes of oscillations with small amplitudes are superposed on the axisymmetric steady disks. The set of eigen-frequency and azimuthal wavenumber of these oscillations are denoted by (ω_1, m_1)

and (ω_2, m_2). The displacement vectors, $\boldsymbol{\xi}_{\rm i}(\boldsymbol{r}, t)$ (where i is 1 or 2), associated with these oscillations are expressed as

$$\boldsymbol{\xi}_{\rm i}(\boldsymbol{\xi}, t) = \Re\left[\hat{\boldsymbol{\xi}}_{\rm i}(\boldsymbol{r})\exp(i\omega_{\rm i}t)\right] = \Re\left[\breve{\boldsymbol{\xi}}_{\rm i}\exp[i(\omega_{\rm i}t - m_{\rm i}\varphi)]\right] \qquad ({\rm i} = 1, 2). \tag{11.26}$$

It is noted that these small amplitude normal modes of oscillations are governed by

$$\rho_0\frac{\partial^2\boldsymbol{\xi}_{\rm i}}{\partial t^2} + 2\rho_0(\boldsymbol{u}_0\cdot\nabla)\frac{\partial\boldsymbol{\xi}_{\rm i}}{\partial t} + \mathscr{L}(\boldsymbol{\xi}_{\rm i}) = 0. \tag{11.27}$$

where i = 1 or 2.

We now assume that the following resonant conditions are present among the two oscillations and the deformation:

$$\omega_1 + \omega_2 + \omega_{\rm D} = \Delta\omega \quad \text{and} \quad m_1 + m_2 + m_{\rm D} = 0, \tag{11.28}$$

where $m_{\rm i}$'s (i $=$ 1, 2, and D) are integers. In order to include in our formulation the cases where three frequencies are slightly deviated from the exact resonant condition, $\Delta\omega$ is introduced in the first relation of equation (11.28), where $|\Delta\omega|$ is assumed to be much smaller than the absolute values of ω_1 and ω_2.[3] In the final analyses of instability, however, we consider only the case of $\Delta\omega = 0$. In order to represent the resonant conditions by a simple form without separately considering such cases as $\omega_2 = \omega_1 + \omega_{\rm D}$ and $\omega_2 = -\omega_1 + \omega_{\rm D}$, we have adopted simple resonant forms as equations (11.28). Hence, ω's and m's can be taken to be positive or negative as long as the conditions (11.28) are satisfied.

In the non-linear stages, the ω_1- and ω_2-oscillations satisfying conditions (11.28) resonantly interact with each other through the disk deformation specified by $\omega_{\rm D}$.[4] Our main concern here is to examine how the amplitudes of the ω_1- and ω_2-oscillations are changed by the quasi-nonlinear interactions through the disk deformation. Quasi-nonlinear changes of the ω_1- and ω_2-oscillations by themselves are not considered, assuming that their amplitudes are smaller than that of the disk deformation. It is further assumed that the amplitudes of the oscillations are so small that the amplitude of the disk deformation is unchanged by the resonant couplings with the oscillations.

---

[3]The cases where $\omega_{\rm D} = 0$ can be included in our analyses.

[4]For oscillations to resonantly interact, additional conditions concerning wave forms in the vertical and radial directions are necessary, as mentioned in Sect. 11.1. These conditions are not considered here in detail. If these conditions are not satisfied, the value of the coupling term, W or $W^{\rm T}$, given by equation (11.48) or (11.49) vanishes.

11.3.2 Quasi-nonlinear Resonant Coupling of Oscillations

In the linear stage, the perturbation, $\boldsymbol{\xi}$, imposed on a deformed disk is simply the sum of two oscillations:

$$\begin{aligned}\boldsymbol{\xi}(\boldsymbol{r},t) &= A_1\boldsymbol{\xi}_1(\boldsymbol{r},t) + A_2\boldsymbol{\xi}_2(\boldsymbol{r},t)\\ &= \Re\left[A_1\hat{\boldsymbol{\xi}}_1\exp(i\omega_1 t) + A_2\hat{\boldsymbol{\xi}}_2\exp(i\omega_2 t)\right]\\ &= \Re\left[A_1\breve{\boldsymbol{\xi}}_1\exp[i(\omega_1 t - m_1\varphi)] + A_2\breve{\boldsymbol{\xi}}_2\exp[i(\omega_2 t - m_2\varphi)]\right],\end{aligned} \tag{11.29}$$

where A_1 and A_2 are amplitudes and are arbitrary constants at the linear stage. In the quasi-nonlinear stage, the oscillations are modified through quasi-nonlinear couplings with disk deformation, $\boldsymbol{\xi}_{\rm D}$, so that they satisfy a quasi-nonlinear wave equation. The quasi-nonlinear wave equation is

$$\rho_0\frac{\partial^2\boldsymbol{\xi}}{\partial t^2} + 2\rho_0(\boldsymbol{u}_0\cdot\nabla)\frac{\partial\boldsymbol{\xi}}{\partial t} + \mathscr{L}(\boldsymbol{\xi}) = \boldsymbol{C}(\boldsymbol{\xi},\boldsymbol{\xi}_{\rm D}), \tag{11.30}$$

where $\boldsymbol{C}$ is the quasi-nonlinear coupling terms and is given by (see equations (11.20) and (11.22))

$$\begin{aligned}C_k(\boldsymbol{\xi},\boldsymbol{\xi}_{\rm D}) = &-\rho_0\xi_i\xi_{{\rm D}j}\frac{\partial^3}{\partial r_i\partial r_j\partial r_k}\psi_0 - \frac{\partial}{\partial r_j}\left(p_0\frac{\partial\xi_i}{\partial r_k}\frac{\partial\xi_{{\rm D}j}}{\partial r_i} + p_0\frac{\partial\xi_{{\rm D}i}}{\partial r_k}\frac{\partial\xi_j}{\partial r_i}\right)\\ &+\frac{\partial}{\partial r_j}\left[(\Gamma_1-1)p_0\left(\frac{\partial\xi_j}{\partial r_k}{\rm div}\boldsymbol{\xi}_{\rm D} + \frac{\partial\xi_{{\rm D}j}}{\partial r_k}{\rm div}\boldsymbol{\xi}\right)\right]\\ &+\frac{\partial}{\partial r_k}\left[(\Gamma_1-1)p_0\frac{\partial\xi_i}{\partial r_j}\frac{\partial\xi_{{\rm D}j}}{\partial r_i}\right] + \frac{\partial}{\partial r_k}\left[(\Gamma_1-1)^2p_0{\rm div}\boldsymbol{\xi}\,{\rm div}\boldsymbol{\xi}_{\rm D}\right],\end{aligned} \tag{11.31}$$

where the subscript k attached to ξ and r, for example, denotes the k-component of vectors $\boldsymbol{\xi}$ and $\boldsymbol{r}$ in the Cartesian coordinates, and we use Einstein's convention of indices (i.e., take summation if a term has the same index variable twice).

It is noted here that in the cases where the disk deformation is due to an external force, additional terms should be added to equation (11.31). If the external force is tidal one, for example, there appears a quasi-nonlinear term in $\delta\nabla\psi$. This is because a general expression for $\delta\nabla\psi$ is given by

$$\delta\nabla\psi = \nabla\psi_{\rm D} + \xi_j\frac{\partial}{\partial r_j}\nabla(\psi_0+\psi_{\rm D}) + \frac{1}{2}\xi_j\xi_k\frac{\partial^2}{\partial r_j\partial r_k}\nabla(\psi_0+\psi_{\rm D}) + \ldots, \tag{11.32}$$

where ψ_0 is the unperturbed gravitational potential in the unperturbed axisymmetric disk. Corresponding to this, the quasi-nonlinear wave equation (11.30) is modified to

$$\rho_0\frac{\partial^2\boldsymbol{\xi}}{\partial t^2}+2\rho_0(\boldsymbol{u}_0\cdot\nabla)\frac{\partial\boldsymbol{\xi}}{\partial t}+\mathscr{L}(\boldsymbol{\xi})=\boldsymbol{C}^{\mathrm{T}}(\boldsymbol{\xi},\boldsymbol{\xi}_{\mathrm{D}}), \tag{11.33}$$

and $\boldsymbol{C}^{\mathrm{T}}$ is

$$\boldsymbol{C}^{\mathrm{T}}(\boldsymbol{\xi},\boldsymbol{\xi}_{\mathrm{D}})=\boldsymbol{C}(\boldsymbol{\xi},\boldsymbol{\xi}_{\mathrm{D}})-\rho_0\xi_j\frac{\partial}{\partial r_j}\nabla\psi_{\mathrm{D}}. \tag{11.34}$$

Here and hereafter, however, the coupling terms are sometimes represented by $\boldsymbol{C}(\boldsymbol{\xi},\boldsymbol{\xi}_{\mathrm{D}})$ without $\boldsymbol{C}(\boldsymbol{\xi},\boldsymbol{\xi}_{\mathrm{D}})$ and $\boldsymbol{C}^{\mathrm{T}}(\boldsymbol{\xi},\boldsymbol{\xi}_{\mathrm{D}})$ being distinguished, because there is no formal difference in the final results except that $\boldsymbol{C}(\boldsymbol{\xi},\boldsymbol{\xi}_{\mathrm{D}})$ is replaced by $\boldsymbol{C}^{\mathrm{T}}(\boldsymbol{\xi},\boldsymbol{\xi}_{\mathrm{D}})$.

The disk oscillations, $\boldsymbol{\xi}(\boldsymbol{r},t)$, resulting from the quasi-nonlinear coupling through disk deformation, $\boldsymbol{\xi}_{\mathrm{D}}(\boldsymbol{r},t)$, will be written generally in the form:

$$\boldsymbol{\xi}(\boldsymbol{r},t)=\Re\sum_{\mathrm{i}=1}^{2}A_{\mathrm{i}}(t)\hat{\boldsymbol{\xi}}_{\mathrm{i}}(\boldsymbol{r})\exp(\mathrm{i}\omega_{\mathrm{i}}t)+\Re\sum_{i}^{2}\biggl(\sum_{\alpha\neq 1,2}A_{i,\alpha}\hat{\boldsymbol{\xi}}_{\alpha}(\boldsymbol{r})\biggr)\exp(i\omega_\alpha \mathrm{t})$$
$$+\text{oscillating terms with other frequencies.} \tag{11.35}$$

The original two oscillations, $\boldsymbol{\xi}_1$ and $\boldsymbol{\xi}_2$, resonantly interact through the disk deformation. By this resonant interaction their amplitudes secularly change with time, which is taken into account in equation (11.35) by taking the amplitudes, A_{i}'s, to be slowly varying functions of time. The terms whose time-dependence is the same as $\exp(i\omega_{\mathrm{i}}t)$ but whose spatial dependence is different from $\hat{\boldsymbol{\xi}}_{\mathrm{i}}$ are expressed by a sum of eigen-functions, $\hat{\boldsymbol{\xi}}_{\alpha}$ ($\alpha\neq 1$ and 2), assuming that the set of eigenfunctions, $\hat{\boldsymbol{\xi}}_{\beta}$ ($\beta=1,2,3,\ldots$), make a complete set. The terms whose time-dependences are different from $\exp(i\omega_{\mathrm{i}}t)$ are not written down explicitly in equation (11.35), because these terms disappear by taking long-term time average when we are interested in phenomena with frequencies of ω_{i} (i $=1$ and 2).

In order to derive equations describing the time evolution of $A_{\mathrm{i}}(t)$ (i $=1$ or 2), we substitute equation (11.35) into the left-hand side of equation (11.30) or (11.33). Then, considering that $\hat{\boldsymbol{\xi}}_{\mathrm{i}}$'s and $\hat{\boldsymbol{\xi}}_{\alpha}$'s are displacement vectors associated with eigenfunctions of the linear wave equation (11.3), we find that the left-hand side of equation (11.30) or (11.33) becomes the real part of

$$2\rho_0\sum_{i=1}^{2}\frac{dA_{\mathrm{i}}}{dt}\biggl[i\omega_{\mathrm{i}}+(\boldsymbol{u}_0\cdot\nabla)\biggr]\hat{\boldsymbol{\xi}}_{\mathrm{i}}\exp(i\omega_{\mathrm{i}}t)$$
$$+\rho_0\sum_{i=1}^{2}\sum_{\alpha}A_{i,\alpha}\biggl[(\omega_\alpha^2-\omega_{\mathrm{i}}^2)-2i(\omega_\alpha-\omega_{\mathrm{i}})(\boldsymbol{u}_0\cdot\nabla)\biggr]\hat{\boldsymbol{\xi}}_{\alpha}\exp(i\omega_{\mathrm{i}}t), \tag{11.36}$$

where d^2A_{i}/dt^2 has been neglected, because $A_{\mathrm{i}}(t)$'s (i $=1$ and 2) are slowly varying functions of time.

Now, the real part of equation (11.36) is integrated over the whole volume of disks after being multiplied by $\boldsymbol{\xi}_1[= \Re\,\hat{\boldsymbol{\xi}}_1\exp(i\omega_1 t)]$.[5] Then, the terms resulting from the second term of equation (11.36) vanish by the orthogonality relations given in Chap. 3 (see equation (3.18)). It is also noted that we are interested only in secularly changing terms, i.e., time-periodic terms are neglected by taking time average. Then, the result of the integration is

$$\Re\, i\frac{dA_1}{dt}\Big\langle \rho_0\hat{\boldsymbol{\xi}}_1^*[\omega_1 - i(\boldsymbol{u}_0\cdot\nabla)]\hat{\boldsymbol{\xi}}_1\Big\rangle, \tag{11.37}$$

where $\langle X\rangle$ denotes the volume integration of $X(\boldsymbol{\xi})$, and the asterisk $*$ denotes the complex conjugate. It is noted that the cross term between $\hat{\boldsymbol{\xi}}_1^*$ and $\hat{\boldsymbol{\xi}}_2$ disappears by the volume integration, because the ω_1- and ω_2-oscillations are different modes and $\omega_2 = \omega_1$ and $m_2 = m_1$ cannot be realized simultaneously.

Equation (11.37) can be further reduced to

$$\Re\, i\frac{2E_1}{\omega_1}\frac{dA_1}{dt}, \tag{11.38}$$

because the terms in the brackets of equation (11.37) are related to the wave energy, E_1, of the oscillation $\boldsymbol{\xi}_1$, which is (see Chap. 3)

$$E_1 = \frac{1}{2}\omega_1\Big[\omega_1\langle\rho_0\hat{\boldsymbol{\xi}}_1^*\hat{\boldsymbol{\xi}}_1\rangle - i\langle\rho_0\hat{\boldsymbol{\xi}}_1^*(\boldsymbol{u}_0\cdot\nabla)\hat{\boldsymbol{\xi}}_1\rangle\Big]. \tag{11.39}$$

It is of importance to note that in the case of the oscillations with small azimuthal wavenumber (which is m) in geometrically thin disks the wave energy can be expressed as (see equation (3.31))

$$E_1 = \frac{\omega_1}{2}\Big\langle(\omega_1 - m_1\Omega)\rho_0(\hat{\xi}_{1,r}^*\hat{\xi}_{1,r} + \hat{\xi}_{1,z}^*\hat{\xi}_{1,z})\Big\rangle, \tag{11.40}$$

where $\Omega(r)$ is the angular velocity of disk rotation. An importance of this expression for wave energy is that the sign of wave energy is related to the value of $\omega - m\Omega$ in the region where the wave exists predominantly. That is, if a wave exists inside the radius of the corotation resonance where $\omega - m\Omega = 0$, the wave energy is negative (when $\omega > 0$), while it is positive if the wave is outside the radius of the resonance (when $\omega > 0$).

[5] The formula

$$\Re(A)\Re(B) = \frac{1}{2}\Re[AB + AB^*] = \frac{1}{2}\Re[AB + A^*B]$$

is used, where A and B are complex variables and B^* is the complex conjugate of B.

Now, after preparations in the above two paragraphs, we multiply $\boldsymbol{\xi}_1(\boldsymbol{r})$ to the real part of equation (11.30) and integrate over the whole volume. Then, we have

$$\Re\, i\frac{2E_1}{\omega_1}\frac{dA_1}{dt} = \frac{1}{2}\Re\left[A_2(t)A_{\rm D}\left\langle \hat{\boldsymbol{\xi}}_1 \cdot \boldsymbol{C}(\hat{\boldsymbol{\xi}}_2, \hat{\boldsymbol{\xi}}_{\rm D})\right\rangle \exp(i\Delta\omega t)\right], \tag{11.41}$$

where $\boldsymbol{C}(\hat{\boldsymbol{\xi}}_2, \hat{\boldsymbol{\xi}}_{\rm D})$ is replaced to $\boldsymbol{C}^{\rm T}(\hat{\boldsymbol{\xi}}_2, \hat{\boldsymbol{\xi}}_{\rm D})$, when tidal deformation is considered. In deriving the right-hand side of equation (11.41), the time periodic terms with high frequencies (i.e., non-resonant terms) have been neglected by time average being taken.

Similarly, we multiply $\boldsymbol{\xi}_2(\boldsymbol{r})$ to the real part of equation (11.30) and integrate over the whole volume to lead to

$$\Re\, i\frac{2E_2}{\omega_2}\frac{dA_2}{dt} = \frac{1}{2}\Re\left[A_1(t)A_{\rm D}\left\langle \hat{\boldsymbol{\xi}}_2 \cdot \boldsymbol{C}(\hat{\boldsymbol{\xi}}_1, \hat{\boldsymbol{\xi}}_{\rm D})\right\rangle \exp(i\Delta\omega t)\right]. \tag{11.42}$$

As mentioned above, in the case of tidal deformation, $\boldsymbol{C}(\hat{\boldsymbol{\xi}}_1, \hat{\boldsymbol{\xi}}_{\rm D})$ is replaced by $\boldsymbol{C}^{\rm T}(\hat{\boldsymbol{\xi}}_1, \hat{\boldsymbol{\xi}}_{\rm D})$. The set of equations (11.41) and (11.42) are simultaneous differential equations describing time evolution of A_1 and A_2.

11.3.3 Commutative Relations Among Coupling Terms

Before studying the simultaneous equations (11.41) and (11.42), we should notice an important and general characteristic of the coupling terms. This characteristic leads the stability criterion to a simple and general form. That is, we consider three arbitrary functions of $\boldsymbol{\xi}$, say $\boldsymbol{\xi}^{(1)}$, $\boldsymbol{\xi}^{(2)}$, and $\boldsymbol{\xi}^{(3)}$, which are continuous and ρ_0 times each of them vanishes at far outside of the disk. Then, for arbitrary exchanges of $\boldsymbol{\xi}^{(1)}$, $\boldsymbol{\xi}^{(2)}$, and $\boldsymbol{\xi}^{(3)}$, we have the following commutative relations (e.g., Kato 2013):

$$\begin{aligned}\int \rho_0\boldsymbol{\xi}^{(3)}\boldsymbol{C}(\boldsymbol{\xi}^{(1)}, \boldsymbol{\xi}^{(2)})d^3r &= \int \rho_0\boldsymbol{\xi}^{(1)}\boldsymbol{C}(\boldsymbol{\xi}^{(3)}, \boldsymbol{\xi}^{(2)})d^3r \\ = \int \rho_0\boldsymbol{\xi}^{(2)}\boldsymbol{C}(\boldsymbol{\xi}^{(1)}, \boldsymbol{\xi}^{(3)})d^3r &= \int \rho_0\boldsymbol{\xi}^{(3)}\boldsymbol{C}(\boldsymbol{\xi}^{(2)}, \boldsymbol{\xi}^{(1)})d^3r = \cdots. \end{aligned} \tag{11.43}$$

These commutative relations can be shown by performing the integrations by parts with using the expression for $\boldsymbol{C}$ given by equation (11.31). Integrations by parts lead integrands to forms symmetric with respect to $\boldsymbol{\xi}^{(1)}$, $\boldsymbol{\xi}^{(2)}$, and $\boldsymbol{\xi}^{(3)}$. In the case where the coupling term is $\boldsymbol{C}^{\rm T}$, however, commutability is less general. Let us regards $\boldsymbol{\xi}^{(3)}$ as $\boldsymbol{\xi}_D$ (the displacement vector associated with tidal force). Then, $\boldsymbol{\xi}^{(1)}$ and $\boldsymbol{\xi}^{(2)}$ are commutative as

$$\int \rho_0\boldsymbol{\xi}^{(1)}\boldsymbol{C}^{\rm T}(\boldsymbol{\xi}^{(2)}, \boldsymbol{\xi}_{\rm D})d^3r = \int \rho_0\boldsymbol{\xi}^{(2)}\boldsymbol{C}(\boldsymbol{\xi}^{(1)}, \boldsymbol{\xi}_{\rm D})d^3r, \tag{11.44}$$

but not commutable with $\boldsymbol{\xi}_{\rm D}$, i.e., we have

$$\int \rho_0 \boldsymbol{\xi}^{(1)} \boldsymbol{C}^{\rm T}(\boldsymbol{\xi}^{(2)}, \boldsymbol{\xi}_{\rm D}) d^3r \neq \int \rho_0 \boldsymbol{\xi}_{\rm D} \boldsymbol{C}(\boldsymbol{\xi}^{(2)}, \boldsymbol{\xi}^{(1)}) d^3r. \tag{11.45}$$

It is noted, however, what we need in the following analyses is relation (11.44) alone.

11.4 Conditions on Growth of Resonant Oscillations

Since the amplitude $A_{\rm D}$ is assumed to be constant, the time evolutions of A_1 and A_2 are determined by solving the simultaneous equations (11.41) and (11.42). What are governed by equations (11.41) and (11.42) are the imaginary part of A_1 and A_2, say, $A_{1,\rm i}$ and $A_{2,\rm i}$. Their real parts are not related to the resonance, and we consider the cases where they can be neglected.

Hereafter, we restrict our attention only to the case of $\Delta\omega = 0$. Then, we have from equations (11.41) and (11.42)

$$-\frac{2E_1}{\omega_1}\frac{A_{1,\rm i}}{dt} = -\frac{1}{2}A_{2,\rm i}\Im(A_{\rm D}W), \tag{11.46}$$

$$-\frac{2E_2}{\omega_2}\frac{A_{2,\rm i}}{dt} = -\frac{1}{2}A_{1,\rm i}\Im(A_{\rm D}W), \tag{11.47}$$

where

$$W \equiv \left\langle \hat{\boldsymbol{\xi}}_1 \cdot \boldsymbol{C}(\hat{\boldsymbol{\xi}}_2, \hat{\boldsymbol{\xi}}_{\rm D}) \right\rangle = \left\langle \hat{\boldsymbol{\xi}}_2 \cdot \boldsymbol{C}(\hat{\boldsymbol{\xi}}_1, \hat{\boldsymbol{\xi}}_{\rm D}) \right\rangle. \tag{11.48}$$

The last equality of equation (11.48) comes from commutability between $\hat{\boldsymbol{\xi}}_1$ and $\hat{\boldsymbol{\xi}}_2$ (see equation (11.43) or (11.44)). In the case where the deformation is due to the tidal force, W's in equations (11.46) and (11.47) are replaced by $W^{\rm T}$ which is

$$W^{\rm T} \equiv \left\langle \hat{\boldsymbol{\xi}}_1 \cdot \boldsymbol{C}^{\rm T}(\hat{\boldsymbol{\xi}}_2, \hat{\boldsymbol{\xi}}_{\rm D}) \right\rangle = \left\langle \hat{\boldsymbol{\xi}}_2 \cdot \boldsymbol{C}^{\rm T}\boldsymbol{\xi}(\hat{\boldsymbol{\xi}}_1, \hat{\boldsymbol{\xi}}_{\rm D}) \right\rangle. \tag{11.49}$$

Eliminating $A_{2,\rm i}$ from equations (11.46) and (11.47), we have

$$\frac{d^2A_{1,\rm i}}{dt^2} = \frac{1}{16}\left(\frac{E_1E_2}{\omega_1\omega_2}\right)^{-1}\left[\Im(A_{\rm D}W)\right]^2 A_{1,\rm i}. \tag{11.50}$$

A similar equation is obtained for $A_{2,\rm i}$ by eliminating $A_{1,\rm i}$ instead of $A_{2,\rm i}$ from equations (11.46) and (11.47).

Equation (11.50) shows that if

$$\frac{E_1 E_2}{\omega_1 \omega_2} > 0, \tag{11.51}$$

the oscillations grow with the growth rate given by

$$\left(\frac{\omega_1 \omega_2}{16 E_1 E_2}\right)^{1/2} |\Im(A_{\rm D} W)|. \tag{11.52}$$

In the case where the disk deformation is due to tidal force, W in equation (11.52) is changed to $W^{\rm T}$.

The resonant interaction of two oscillations through disk deformation are schematically shown in Fig. 11.1.

11.5 Cause of Instability and Three Wave Interaction

In this section we discuss the cause of the resonant instability. In the case where $|\omega_{\rm D}|$ is small, the resonance occurs between two oscillations with opposite signs of ω's, i.e., $\omega_1 \omega_2 < 0$. In this case the instability condition (11.51) is written as

$$E_1 E_2 < 0. \tag{11.53}$$

This might suggest that the instability is a result of energy exchange between two oscillations through a catalytic action of disk deformation. Positive energy flow from a negative-energy oscillation to a positive-energy oscillation leads to growth of both oscillations.

This interpretation is not always correct. Energy exchange between deformation and oscillations is of importance in the case where $|\omega_{\rm D}|$ is large compared with $|\omega_1|$ and $|\omega_2|$.[6] In this case the resonance ($\omega_1 + \omega_2 + \omega_{\rm D} = 0$) occurs for oscillations with the same signs of ω_1 and ω_2, and the instability condition (11.51) is written as

$$E_1 E_2 > 0. \tag{11.54}$$

Equation (11.54) clearly shows that in the present case of $|\omega_{\rm D}|$ being large, the instability cannot be interpreted as energy exchange between two oscillations. That is, both oscillations grow by getting energy from the deformation (in the case of $E_1 > 0$ and $E_2 > 0$), or by lossing energy to the deformation (in the case of $E_1 < 0$ and $E_2 < 0$). In other words, energy exchange between oscillations and deformation is essential for growth of the oscillations.

[6]The tidal instability in dwarf novae, which is discussed in Sect. 12.2, corresponds to this case.

To understand this situation, it is helpful to examine the case where the disk deformation is not fixed, but subject to changes of amplitudes of $\boldsymbol{\xi}_1$ and $\boldsymbol{\xi}_2$. That is, we consider the case where the ω_D-deformation is an another disk oscillation and has resonant interactions with ω_1- and ω_2-oscillations (i.e., a three-modes interaction). In this case, we can study the amplitude variation of the ω_D-oscillation by performing similar analyses used to derive equation (11.41) and (11.42). Then, as the results we have

$$\Re i \frac{2E_D}{\omega_D}\frac{dA_D}{dt} = \frac{1}{2}\Re\left[A_1(t)A_2(t)\left\langle \hat{\boldsymbol{\xi}}_D \cdot \boldsymbol{C}(\hat{\boldsymbol{\xi}}_1, \hat{\boldsymbol{\xi}}_2)\right\rangle \exp(i\Delta\omega t)\right], \tag{11.55}$$

where E_D is the wave energy of disk deformation. In this case we have the commutative relation:

$$\left\langle \hat{\boldsymbol{\xi}}_D \cdot \boldsymbol{C}(\hat{\boldsymbol{\xi}}_1, \hat{\boldsymbol{\xi}}_2)\right\rangle = \left\langle \hat{\boldsymbol{\xi}}_1 \cdot \boldsymbol{C}(\hat{\boldsymbol{\xi}}_2, \hat{\boldsymbol{\xi}}_D)\right\rangle = \left\langle \hat{\boldsymbol{\xi}}_2 \cdot \boldsymbol{C}(\hat{\boldsymbol{\xi}}_1, \hat{\boldsymbol{\xi}}_D)\right\rangle \equiv W. \tag{11.56}$$

Hence, if $\Delta\omega = 0$, we have the following set of equations (see equations (11.46) and (11.47)):

$$-\frac{2E_1}{\omega_1}\frac{dA_{1,i}}{dt} = -\frac{1}{2}A_{2,i}A_{D,i}\Re W, \tag{11.57}$$

$$-\frac{2E_2}{\omega_2}\frac{dA_{2,i}}{dt} = -\frac{1}{2}A_{1,i}A_{D,i}\Re W, \tag{11.58}$$

$$-\frac{2E_D}{\omega_D}\frac{dA_{D,i}}{dt} = -\frac{1}{2}A_{1,i}A_{2,i}\Re W, \tag{11.59}$$

where $\Re W$ is the real part of W. That is, in addition to equations describing time evolutions of $A_{1,i}$ and $A_{2,i}$, we have an equation describing time evolution of $A_{D,i}$ and time evolutions of $A_{1,i}$, $A_{2,i}$, and $A_{D,i}$ are coupled each other.

In order to examine how the amplitude, $A_{D,i}(t)$, changes with time, we take the time derivative of equation (11.59) and substitute equations (11.57) and (11.58) into the resulting equation. Then, we have

$$\frac{d^2A_{D,i}}{dt^2} = \frac{1}{16}\left(\frac{E_2E_D}{\omega_2\omega_D}\right)^{-1}\left[A_{1,i}^2 + \frac{E_2/\omega_2}{E_1/\omega_1}A_{2,i}^2\right](\Re W)^2 A_{D,i}. \tag{11.60}$$

This equation shows that in the case of $(E_1/\omega_1)(E_2/\omega_2) > 0$, $A_{D,i}$ oscillates with time unless $(E_D/\omega_D)(E_2/\omega_2) > 0$. However, $(E_1/\omega_1)(E_2/\omega_2) > 0$ and $(E_D/\omega_D)(E_2/\omega_2) > 0$ cannot be realized at the same time at resonance. This comes from the following situation. The sign of E_D/ω_D is equal to the sign of $\langle\omega_D - m_D\Omega\rangle$, where angle $\langle\ \rangle$ denotes a spatial average in the region where the wave exists (remember the definition of wave energy, i.e., see equation (11.40)). When the resonant conditions are satisfied, we have

$$\mathrm{Sign}\ \langle \omega_D - m_D\Omega \rangle = -\mathrm{Sign}\ [\langle \omega_1 - m_1\Omega \rangle + \langle \omega_2 - m_2\Omega \rangle]. \tag{11.61}$$

Roughly speaking, the signs of $\langle \omega_1 - m_1\Omega \rangle$ and $\langle \omega_2 - m_2\Omega \rangle$ are equal, respectively, to the signs of E_1/ω_1 and E_2/ω_2 (see equation (11.40)). This means

$$\mathrm{Sign}\ \frac{E_D}{\omega_D} < 0, \qquad \text{if}\ \frac{E_1}{\omega_1} > 0\ \text{and}\ \frac{E_2}{\omega_2} > 0, \tag{11.62}$$

$$\mathrm{Sign}\ \frac{E_D}{\omega_D} > 0, \qquad \text{if}\ \frac{E_1}{\omega_1} < 0\ \text{and}\ \frac{E_2}{\omega_2} < 0. \tag{11.63}$$

In both cases we have $(E_D/\omega_D)(E_2/\omega_2) < 0$. In other words, when ω_1- and ω_2-oscillations grow by $(E_1/\omega_1)(E_2/\omega_2) > 0$, the sign of $(E_2/\omega_2)(E_{\rm D}/\omega_{\rm D})$ is negative and the amplitude of ω_D-oscillation changes with time (oscillates with time).

This situation can be also demonstrated clearly from the following considerations. Equations (11.57), (11.58), and (11.59) are a set of simultaneous nonlinear differential equations with respect to $A_{1,\rm i}$, $A_{2,\rm i}$, and $A_{\rm D,i}$. By multiplying $A_{1,\rm i}$ and $A_{2,\rm i}$, respectively, to equation (11.57) and (11.58), and taking the difference of the resulting two equations, we have

$$\frac{E_1}{\omega_1}A_{1,\rm i}^2 - \frac{E_2}{\omega_2}A_{2,\rm i}^2 = \text{const.} \tag{11.64}$$

By the similar procedures, we also have

$$\frac{E_1}{\omega_1}A_{1,\rm i}^2 - \frac{E_{\rm D}}{\omega_{\rm D}}A_{\rm D,i}^2 = \text{const.} \tag{11.65}$$

$$\frac{E_2}{\omega_2}A_{2,\rm i}^2 - \frac{E_{\rm D}}{\omega_{\rm D}}A_{\rm D,i}^2 = \text{const.}, \tag{11.66}$$

where the right-hand sides of equations (11.64), (11.65), and (11.66) are time-independent constants. Equation (11.64) shows that when $(E_1/\omega_1)(E_2/\omega_2) > 0$, $A_{1,\rm i}$ and $A_{2,\rm i}$ can increase with time. Equation (11.65), however, demonstrates that in this case $A_{D,\rm i}$ decreases because the signs of E_2/ω_2 and E_D/ω_D are opposite.

More generally speaking, let us consider resonant interactions among three oscillations. Then, two of them can grow by consumption of the other oscillation, but all oscillations can not grow.

Finally, it is noticed that the excitation condition, $(E/\omega)_1(E/\omega)_2 > 0$, is unchanged even in MHD disks, if as E wave energy in magnetized disks is adopted. This is shown in Sect. 11.6.

11.6 Generalization of Stability Criterion to MHD Systems

Analyses in Sects. 11.1, 11.2, 11.3, and 11.4 have shown that oscillations resonantly coupled through disk deformation are amplified if $(E/\omega)_1(E/\omega)_2 > 0$. This was shown for hydrodynamical perturbations. The procedure of the analyses is rather general and suggests that it will be extended to the cases of MHD disks. Really, the criterion is applicable, without changing the expression, even in cases of MHD oscillations in magnetized disks, if E_1 and E_2 are wave energies generalized to MHD oscillations (Kato 2014). The necessary conditions for derivation of the criterion, $(E/\omega)_1(E/\omega)_2 > 0$, are that perturbations associated with the oscillations vanish on the boundary of the system. In the followings we will present the outlines of the derivation of the criterion, $(E/\omega)_1(E/\omega)_2 > 0$, in MHD oscillations.

11.6.1 *Quasi-nonlinear Wave Equation Describing Couplings*

The unperturbed disk which we consider is steady and axisymmetric. In the Lagrangian formulation, hydromagnetic perturbations superposed on the unperturbed disks can be described by extending the hydrodynamical formulation by Lynden-Bell and Ostriker (1967) to hydromagnetic cases as (cf., equation (3.1))

$$\frac{D_0^2\boldsymbol{\xi}}{Dt^2} = \delta\left(-\nabla\psi - \frac{1}{\rho}\nabla p + \frac{1}{4\pi\rho}\mathrm{curl}\boldsymbol{B}\times\boldsymbol{B}\right), \tag{11.67}$$

where δ represents the Lagrangian variation and $\boldsymbol{\xi}(\boldsymbol{r},t)$ is a displacement vector associated with the perturbations, $\boldsymbol{B}$ is the magnetic field flux density, and other notations are the same as those in Sect. 11.1.

In the case where the perturbations have small amplitudes and are non-dissipative, equation (11.67) is written as

$$\rho_0\frac{\partial^2\boldsymbol{\xi}}{\partial t^2} + 2\rho_0(\boldsymbol{u}_0\cdot\nabla)\frac{\partial\boldsymbol{\xi}}{\partial t} + \mathscr{L}(\boldsymbol{\xi}) = 0, \tag{11.68}$$

where $\mathscr{L}(\boldsymbol{\xi})$ consists of hydrodynamic and hydromagnetic parts as

$$\mathscr{L}(\boldsymbol{\xi}) = \mathscr{L}^{\mathrm{G}}(\boldsymbol{\xi}) + \mathscr{L}^{\mathrm{B}}(\boldsymbol{\xi}). \tag{11.69}$$

A detailed expression for $\mathscr{L}^{\mathrm{G}}(\boldsymbol{\xi})$ is given by equation (11.23), but that of $\mathscr{L}^{\mathrm{B}}(\boldsymbol{\xi})$ is not presented here (for a detailed expression, see Bernstein et al. 1958; see also Khalzov et al. 2008), because its detailed expression is not needed here. What we need here is that both of them are Hermitian in the following sense:

$$\int \boldsymbol{\eta}\cdot\mathscr{L}^{\rm G}(\boldsymbol{\xi})d^3r = \int \boldsymbol{\xi}\cdot\mathscr{L}^{\rm G}(\boldsymbol{\eta})d^3r, \quad \int \boldsymbol{\eta}\cdot\mathscr{L}^{\rm B}(\boldsymbol{\xi})d^3r = \int \boldsymbol{\xi}\cdot\mathscr{L}^{\rm B}(\boldsymbol{\eta})d^3r, \tag{11.70}$$

where $\boldsymbol{\xi}$ and $\boldsymbol{\eta}$ are any non-singular functions of $\boldsymbol{r}$ defined in the unperturbed volume of the disk and having continuous first and second derivatives everywhere. The integration is performed over the whole volume of the system, assuming that the surface integrals vanish.

Now, we consider two normal modes of oscillations, $\boldsymbol{\xi}_1$ and $\boldsymbol{\xi}_2$, with (ω_1, m_1) and (ω_2, m_2), respectively. In the linear stage, the perturbation, $\boldsymbol{\xi}$, imposed on a deformed disk is simply the sum of these two oscillations:

$$\begin{aligned}\boldsymbol{\xi}(\boldsymbol{r},t) &= A_1\boldsymbol{\xi}_1(\boldsymbol{r},t) + A_2\boldsymbol{\xi}_2(\boldsymbol{r},t)\\ &= \Re\Big[A_1\breve{\boldsymbol{\xi}}_1\exp[i(\omega_1 t - m_1\varphi)] + A_2\breve{\boldsymbol{\xi}}_2\exp[i(\omega_2 t - m_2\varphi)]\Big],\end{aligned} \tag{11.71}$$

where A_1 and A_2 are amplitudes and are arbitrary constants. In the quasi-nonlinear stage, the oscillations are coupled through disk deformation, $\boldsymbol{\xi}_{\rm D}$, so that they satisfy a quasi-nonlinear wave equation. The nonlinear wave equation is

$$\rho_0\frac{\partial^2\boldsymbol{\xi}}{\partial t^2} + 2\rho_0(\boldsymbol{u}_0\cdot\nabla)\frac{\partial\boldsymbol{\xi}}{\partial t} + \mathscr{L}(\boldsymbol{\xi}) = \boldsymbol{C}(\boldsymbol{\xi},\boldsymbol{\xi}_{\rm D}), \tag{11.72}$$

where $\boldsymbol{C}$ is the quasi-nonlinear coupling terms and consists of hydrodynamic and hydromagnetic terms:

$$\boldsymbol{C}(\boldsymbol{\xi},\boldsymbol{\xi}_{\rm D}) = \boldsymbol{C}^{\rm G}(\boldsymbol{\xi},\boldsymbol{\xi}_{\rm D}) + \boldsymbol{C}^{\rm B}(\boldsymbol{\xi},\boldsymbol{\xi}_{\rm D}), \tag{11.73}$$

where $\boldsymbol{C}^{\rm G}(\boldsymbol{\xi},\boldsymbol{\xi}_{\rm D})$ is the same as $\boldsymbol{C}(\boldsymbol{\xi},\boldsymbol{\xi}_{\rm D})$ or $\boldsymbol{C}^{\rm T}(\boldsymbol{\xi},\boldsymbol{\xi}_{\rm D})$ in Sect. 11.3.2, and $\boldsymbol{C}^{\rm B}(\boldsymbol{\xi})$ is a new term appeared here due to effects of magnetic fields.

A detailed expression for $\boldsymbol{C}^{\rm B}(\boldsymbol{\xi})$ is very complicated (see Appendix 2 by Kato 2014), but we can show after very long manipulations the following commutative relations (see Appendix 3 by Kato 2014)

$$\int \rho_0\hat{\boldsymbol{\xi}}_1\cdot\boldsymbol{C}^{\rm B}(\hat{\boldsymbol{\xi}}_2,\hat{\boldsymbol{\xi}}_{\rm D})d^3r = \int \rho_0\hat{\boldsymbol{\xi}}_2\cdot\boldsymbol{C}^{\rm B}(\hat{\boldsymbol{\xi}}_1,\hat{\boldsymbol{\xi}}_{\rm D})d^3r. \tag{11.74}$$

Hence, we have finally

$$W \equiv \int \rho_0\hat{\boldsymbol{\xi}}_1\cdot\boldsymbol{C}(\hat{\boldsymbol{\xi}}_2,\hat{\boldsymbol{\xi}}_{\rm D})d^3r = \int \rho_0\hat{\boldsymbol{\xi}}_2\cdot\boldsymbol{C}(\hat{\boldsymbol{\xi}}_1,\hat{\boldsymbol{\xi}}_{\rm D})d^3r. \tag{11.75}$$

Then, by similar procedures used in Sects. 11.2, 11.3, and 11.4 for hydrodynamical oscillations, we can show that the condition of resonant amplification is

$$\frac{E_1}{\omega_1}\frac{E_2}{\omega_2} > 0, \tag{11.76}$$

where E_1 and E_2 are wave energies of two oscillations coupling through disk deformation, and their formal expressions are the same as those in the case of hydrodynamical waves, if $\mathscr{L}(\boldsymbol{\xi})$ in hydrodynamical cases is extended by adding $\mathscr{L}^{\rm B}(\boldsymbol{\xi})$ as shown in equation (11.69). That is, E_i ($i = 1, 2$) is given by (see also Sect. 3.1.2)

$$E_{\rm i} = \frac{1}{4}\biggl[\omega_{\rm i}^2\int\rho_0\hat{\boldsymbol{\xi}}_{\rm i}^*\hat{\boldsymbol{\xi}}_{\rm i}d^3r + \int\hat{\boldsymbol{\xi}}_{\rm i}^*\cdot\mathscr{L}(\hat{\boldsymbol{\xi}}_{\rm i})d^3r\biggr] \quad (i = 1, 2), \tag{11.77}$$

which is rearranged in the form given by equation (11.39) (see also Sect. 3.1.2). Of course, the necessary conditions for the resonance are $m_1 + m_2 + m_{\rm D} = 0$, and $\omega_1 + \omega_2 + \omega_{\rm D} = 0$. Furthermore, the oscillations must be trapped in disks and their trapped regions must be overlapped so that the coupling occurs.

References

Bernstein, I. B., Frieman, E. A., Kruskal, M. D., & Kulsrud, R. M. 1958, Proc. R. Soc. A-Math. Phys. Eng.Sci., A244, 17

Ferreria, B. T., & Ogilvie, G. I. 2008, Mon. Not. R. Astron. Soc., 386, 2297

Kato, S. 2004, Publ. Astron. Soc. Jpn., 56, 905

Kato, S. 2008, Publ. Astron. Soc. Jpn., 60, 111

Kato, S. 2013, Publ. Astron. Soc. Jpn., 65, 75

Kato, S. 2014, Publ. Astron. Soc. Jpn., 66, 24

Kato, S. & Unno, W. 1967, Publ. Astron. Soc. Jpn., 19, 1

Kato, S., Okazaki, A. T., & Oktariani, F. 2011, Publ. Astron. Soc. Jpn., 63, 363

Khalzov, I. V., Smolyakov, A. I., & Hgisonis, V. I. 2008, Phys. Plasmas, 15, 4501

Lynden-Bell, D., & Ostriker, J. P. 1967, Mon. Not. R. Astron. Soc., 136, 293

Oktariani, F., Okazaki, A. T., & Kato, S. 2010, Publ. Astron. Soc. Jpn., 62, 709

Chapter 12
Wave-Wave Resonant Instability in Deformed Disks: Applications

Abstract In Chap. 11 we have shown that if certain resonant conditions are satisfied, a set of two oscillation modes are resonantly excited in deformed disks. This wave-wave resonant process is the cause of positive superhumps observed in dwarf novae, and perhaps will be one of possible causes of negative superhumps. Furthermore, some oscillatory phenomena in X-ray sources might be attributed to this process. In this context we present here attempts to apply the process to describing high-frequency QPOs in X-ray sources, in addition to superhumps.

Keywords High-frequency QPOs • Superhumps • Tidal deformation • Tilt • Warp

12.1 Types of Tidal Waves

In tidally deformed disks, $m_{\rm D}$ and $\omega_{\rm D}$ are related by $\omega_{\rm D} = m_{\rm D}\Omega^*_{\rm orb}$, when the orbit of the secondary star around the primary star is circular ($e = 0$) and the orbital plane is aligned with the disk plane ($\delta = 0$), where $\Omega^*_{\rm orb}$ is the orbital frequency of the secondary star around the primary star, observed from the primary star,[1] and $m_{\rm D}$ is an integer. If the secondary star has a circular orbit (i.e., $e = 0$) around the primary and is far outside the disk, the tidal deformation with $m_{\rm D} = 2$ much dominates over other deformations. In the cases where $e \neq 0$ and $\delta \neq 0$, various types of tide waves appear.

Before examining trapped oscillations satisfying the resonant conditions, $m_1 + m_2 + m_{\rm D} = 0$ and $\omega_1 + \omega_2 + \omega_{\rm D} = 0$, we must know what sets of $(m_{\rm D}, \omega_{\rm D}/\Omega^*_{\rm orb})$ are allowed as tidal waves. Detailed studies on these sets are presented in Appendix D. Based on the results in Appendix D, we summarize here the possible sets of ($m_{\rm D}$, $\omega_{\rm D}/\Omega^*_{\rm orb}$). There are three factors determining the combination of $m_{\rm D}$ and $\omega_{\rm D}/\Omega^*_{\rm orb}$. They are (i) binary separation (i.e., R/a), where R is the distance of the observational point on the disk from the disk center and a is the mean separation between the

[1] To distinguish $\Omega^*_{\rm orb}$ from the orbital frequency, $\Omega_{\rm orb}$, observed from the inertial frame, the asterisk has been attached. The relation between $\Omega^*_{\rm orb}$ and $\Omega_{\rm orb}$ is $\Omega_{\rm orb} = (1+q)^{1/2}\Omega^*_{\rm orb}$, where $q = M_{\rm s}/M$.

S. Kato, *Oscillations of Disks*, Astrophysics and Space Science Library 437,
DOI 10.1007/978-4-431-56208-5_12

primary and secondary, (ii) eccentricity (i.e., e) of the orbit of secondary star, and (iii) inclination (i.e., δ) of the orbital plane from the disk plane.

A general form of tidal potential, $\psi_{\rm D}$, is given in Appendix D (equation (D.1)). The potential is expanded by power series of R/a, e and δ. Then, as is well-known, in the limit where the binary orbit is circular ($e = 0$) and coplanar with the disk plane($\delta = 0$) and the secondary star is far outside ($R/a \ll 1$), the tidal potential has a form proportional to $(R/a)^2 \exp[i(2\Omega^*_{\rm orb}t - 2\varphi)]$, i.e., $(m_{\rm D}, \omega_{\rm D}/\Omega^*_{\rm orb}) = (2, 2)$ (see Table 12.1). This is the result in the lowest order approximation when the tidal potential is expanded by a power series of R/a, e, and the disk inclination angle δ. Table 12.1 shows the set of $\omega_{\rm D}$ and $m_{\rm D}/\Omega^*_{\rm orb}$ which appear in each stage of expansion of the tidal potential by power series of R/a and e. No inclination of the orbital plane of secondary star has been assumed. If $e = 0$ and $\delta = 0$, but unless R/a is sufficiently small, for example, the tidal potential has terms proportional to $(R/a)^3$ in addition to those of $(R/a)^2$. These additional terms consist of terms proportional to $\exp[i(\Omega^*_{\rm orb}t - \varphi)]$ and to $\exp[i(3\Omega^*_{\rm orb}t - 3\varphi)]$ (see Table 12.1), i.e., we have additional tidal waves of $m_{\rm D} = 1$, $\omega_{\rm D} = \Omega^*_{\rm orb}$ and $m_{\rm D} = 3$, $\omega_{\rm D} = 3\Omega^*_{\rm orb}$. It is noted that the tidal wave with $m_{\rm D} = 3$ and $\omega_{\rm D} = 3\Omega^*_{\rm orb}$ is what is considered in the studies of the superhump phenomena.

In the cases where the secondary star is far from the disk and the orbit is slightly eccentric, by expanding the tidal potential in terms of a power series of e, we have additional tidal waves proportional to $\exp[i(\Omega^*_{\rm orb}t - 2\varphi)]$ and $\exp[i(3\Omega^*_{\rm orb}t - 2\varphi)]$ in the order of e^1. The results till the orders of $(R/a)^4$ and e^2 are summarized in Table 12.1 in coplanar cases.

It is noted that in the coplanar cases, the velocity fields associated with the tidal potential are plane-symmetric with respect to the equatorial plane. This means that $n_{\rm D}$ is even (not odd) and in the lowest order of approximations $n_{\rm D} = 0$.

Table 12.1 The set of ($m_{\rm D}$, $\omega_{\rm D}/\Omega^*_{\rm orb}$) of the tidal waves which appear by expanding the tidal potential by power series of small dimensionless parameters, R/a and e in coplanar ($\delta = 0$) case.

	The set of $(m_{\rm D}, \omega_{\rm D}/\Omega^*_{\rm orb})$		
Separation	e^0	e^1	e^2
$(R/a)^2$	(2, 2)	(2,1)	(2,0)
		(2,3)	(2,2)
			(2.4)
$(R/a)^3$	(1,1)	(1,0)	(1, −1)
		(1,2)	(1,1)
			(1,3)
	(3,3)	(3,2)	(3,1)
		(3,4)	(3,3)
			(3,5)
$(R/a)^4$	(2,2)	(2,1)	(2,0)
		(2,3)	(2,2)
			(2,4)
	(4,4)	(4,3)	(4,2)
		(4,5)	(4,4)

Table 12.2 The set of (m_D, ω_D/Ω^*_{orb}) of the tidal waves ($n_D = 1$) which additionally appear in the cases where the orbital plane is weakly misaligned from the disk plane.

	The set of (m_D, ω_D/Ω^*_{orb})		
Separation	e^0	e^1	e^2
$(R/a)^2$	(1,0)	(1,−1)	(1,−2)
		(1,1)	(1,0)
			(1,2)
	(1,2)	(1,1)	(1,0)
		(1,3)	(1,2)
			(1,4)
$(R/a)^3$	(2,1)	(2.0)	(2,−1)
		(2,2)	(2,1)
			(2,3)
	(2,3)	(2,2)	(2,0)
		(2,4)	(2,3)
			(2,5)
$(R/a)^4$	(1,2)	(1,1)	(1.0)
		(1,3)	(1,2)
			(1,4)
	(3,2)	(3,1)	(3,0)
		(3,3)	(3,2)
			(3,4)
	(3,4)	(3,3)	(3,2)
		(3,5)	(3,4)
			(3,6)

If the orbital plane is misaligned with the disk plane ($\delta \neq 0$), another types of tidal waves appear, which are asymmetric with respect to disk plane, i.e., n_D is odd and in the lowest order of approximations we have tidal waves with $n_D = 1$. The set of (m_D, ω_D/Ω^*_{orb}) for these tidal waves are summarized in Table 12.2 (see Appendix D for detailed calculations).

12.2 Applications to (Positive) Superhumps in Dwarf Novae

Typical objects with disks externally deformed by tidal forces are dwarf novae, where the gravitational potential in disks is tidally deformed by a secondary star. Whitehurst (1988) found by numerical simulations that under a certain condition tidally deformed disks become unstable to an eccentric deformation with slow precession. This instability is now called the tidal instability and is known to be due to a so-called 3 : 1 resonance resulting from mode-mode coupling processes (Lubow 1991, see also Hirose and Osaki 1990). This disk precession is a result of excitation of one-armed oscillation by the tidal instability and it brings about superhump phenomena (Osaki 1985). (See Osaki (1996) for review of superoutburst and superhump of dwarf novae.)

Lubow (1992) further shows that a tilt is also excited at the 3 : 1 resonance in dwarf novae disks by a mode-mode coupling process similar to that in the case of superhumps.

The wave-wave resonant instability process presented in Chap. 11 can be regarded as a generalization of Lubow's mode-mode coupling process from a slightly different viewpoint. This generalization allows us to study resonant instabilities in deformed disks in a perspective way from a wide viewpoint, and allows us to regard the tidal instability in dwarf novae as one of examples of wave-wave resonant instabilities in deformed disks (Kato 2013).

In this section (Sect. 12.2) the tidal excitation of eccentric precession mode (one-armed p-mode) is argued in terms of a picture of wave-wave resonant instability. The excitation of tilt mode in deformed disks is investigated in the next section (Sect. 12.3), and in the subsequent sections (Sects. 12.4 and 12.5), possible applications to high-frequency QPOs are investigated.

12.2.1 Precession Mode (ω_1-Mode) and Its Counter-Mode (ω_2-Mode)

As the ω_1-oscillation, we take one-armed ($m_1 = 1$) low-frequency p-mode oscillation with a positive frequency. This is the low-frequency eccentric precession mode. The propagation regions of this mode is specified by $\omega - \Omega < -\kappa$ (see Chap. 6). This is one of the propagation regions specified by $(\omega - \Omega)^2 - \kappa^2 > 0$ for one-armed p-mode oscillations (see Chap. 6). In tidally deformed disks, $\Omega(r) - \kappa(r)$ is positive and increases outwards as (see Chap. 2),

$$\Omega - \kappa = \frac{3}{4} q \Omega_{\rm K} \left(\frac{r}{a}\right)^3, \tag{12.1}$$

in the lowest order of approximations, where q is the radio of mass of the secondary to that of the primary, i.e., $q = M_{\rm s}/M$, and a is the separation between the primary and secondary stars. Hence, an oscillation with a positive frequency, say ω_1, is trapped between the radius, $r_{\rm c}$, where $\omega_1 = (\Omega - \kappa)_{\rm c}$ (the subscript c denotes the value at $r_{\rm c}$) and the radius, $r_{\rm t}(>r_{\rm c})$, where the disk is truncated by the tidal effects (see Fig. 12.1).[2] This oscillation has $E_1/\omega_1 < 0$, since in its propagation region it has $\omega_1 - \Omega \le -\kappa$ and thus $\omega_1 - \Omega < 0$.

Next, let us consider a p-mode oscillation with ω_2 and m_2. Here, ω_2 and m_2 are taken to be positive. The local dispersion relation shows that one of the propagation region of the ω_2-oscillation is specified by $\omega_2 - m_2\Omega < -\kappa$ (see Chap. 6). Since

[2] In Chaps. 6 and 7, $r_{\rm t}$ is denoted by $r_{\rm D}$. Both of them are the same.

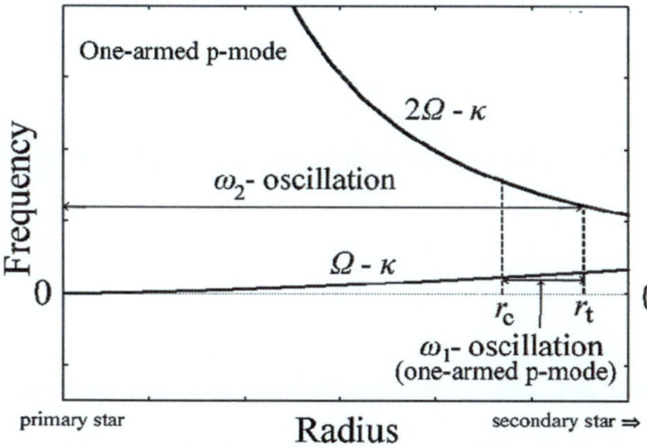

Fig. 12.1 Schematic diagram showing frequencies and propagation regions of the ω_1- and ω_2-oscillations in the case where the ω_1-oscillation is one-armed low-frequency p-mode oscillation, and the ω_2-oscillation is two-armed p-mode oscillation. The scales of coordinates are arbitrary, and are not linear. The one-armed p-mode oscillation is trapped between r_c and r_t. The inside of r_c is the evanescent region. The ω_2-oscillation can propagate inside r_t.

$m_2\Omega - \kappa$ is positive and decreases outwards, the propagation region of the ω_2-oscillation with $\omega_2 > 0$ is bound from outside. Figure 12.1 schematically shows the case of $m_2 = 2$. The outer edge of the propagation region is taken at r_t, i.e., $\omega_2 = (2\Omega - \kappa)_t$, where the subscript t denotes the value at r_t. This oscillation has $E_2/\omega_2 < 0$, since $\omega_2 - m_2\Omega < 0$.

A comment on the outer edge of the propagation region of the ω_2-oscillation is necessary. This outer edge needs not always to be at r_t. We adopt, however, the outer edge at r_t, because the growth rate of resonant instability will be large when the outer boundaries of ω_1- and ω_2-oscillations are close (i.e., W^T given by equation (11.49) is large in this case). We should remember that in the limit of pressureless disks, the ω_1- and ω_2-oscillations are localized at the radii where $\omega_1 = \Omega - \kappa$ and $\omega_2 = 2\Omega - \kappa$, respectively. Hence, for the resonant interaction to occur, both of these radii needs to coincide.

In summary, we have $(E_1/\omega_1)(E_2/\omega_2) > 0$. Furthermore, the propagation regions of the ω_1- and ω_2-oscillations are overlapped (see Fig. 12.1). It is noted that this condition of overlapping of propagation regions is necessary for instability. Otherwise, the value of coupling term, W^T (see, e.g., equation (11.49)), is small, and growth rate is small even if resonant conditions concerning ω's and m's are satisfied.

Let us next consider the resonant condition, $\omega_1 + \omega_2 + \omega_D = 0$. Considering that there are various types of tidal waves, we write here

$$\omega_D = n^* \Omega^*_{orb}, \tag{12.2}$$

where n^* is a positive or negative integer, depending on the type of tidal wave. Since $\omega_1 = (\Omega - \kappa)_c$ and $\omega_2 = (m_2\Omega - \kappa)_t$, the resonant condition, $\omega_1 + \omega_2 + \omega_D = 0$, is written as

$$(\Omega - \kappa)_c + (m_2\Omega - \kappa)_t + n^*\Omega^*_{\rm orb} = 0. \tag{12.3}$$

For simplicity, if the difference between Ω and κ is neglected, the above equation leads to

$$\Omega_t = \frac{n^*}{1 - m_2}\Omega^*_{\rm orb}. \tag{12.4}$$

First, let us consider the case of $m_2 = 2$, which means $m_D = -3$ because one of the resonant condition is $m_1 + m_2 + m_D = 0$. In this case, the resonant condition, equation (12.4), is written as

$$\Omega_t = -n^*\Omega^*_{\rm orb} = -\omega_D. \tag{12.5}$$

If we adopt $n^* = -3$ (i.e., $\omega_D = -3\Omega^*_{\rm orb}$), equation (12.5) gives $\Omega_t = 3\Omega^*_{\rm orb}$. The set of $m_D = -3$ and $\omega_D = -3\Omega^*_{\rm orb}$ is equivalent with $m_D = 3$ and $\omega_D = 3\Omega^*_{\rm orb}$, because the signs of m_D and ω^*_D can be changed simultaneously. This is the case of the so-called 3 : 1 resonance, i.e., $\Omega_t : \Omega^*_{\rm orb} = 3 : 1$ (see equation (12.4)), and the set of $(m^*_D, \omega_D/\Omega^*_{\rm orb})$ of the tidal wave contributing this resonance is (3, 3). We see that this tidal wave really exists by expanding the tidal potential till the order of $(R/a)^3$ with $e = 0$ (see Table 12.1). This is the largest term in tidal potential except for that of the set of (2, 2) which exist in the stage of $(R/a)^2$ (see Table 12.1).

The above arguments show that the wave-wave resonant instability is not restricted to the case of the above 3 : 1 resonance. If we consider, for example, $n^* = -2$, the set of $(m_D, \omega_D/\Omega^*_{\rm orb})$ is $(-3, -2)$, i.e., (3, 2) and $\Omega_t = 2\Omega^*_{\rm orb}$, i.e., 2 : 1 resonance. This set of $(m_D, \omega_D/\Omega^*_{\rm orb})$ exists in the tidal waves in the term of $(R/a)^3$, although eccentricity of the orbit is required (see Table 12.1). If disks are truncated by the 3 : 1 resonance, this 2 : 1 resonance will not be so important. We can consider other cases such as $n^* = -4$. The results are summarized in Table 12.3, including the cases of $m_2 = 3$.

Table 12.3 shows that the one-armed precession mode can be excited at various stages of disk expansion. The most important resonance is due to $m_D = -3$ and

Table 12.3 Characteristic quantities required for resonant excitation of low-frequency, precession mode ($m_1 = 1$ and $n_1 = 0$).

(m_2, n_2)	$(m_D, \omega_D/\Omega^*_{\rm orb})$	$\Omega_t/\Omega^*_{\rm orb}$	r_t/a	r/a	Eccentricity
$(2, 0)$	$(-3, -2)$	2	0.63	$(r/a)^3$	e^1
	$(-3, -3)$	3	0.48	$(r/a)^3$	e^0
	$(-3, -4)$	4	0.40	$(r/a)^3$	e^1
	$(-3, -5)$	5	0.34	$(r/a)^3$	e^2
$(3, 0)$	$(-4, -2)$	1	1.00	$(r/a)^4$	e^2
	$(-4, -4)$	2	0.63	$(r/a)^4$	e^0

$n^* = -3$.[3] The next one will be the resonance resulting from the tidal deformation of $(m_D, n^*) = (-3, -4)$, which occurs at $\Omega_t = 4\Omega^*_{orb}$. Hereafter, however, we restrict our attention to the former 3 : 1 resonance, i.e., the coupling of one-armed precession mode and two-armed p-mode with disk deformation with $m_D = 3$ and $\omega_D/\Omega^*_{orb} = 3$, which occurs at $\Omega_t : \Omega^*_{orb} = 3 : 1$. Hirose and Osaki (1990) derived this 3 : 1 resonant instability as a parametric resonance of particles, and Lubow (1991) derived this as a resonant instability due to mode-mode coupling in fluid disks. The present derivation of the instability is a generalization of Lubow's picture of mode-mode coupling to a picture of a wave-wave resonant instability of two disk oscillations.

It should be noted here that there still remains an issue to be addressed. This is whether the ω_2-oscillation satisfying the resonant condition (12.3) is really a trapped oscillation. The frequency required to the ω_2-oscillation is on the order of Ω_t (see that $\omega_2 \sim (2\Omega - \kappa)_t$). Hence, this problem is, in other words, whether a two-armed p-mode oscillation with frequency of $\sim\Omega_t$ is really trapped between the inner edge of disks and the outer truncation radius, r_t. The required frequency, $\sim\Omega_t$, is much smaller than $2\Omega - \kappa$ in the inner region of disks. This means that the ω_2-oscillation is a higher overtone in the radial direction, and has a short radial wavelength in the inner region. In other words, a trapped oscillation whose frequency is close to ω_2 always exist practically.

It is important to note here that in application of the present wave-wave resonant instability to excitation of quasi-periodic oscillations in X-ray binaries, examination whether there is ω_2-oscillation satisfying the condition of $\omega_1 + \omega_2 + \omega_D = 0$ is not trivial, as will be mentioned later (see Sects. 12.4 and 12.5).

12.2.2 Comparison with Observations of Superhumps

As described in Introduction (Chap. 1), in the superoutburst stage of dwarf novae, oscillations whose period is slightly longer than the observed binary orbital period are observed. These oscillations are called superhumps and are understood to be due to excitation of slowly prograding precession mode (Osaki 1985). In the present wave-wave resonant instability model, the ω_1-oscillation corresponds to the precession mode.

If the precession mode is not superposed on the tidally deformed disks, the luminosity variation observed in the inertial frame is the binary period, Ω_{orb}. If the eccentric disk has a prograde precession, the luminosity variation takes a longer time than the orbital period, since the eccentric position of the disk proceeds. That

[3] The value of m_D is negative. This means that ω_D is also negative, since $\omega_D = m_D\Omega^*_{orb}$. Since both of m_D and ω_D are negative, this is equivalent with the case where both are positive.

is, the superhump frequency, $\omega_{\rm sh}$, is lower than $\Omega_{\rm orb}$ by ω_1:

$$\omega_{\rm sh} = \Omega_{\rm orb} - \omega_1. \tag{12.6}$$

It is noted that $\Omega_{\rm orb}$ and $\Omega^*_{\rm orb}$ are related by

$$\Omega_{\rm orb} = (1+q)^{1/2}\Omega^*_{\rm orb}. \tag{12.7}$$

The purpose here is to check whether the frequency, $\omega_{\rm sh}$, can really describe the frequency of the observed superhumps. To examine this issue, we must know the frequency ω_1. This frequency has been calculated in Chap. 7 (Sect. 7.4.1) by using the WKB method. The results (equation (7.34)) show that the width of the trapped region , i.e., $r_{\rm t} - r_{\rm c}$, is given by

$$\frac{r_{\rm t} - r_{\rm c}}{a} = \frac{\pi}{6q^{1/2}}\frac{c_{\rm s}}{(GM/a)^{1/2}}. \tag{12.8}$$

This is a relation between $r_{\rm t}/a$ and $r_{\rm c}/a$ with two dimensionless parameters q and $c_{\rm s}/(GM/a)^{1/2}$.

Since ω_1 is given by

$$\omega_1 = (\Omega - \kappa)_{\rm c} = \frac{3}{4}q\Omega_{\rm K}\left(\frac{r_{\rm c}}{a}\right)^3, \tag{12.9}$$

we have (see equation (7.35))

$$\frac{\omega_1}{\Omega_{\rm orb}} \sim \frac{3}{4}\frac{q}{(1+q)^{1/2}}\left(\frac{r_{\rm c}}{a}\right)^{3/2}, \tag{12.10}$$

which gives $\omega_1/\Omega_{\rm orb}$ as a function of q and $c_{\rm s}/(GM/a)^{1/2}$, because $r_{\rm c}$ and $r_{\rm t}$ are related by the trapping condition (12.8) and $r_{\rm t}$ is related to $\Omega_{\rm orb}$ by the resonant condition (equation (12.3) or (12.4)).

Now, we now introduce the dimensionless observable quantity, ϵ, defined by $\epsilon = \Omega_{\rm orb}/\omega_{\rm sh} - 1$. Then, we have

$$\epsilon = \frac{\omega_1/\Omega_{\rm orb}}{1 - \omega_1/\Omega_{\rm orb}}. \tag{12.11}$$

This equation gives ϵ as a function of q and $c_{\rm s}/(GM/a)^{1/2}$, because $\omega_1/\Omega_{\rm orb}$ has been found to be a function of q and $c_{\rm s}/(GM/a)^{1/2}$ in the previous paragraph. Hence, we have a relation between ϵ and q with parameter $c_{\rm s}/(GM/a)^{1/2}$. The results of calculation are shown in Fig. 12.2 for three cases of $c_{\rm s}/(GM/a)^{1/2} = 0.01$, 0.02, and 0.04. In order to compare with observations, the calculated curves are superposed on a $q - \epsilon$ plot by Kato et al. (2009) for recently observed sources. The calculated curve in the case of $c_{\rm s}/(GM/a)^{1/2} = c_{\rm s}/(a\Omega^*_{\rm orb}) = 0.02$ seems to be

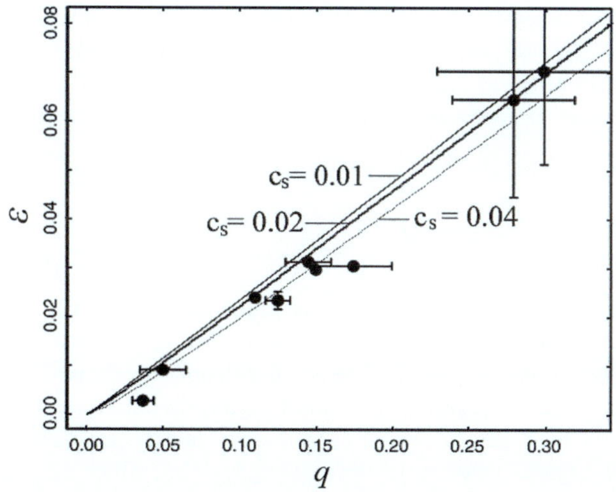

Fig. 12.2 The $q - \epsilon$ relation calculated in three cases of $c_s/(GM/D)^{1/2} = 0.01, 0.02$, and 0.04. These curves are superposed on the diagram plotting observed $q - \epsilon$ relation (figure 34 of T. Kato et al. 2009). The names of nine sources in this figure are listed in table 6 of T. Kato et al. (2009) (After S. Kato 2013, Publ. Astron. Soc. Japan, 65, 56, Publ. Astron. Soc. Japan ©).

close to the observational results. The dimensionless value of $c_s/(GM/a)^{1/2} = 0.02$ will be relevant observationally, because in accretion disks in cataclysmic variables, $a\Omega_{orb} \sim 500$ and $c_s \sim 10\,\mathrm{km\,s^{-1}}$.

12.3 Application to Negative Superhumps in Dwarf Novae

Recent observations show that, in addition to superhumps, another kind of superhumps whose periods are slightly shorter than the orbital period are observed in various phases of outbursts of dwarf novae (e.g., Patterson et al. 1995; Ohshima et al. 2012, see also Fig. 1.5). To distinguish them from "normal" superhumps with period slightly longer than the orbital one, they are called *negative superhumps*. The "normal" superhumps are called *positive superhumps*. Negative superhumps are supposed to be due to tilts of disks, although their origin is yet under debate. See Osaki and Kato (2013a,b) and Kato and Osaki (2013, 2014) for extensive analyses of observational data obtained by the Kepler telescope.

As one of models of negative superhumps, Lubow (1992) has pointed out that the 3 : 1 tidal resonance also excite the tilt mode by a mode-mode coupling process. This means that the present wave-wave resonant coupling process can also describe the 3 : 1 excitation of the tilt. In addition to this, examination of the wave-wave resonant process shows that the process can excite the tilt mode by other sets of wave-wave resonances (Kato 2014). The purpose of this section is to demonstrate this situation. To do so, we need to know what tidal waves exist when the binary orbit is eccentric ($e \neq 0$) and the orbital plane is misaligned with the disk plane ($\delta \neq 0$). This issue has been argued in Sect. 12.1.

12.3.1 Tilt Mode Trapped

Here and hereafter, the ω_1-oscillation is taken to be the tilt mode (in other terminology, it is the mode of the corrugation wave (e.g., see Chap. 6)). That is, we adopt $m_1 = 1$ and $n_1 = 1$. The local dispersion relation[4] shows that this mode has a propagation region in the radial region specified by $\omega_1 - \Omega < -\Omega_\perp$, where $\Omega_\perp(r)$ is the vertical epicyclic frequency and larger in tidally deformed disks than the angular velocity of disk rotation, $\Omega(r)$. Since $\Omega - \Omega_\perp < 0$ and tends to zero in the innermost region of the disks, the tilt mode has a negative frequency, i.e., $\omega_1 < 0$, and is trapped inside a radius, say $r_{\rm L_1}$, where ω_1 becomes equal to $\Omega - \Omega_\perp$, i.e., $\omega_1 = (\Omega - \Omega_\perp)_{\rm L_1}$, the subscript $\rm L_1$ denoting the value at $r_{\rm L_1}$. This is schematically shown in Fig. 12.3.[5] It should be noted that this ω_1-oscillation has negative value of E_1/ω_1, i.e., $E_1/\omega_1 < 0$, since $\omega_1 - \Omega < 0$.

Next, let us consider the trapping condition of the tilt mode. This issue has been studied in Chap. 7 by using the WKB method (Sect. 7.4.2). If the inner edge of the trapped region is taken to be zero, the trapping condition gives $r_{\rm L_1}$ as a function of q and $c_{\rm s}/(GM/a)^{1/2}$. Since the outer edge of the trapping region, $r_{\rm L_1}$, is the radius where $\omega - \Omega = -\Omega_\perp$, we have (see equation (7.39))

$$\omega_1 = (\Omega - \Omega_\perp)_{\rm L_1} = -\frac{3}{4}\frac{q}{(1+q)^{1/2}}\Omega_{\rm orb}\left(\frac{r_{\rm L_1}}{a}\right)^{3/2}. \qquad (12.12)$$

[4]The local dispersion relation for isothermal perturbations in vertically isothermal disks is (e.g., Okazaki et al. 1987, Kato 2001, and Kato et al. 2008)

$$[(\omega - m\Omega)^2 - \kappa^2][(\omega - m\Omega)^2 - n\Omega_\perp^2] = c_{\rm s}^2 k^2 (\omega - m\Omega)^2,$$

where κ and $\Omega_\perp$ are, respectively, the horizontal and vertical epicyclic frequencies, and k is the radial wavenumber of the oscillations. This local dispersion relation shows the presence of oscillation modes propagating in the region of $(\omega - m\Omega)^2 > n\Omega_\perp^2$ $(n = 1, 2, 3 \ldots)$, since $\Omega_\perp$ is always larger than κ.

[5]If ω_1 is smaller than the value of $\Omega - \Omega_\perp$ at the disk outer edge, the propagation region is terminated at the disk edge.

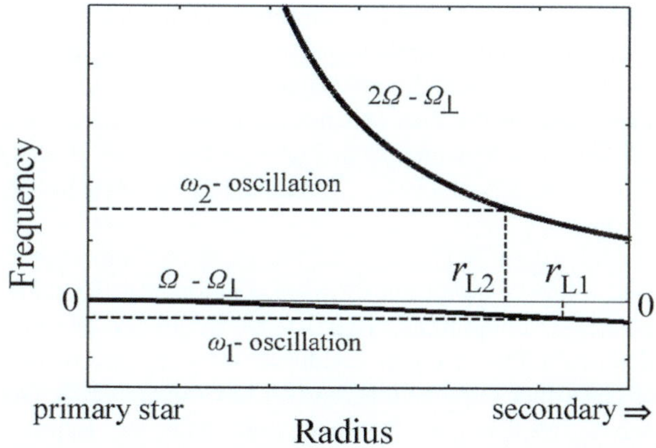

Fig. 12.3 Schematic diagram showing propagation regions of ω_1-oscillation (tilt mode) and ω_2-oscillation. The scales of coordinates are arbitrary, and are not linear. The set (m_1, n_1) of the ω_1-oscillation is (1, 1) and that adopted for ω_2-oscillation is (2,1). The outer edges of the propagation regions of ω_1- and ω_2-oscillations are denoted by $r_{\rm L_1}$ and $r_{\rm L_2}$, respectively (After S. Kato 2014, Publ. Astron. Soc. Japan, 66, 21, Publ. Astron. Soc. Japan ©).

In summary, the trapping condition, equation (12.12), gives the width of the trapped region, $r_{\rm L_1}$, and the frequency of the trapped oscillation, $\omega_1/\Omega_{\rm orb}$, as functions of q and $c_{\rm s}/(GM/a)^{1/2}$. The relation between $r_{\rm L_1}/a$ and $c_{\rm s}/(GM/a)^{1/2}$ is given in Fig. 7.11.[6]

12.3.2 Counter-Mode of Tilt Mode, i.e., ω_2-Mode

For the ω_1-oscillation (tilt) described above to be excited by the wave-wave resonant process, the counter-mode (i.e., ω_2-oscillation) must also have a negative value of E_2/ω_2 so that the excitation condition $(E_1/\omega_1)(E_2/\omega_2) > 0$ is satisfied. Here and hereafter, as ω_2-oscillations, we consider c-mode or vertical p-mode oscillations (i.e., $n_2 \neq 0$). Their propagation regions are specified by $(\omega_2 - m_2\Omega)^2 - n_2\Omega_\perp^2 > 0$ (see the local dispersion relation for disk oscillations), and are separated into two regions of $\omega_2 - m_2\Omega > \sqrt{n_2}\Omega_\perp$ and $\omega_2 - m_2\Omega < -\sqrt{n_2}\Omega_\perp$. We are interested in the oscillations propagating in the region of $\omega_2 - m_2\Omega < -\sqrt{n_2}\Omega_\perp$. The propagation region of this ω_2-oscillation is schematically shown in Fig. 12.3 in the case of $m_2 = 2$ and $n_2 = 1$. This oscillation has $E_2/\omega_2 < 0$, because $\omega_2 - m_2\Omega < 0$ in the propagation region.

[6]In Fig. 7.11 the outer capture radius of oscillations is shown by $r_{\rm out}$. In the present chapter, however, the radius is shown by $r_{\rm L_1}$. Both of them are the same.

Among many c-mode and vertical p-mode oscillations, we concentrate our attention particularly on those with $n_2 = 1$, or $n_2 = 2$, or $n_2 = 3$. Then, if $m_2 \geq 2$, ω_2 is positive, and the propagation regions of these oscillations are between the inner edge of the disk and the radius where $\omega_2 = m_2\Omega - \sqrt{n_2}\Omega_\perp$ is realized, as is shown schematically in Fig. 12.3 for $m_2 = 2$ and $n_2 = 1$. The radius where $\omega_2 = m_2\Omega - \sqrt{n_2}\Omega_\perp$ is realized is denoted by r_{L_2}, i.e., $\omega_2 = (m_2\Omega - \sqrt{n_2}\Omega_\perp)_{\mathrm{L}_2}$.

Like the ω_2-oscillations in the case of the (positive) superhumps, we do not examine here the trapping condition of ω_2-oscillations by the same reason as in the case of (positive) superhumps. That is, the frequency of the ω_2-oscillation which is needed from the resonant condition is $\sim\omega_{\mathrm{D}}$. As is shown later, the resonant condition further requires that ω_{D} is a few times of Ω^*_{orb} (see Tables 12.1 and 12.2). Such a frequency is low compared with the Keprerian frequency in the innermost region of the disks. This means that the trapped oscillations are higher overtones in the radial direction, i.e., they have many nodes in the radial direction. In other words, for any frequency ω_2 required from the resonant condition, there will always be trapped oscillations whose frequencies are close to ω_2. Hence, practically no additional condition is required for trapping of the ω_2-oscillation.

12.3.3 Resonant Conditions and Radii Where Resonances Begin

Next, let us consider the resonant conditions. The frequency ω_1 and the outer trapping radius (capture radius) of the ω_1-oscillation, r_{L_1}, are related by $\omega_1 = (\Omega - \Omega_\perp)_{\mathrm{L}_1}$, and also ω_2 and r_{L_2} are related by $\omega_2 = (m_2\Omega - \sqrt{n_2}\Omega_\perp)_{\mathrm{L}_2}$, as mentioned before. As discussed in Sect. 12.1, the frequency of tidal waves is a multiple of Ω^*_{orb} and thus we write ω_{D} as $\omega_{\mathrm{D}} = n^*\Omega^*_{\mathrm{orb}}$, and examine what sets of n^* and m_{D} satisfy the resonant condition and whether they are expected as a set of tidal waves.

The resonant condition, $\omega_1 + \omega_2 + \omega_{\mathrm{D}} = 0$, is written as

$$(\Omega - \Omega_\perp)_{\mathrm{L}_1} + (m_2\Omega - \sqrt{n_2}\Omega_\perp)_{\mathrm{L}_2} + n^*\Omega^*_{\mathrm{orb}} = 0, \tag{12.13}$$

and the condition of $m_1 + m_2 + m_{\mathrm{D}} = 0$ is

$$m_{\mathrm{D}} = -(1 + m_2). \tag{12.14}$$

Since $(\Omega - \Omega_\perp)_{\mathrm{L}_1}$ is much smaller than other terms in equation (12.13), equation (12.13) gives[7]

$$\Omega_{\mathrm{L}_2} \simeq -\frac{n^*}{m_2 - \sqrt{n_2}} \Omega^*_{\mathrm{orb}}. \tag{12.15}$$

Hereafter, we do not distinguish L_1 and L_2, and they are simply denoted by L (see similar arguments in Sect. 12.2.1). It is important to note here that r_L is not always necessary to be the outer edge of disks, different from the case of trapping of one-armed p-mode oscillation in Sect. 12.2.

Equation (12.15) shows that n^* needs to be a negative integer, because we are interested here in cases where $m_2 - \sqrt{n_2}$ is positive. In the case of dwarf novae, the disk is known to be roughly truncated by the tidal instability of the so-called 3 : 1 resonance at the radius where $\Omega/\Omega^*_{\mathrm{orb}} \sim 3:1$ (Whitehurst 1988). Hence, hereafter, we restrict our attention only to resonances which occur inside the radius of $\Omega/\Omega^*_{\mathrm{orb}} = 3$, i.e., to the cases where $\Omega_\mathrm{L}/\Omega^*_{\mathrm{orb}} > 3.0$. Furthermore, resonances in coplanar systems are considered first, and those which occur only in misaligned systems ($\delta \neq 0$) are examined next.

12.3.3.1 Coplanar Cases

Two cases of $(m_2, n_2) = (2, 1)$ and (2,3) are considered. In these cases, $m_\mathrm{D} = -3$, because one of resonant conditions is $m_1 + m_2 + m_\mathrm{D} = 0$.

(i) The case of $(m_2, n_2) = (2, 1)$

In the present cases, the resonant condition (12.15) is realized at

$$\Omega_\mathrm{L} \sim -n^* \Omega^*_{\mathrm{orb}}. \tag{12.16}$$

Since we are interested in resonances which occurs inside the radius of $\Omega/\Omega^*_{\mathrm{orb}} = 3.0$ (i.e., $\Omega_\mathrm{L}/\Omega^*_{\mathrm{orb}} > 3$), we restrict our attention to cases of $n^* = -3, -4$, and -5. If we adopt $n^* = -3$, equation (12.16) leads to $\Omega_\mathrm{L} = 3\Omega^*_{\mathrm{orb}}$. This means that if there is a tidal wave with $\omega_\mathrm{D} = -3\Omega^*_{\mathrm{orb}}$ and $m_\mathrm{D} = -3$ (this is equivalent to $\omega_\mathrm{D} = 3\Omega^*_{\mathrm{orb}}$ and $m_\mathrm{D} = 3$), a resonance excitation of tilt mode occurs at the stage when the disk size, r_D, extends to the radius of $\Omega_\mathrm{D} \simeq 3\Omega^*_{\mathrm{orb}}$, where Ω_D is Ω at $r = r_\mathrm{D}$. Such tidal waves as $\omega_\mathrm{D} = 3\Omega^*_{\mathrm{orb}}$ and $m_\mathrm{D} = 3$ are really present, if tidal potential is expanded

---

[7] It is noted that as far as equation (12.13) is concerned, there is no restriction on r_{L_1}. If we consider growth rate of the oscillations, the tilt oscillation whose r_{L_1} is close to r_{L_2} will be excited in large amplitude. The growth rate is given by equation (11.52) and can be calculated by using the coupling constant W, with eigen-functions of ω_1- and ω_2-oscillations and the form of tidal wave. Considerations concerning the limiting case of zero-temperature disks, however, suggest that the growth rate will be high when r_{L_1} and r_{L_2} are close, i.e., $r_{\mathrm{L}_1} \sim r_{\mathrm{L}_2}$. (see Kato 2014 for more discussions.)

till the terms of the order of $(r/a)^3$ (see Table 12.1). This is nothing but the 3 : 1 resonant excitation of tilt considered by Lubow (1992) (see also Kato 2014).

Next, let us consider the case of $n^* = -4$. Equation (12.16) shows that the resonant condition is realized at $\Omega_L/\Omega^*_{orb} = 4$. For this resonance to occur, tidal waves with $\omega_D = -4\Omega^*_{orb}$ (i.e., $n^* = -4$) and $m_D = -3$ must be really present. In other words, the presence of tidal waves with $\omega_D = 4\Omega^*_{orb}$ and $m_D = 3$ are required. Table 12.1 shows that such tidal waves are really present, if the orbit of the secondary is slightly eccentric, even in the framework till the order of $(r/a)^3$.

It is noted that the radius r_L specified by $\Omega_L = 4\Omega^*_{orb}$ is about $0.40a$ (see Table 12.4) and smaller than the radius $\sim 0.48a$ where the 3 : 1 resonance occurs. That is, this resonant excitation of tilt by the 4 : 1 resonance begins to occur at the evolutional stage of disks when the disk radius is $\sim 0.40a$ and continues to work during further expansion of disks.

In the case of $n^* = -5$, the resonant condition is satisfied for tidal waves with $\omega_D = -5\Omega^*_{orb}$ and $m_D = -3$, which is equivalent to $\omega_D = 5\Omega^*_{orb}$ and $m_D = 3$. Such tidal waves are really present, if the orbit is eccentric and the terms of the order of e^2 are considered (see Table 12.1). This resonance begins to occur when $r_D \sim 0.34a$ (see Table 12.4) and continue to work during further expansion of disks.

(ii) The case of $(m_2, n_2) = (2, 3)$

In the case of $(m_2, n_2) = (2, 3)$, the resonant condition (12.15) is written as

$$\Omega_L \sim -\frac{n^*}{2-\sqrt{3}}\Omega^*_{orb}, \tag{12.17}$$

and the ratio Ω_L/Ω^*_{orb} becomes larger than 3.0 for n^* which is negative and $|n^*| \geq 1$. Some characteristic quantities related to the resonant radius and the mode of tidal wave required are shown in Table 12.4.

It is important to note here that excitation of the tilt mode starts from an early stage of disk expansion. As a demonstration, let us consider the case where $n^* \equiv \omega_D/\Omega^*_{orb} = -3$. Then, the resonant condition (12.17) gives $\Omega_L \sim 11.2\Omega^*_{orb}$ (see

Table 12.4 Characteristic quantities required for resonant excitation of tilt mode ($m_1 = 1$ and $n_1 = 1$).

Inclination	(m_2, n_2)	$(m_D, \omega_D/\Omega^*_{orb})$	Ω_L/Ω^*_{orb}	r_L/a	R/a	Eccentricity
$\delta = 0$	(2,1)	$(-3, -3)$	3	0.48	$(R/a)^3$	e^0
		$(-3, -4)$	4	0.40	$(R/a)^3$	e^1
		$(-3, -5)$	5	0.34	$(R/a)^3$	e^2
	(2,3)	$(-3, -1)$	3.73	0.41	$(R/a)^3$	e^2
		$(-3, -2)$	7.46	0.26	$(R/a)^3$	e^1
		$(-3, -3)$	11.19	0.20	$(R/a)^3$	e^0
$\delta \neq 0$	(2,2)	$(-3, -2)$	3.41	0.44	$(R/a)^4$	e^0
		$(-3, -3)$	5.12	0.34	$(R/a)^4$	e^1
		$(-3, -4)$	6.83	0.28	$(R/a)^4$	e^0

also Table 12.4), which leads to $r_L \sim 0.20a$ (see also Table 12.4). The required tidal wave for this resonance is $m_D = -3$ and $\omega_D = -3\Omega^*_{orb}$. In other words, $m_D = 3$ and $\omega_D = 3\Omega^*_{orb}$, and no eccentricity is required. That is, this resonant excitation of tilt is due to the same tidal wave as in the case of the 3 : 1 resonance,[8] but starts from a much earlier stage of disk expansion and continue to work during further expansion of disks.

12.3.3.2 Misaligned Cases

If the disk and orbital planes are misaligned (i.e., $\delta \neq 0$), the tilt mode ($m_1 = 1$ and $n_1 = 1$) can be resonantly excited even when the ω_2-oscillation is plane-symmetric (i.e., n_2 = even), since the tidal force has a plane-asymmetric term. Here, we consider, in particular, the case of $(m_2, n_2) = (2, 2)$, where $m_D = -3$. The resonant condition (12.15) requires

$$\Omega_L \simeq -\frac{n^*}{2-\sqrt{2}}\Omega^*_{orb}. \tag{12.18}$$

Here, three cases of $n^* = -2, -3$, and -4 are mentioned briefly. Some characteristic quantities related to the resonant radii and the modes of tidal wave required are summarized in Table 12.2.[9] The case of $n^* = -2$ means $\omega_D = -2\Omega^*_{orb}$. This tidal waves are equivalent to those with $\omega_D = 2\Omega^*_{orb}$ and $m_D = 3$. Such tidal waves are really present if we consider the tidal potential till the order of $(r/a)^4$, even if the orbit is circular ($e = 0$) (see Table 12.4). The radius r_L for this resonance is found, from equation (12.18), to be $\Omega_L/\Omega^*_{orb} = 2/(2-\sqrt{2}) = 2+\sqrt{2} = 3.41$ (see Table 12.4), and is really inside the radius of the 3 : 1 resonance.

In the case of $n^* = -3$, we have $\omega_D = -3\Omega^*_{orb}$ and $m_D = -3$. This is equivalent to $\omega_D = 3\Omega^*_{orb}$ and $m_D = 3$. Such tidal waves really exist if the orbit is eccentric (see Table 12.2) and appear in the order of e^1. The radius, r_L, for this resonance is given by, from equation (12.18), $\Omega_L/\Omega^*_{orb} = 3/(2-\sqrt{2}) = 1.5(2+\sqrt{2}) = 5.12$ (see Table 12.4). Similarly, in the case of $n^* = -4$, we have $\omega_D = -4\Omega^*_{orb}$ and $m_D = -3$, or $\omega_D = 4\Omega^*_{orb}$ and $m_D = 3$. These tidal waves really appear if we consider terms till $(r/a)^4$ even if $e = 0$ (see Table 12.2). The radius r_L for this resonance is given by $\Omega_L/\Omega^*_{orb} = 4/(2-\sqrt{2}) = 2(2+\sqrt{2}) = 6.83$ (see Table 12.4).

[8]The statement of this sentence is not rigorous. In the case of the 3 : 1 resonance for positive superhumps, $n_1 = 0$, $n_2 = 0$ and thus n_D required is $n_D = 0$. Furthermore, in the case of the 3 : 1 resonance for negative superhumps, $n_1 = 1$, $n_2 = 1$, and thus the tide wave with $n_D = 0$ can have resonance. In the present case, however, $n_1 = 1$, $n_2 = 3$, and thus the tidal wave which can contribute to the resonance is the tidal wave with $n_D = 2$. Amplitude of this tidal wave will be smaller than that of n_D.

[9]It is noted that the resonant excitation treated in the coplanar cases are still present in misaligned systems. The resonant processes to be mentioned below are additions to them.

Summing the above results, we list some characteristic quantities related to excitation of tilts in Table 12.4. Although tilts are excited by various couplings processes inside the tidal truncation radius, this does not mean that all of them are excited to an observable level in dwarf novae. In close binary systems such as dwarf novae, the eccentricity of the orbits of the secondary star is small, i.e., $e \sim 0$. Hence, in the resonances labelled by e^2 in the last column of Table 12.4, the amplitude of the tidal deformation will be small, and thus the growth will be weak. The tilts which are excited in the inner region of disks will also not grow because the amplitude of disk deformation is small by the factor of $(R/a)^3$. In addition, in the case of dwarf novae, the inclination of the orbital plane to the disk plane is small, i.e., $\delta \sim 0$. Hence, the tilts resulting from $\delta \neq 0$ are also not expected with observable amplitude. Considerations of the above situations lead us to the conclusions that excitation processes important in the case of tilt will be those of the second line ($\Omega_{\rm L}/\Omega_{\rm orb} = 3$ and $r_{\rm L}/a = 0.48$), the third line ($\Omega_{\rm L}/\Omega_{\rm orb} = 4$ and $r_{\rm L}/a = 0.40$), and the sixth line ($\Omega_{\rm L}/\Omega_{\rm orb} = 7.46$ and $r_{\rm L}/a = 0.26$) in Table 12.4.

It is of importance to note that the tilt excitation by resonance is continued after the disk radius reached at $r_{\rm L}$. That is, the excitation continues in the expansion stage of disks where $r_{\rm L} < r_{\rm D}$. This might be one of causes why the negative superhumps appear like clouds in the two-dimensional power spectrum diagram of Fig. 1.5.

In Be/X-ray systems, the Be-star disk and the orbital plane of a compact star (neutron star in general) is generally misaligned, and further a secondary star has a large excentric orbit. In colliding or merging black-hole binaries, each black hole will have its own disk, although the system as a whole will be surrounded by a common envelope. In Be/X-ray stars, for example, in addition to normal outbursts, giant outbursts which occur less frequently are observed. The giant outbursts are supposed to be due to an interaction between warped precessing Be-star disk and secondary star which has a large eccentric orbit (Moritani et al. 2013). The origin of giant outbursts is still under debate. It might be important to examine the wave-wave resonant process in disks whose plane is inclined from the orbital plane of an eccentric binary system. In such systems resonant excitation of tilt at various radii listed in Tables 12.1 and 12.2 may have comparable growth rates.

12.4 Possible Excitation of High-Frequency QPOs in X-Ray Binaries: I Warped Disks

In Sects. 12.1, 12.2, and 12.3, we have examined resonant excitation of oscillations in tidally deformed disks. In neutron-star and black-hole binaries, tidal deformation of the innermost region of relativistic disks will not be expected. This does not mean, however, that deformations are not expected in such disks. For example, warps may be present even in the innermost part of relativistic disks (e.g., Ferreira and Ogilvie 2009). Furthermore, a spiral pattern may be expected in small inner tori which are formed in the very high state (the steep power-law state) of black-hole sources (e.g., Machida and Matsumoto 2008).

In Sects. 12.4 and 12.5, we consider possible excitations of high-frequency QPOs in low-mass X-ray binaries by disk deformation. We assume two kinds of low-frequency disk deformations. One is a deformation by warp ($m_D = 1$, $n_D = 1$ and $\omega_D \sim 0$). The other is a deformation by two-armed low-frequency pattern ($m_D = 2$, $n_D = 0$ or 1, and $\omega_D \sim 0$). First, we consider the former case, and the latter one is examined in the next section (Sect. 12.5).

It is noted that the disk is taken to be relativistic, and thus in the inner region of disks κ is much smaller than Ω.

12.4.1 *Warped Disks with* $m_D = -1, n_D = 1$ *and* $\omega_D = 0$

If central source is a rotating compact object, its rotation axis might not be always perpendicular to the plane of binary orbit where accretion disk is formed. That is, the disk is misaligned or tilted. Misalignment between the axis of orbital angular momentum of disk and the spin axis of central black hole brings about deformations of the inner disk (Lense-Thirring precession) (Bardeen and Petterson, 1975). This differential precession tends to twist the disk, leading to a stationary warped disk (Ferreira and Ogilvie 2009). In this context, we adopt disk deformation characterized by[10]

$$m_D = -1,\ n_D = 1 \text{ and } \omega_D = 0. \tag{12.19}$$

12.4.2 *Excitation of p-Mode and g-Mode Oscillations by Their Coupling*

As the ω_1-oscillation we consider one-armed p-mode oscillation (i.e., $m_1 = 1$ and $n_1 = 0$). Its propagation regions are specified by $(\omega_1 - \Omega)^2 < \kappa^2$, which are separated into two regions; one is the region where $\omega_1 > \Omega + \kappa$ and the other is the region where $\omega_1 < \Omega - \kappa$. The former propagation region extends outward without boundary, and no trapped oscillation is expected. Hence, our attention is on an oscillation in the latter propagation region. Trapping of a p-mode oscillation in this region is schematically shown by the upper wavy line in Fig. 12.4 (the line is drawn arbitrary and the trapping condition is not really examined). Some characteristics of

[10]It is noted that the minus sign of m_D is related to the fact that the resonant condition we adopt is $m_1 + m_2 + m_D = 0$. If $m_1 + m_2 = m_D$ is adopted as the resonant condition, what we need is $m_D = 1$.

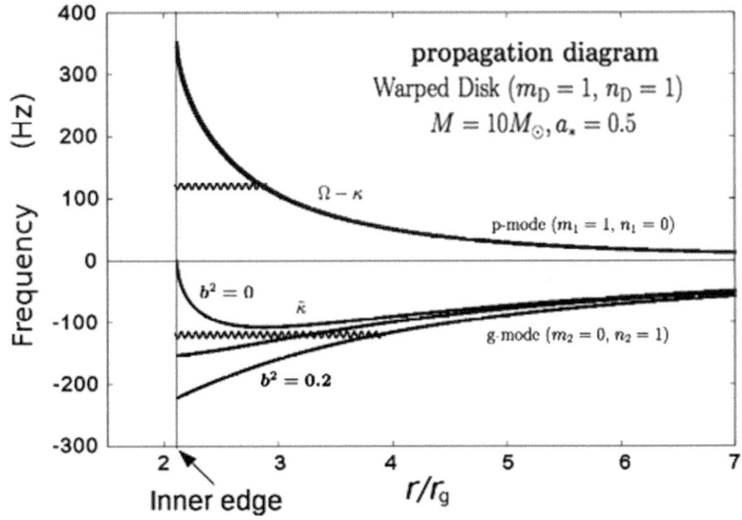

Fig. 12.4 Propagation diagram showing propagation regions of one-armed p-mode oscillation ($m_1 = 1, n_1 = 0$) and axisymmetric g-mode oscillation ($m_2 = 0, n_2 = 1$) in the case where the mass of the central source is $10\,M_\odot$ with spin parameter $a_* = 0.5$. The *upper curve* is the radial distribution of $\Omega - \kappa$. The *lower three curves* show the radial distribution of $-\tilde{\kappa}$ in three cases of $b^2 = 0, 0.1$, and 0.2. Here, $\tilde{\kappa}$ is a modified κ given by equation (12.22) and $b^2 \equiv \tilde{c}_A^2/c_s^2$, where $\tilde{c}_A$ is Alfvén speed defined by poloidal magnetic field, B_p and acoustic speed in disks, c_s. The horizontal wavy lines show qualitatively positions of vertical p-mode oscillation and g-mode oscillation ($b^2 = 0.2$). The positions of trapped oscillations are not results derived from quantitative calculation of the trapping condition, but are shown only for presentation of qualitative image.

this trapped one-armed p-mode oscillation are

$$m_1 = 1,\ n_1 = 0,\ \omega_1 > 0,\ \text{and } E_1/\omega_1 < 0 \quad \text{(one armed p mode)}. \tag{12.20}$$

It is noted that this oscillation has $E_1/\omega_1 < 0$, because $\omega_1 - \Omega < -\kappa < 0$.

Next, the counter-mode of the ω_1-oscillation, i.e., ω_2-oscillation, is considered. The resonant conditions, $m_1 + m_2 + m_D = 0$ and $\omega_1 + \omega_2 + \omega_D = 0$, require that $m_2 = 0$ and $\omega_2 = -\omega_1 < 0$. Furthermore, for ω_2-oscillation to coupling with ω_1-oscillation ($n_1 = 0$) and warp ($n_D = 1$), $n_2 = 1$ is required. That is, the ω_2-oscillation is an axisymmetric g-mode oscillation ($m_2 = 0$, $n_2 = 1$, and $\omega_2 < 0$).

The next problem is to examine the propagation region of the ω_2-oscillation mentioned above (i.e., axisymmetric g-mode oscillation). Some considerations are necessary to examine this issue. In the case where there is no global magnetic field in disks, the propagation region is specified by $-\kappa < \omega < \kappa$ (see Chap. 6), where $\kappa(r)$ is radial epicyclic frequency. Because the radial epicyclic frequency has a maximum, say κ_{max}, at a radius near to the inner edge of the disks, waves with $|\omega| < \kappa_{max}$ are self-trapped in a region close to the inner edge of disks (see Chap. 6).

Fu and Lai (2009), however, showed that this self-trapping of g-mode oscillations is strongly affected and destroyed by the presence of global poloidal magnetic fields, even if the fields are weak.[11] Their results show that the propagation region is changed to

$$-\tilde{\kappa} < \omega < \tilde{\kappa}, \tag{12.21}$$

where $\tilde{\kappa}^2$ is given by

$$\tilde{\kappa}^2 = \frac{1}{2}\left[\kappa^2 + 2\Omega_\perp^2 b^2 + (\kappa^4 + 16\Omega_\perp^2\Omega^2 b^2)^{1/2}\right], \tag{12.22}$$

where $b \equiv \tilde{c}_A/c_s$, $\tilde{c}_A$ being the Alfvén speed defined by the poloidal component of the magnetic fields. The radial distribution of $-\tilde{\kappa}(r)$ is shown in Fig. 12.4 for three cases of $b^2 = 0$, $b^2 = 0.1$, and $b^2 = 0.2$. In the case where $b^2 = 0.1$, for example, self-trapping of axisymmetric g-mode oscillations already disappears, because there is no minimum of $-\tilde{\kappa}(r)$. If the inner edge of disks acts as a reflecting boundary of waves, however, trapping of g-mode oscillations will be possible between the inner edge of the disks and the radius where $\omega = -\tilde{\kappa}$ is realized. Such trapping of g-mode oscillations is schematically shown in the case of $b^2 = 0.2$ in Fig. 12.4. It should be noted that the wavy line demonstrating trapping of oscillations is only for demonstration, and is not drawn so that the trapping condition is really satisfied.

In summary, we have axisymmetric trapped g-mode oscillations characterized by

$$m_2 = 0,\ n_2 = 1,\ \omega_2 < 0,\ \text{and}\ E_2/\omega_2 < 0 \quad \text{(axisymmetric g mode)}. \tag{12.23}$$

It is noted that $E_2/\omega_2 < 0$, because $E_2 > 0$ (axisymmetric oscillation) and $\omega_2 < 0$.

The above two oscillations (oscillations specified by equations (12.20) and (12.23)) satisfy the excitation condition, $(E_1/\omega_1)(E_2/\omega_2) > 0$. In addition, their propagation regions are overlapped as is shown in Fig. 12.4. Hence, if the above two trapped oscillations have frequencies really close to the resonant condition $\omega_1 + \omega_2 = 0$, they are excited simultaneously.

To examine what case the condition of $\omega_1 + \omega_2 = 0$ is satisfied, we must calculate the eigenfrequencies of trapped oscillations, for example, by the WKB method. The eigenfrequencies depend on various quantities characterizing disk structure such as metric (spin of of the central source), disk temperature, magnetic fields, and the node number of oscillations in the radial direction. Although what case the condition of $\omega_1 + \omega_2 = 0$ is satisfied is not examined here in details, Fig. 12.4 suggests that the condition will be realized around 100 Hz in cases of $M = 10\,M_\odot$ and $a_* = 0.5$.

It is noted that the resonant excitation of the above two coupled oscillations (one-armed p-mode oscillation and axisymmetric g-mode oscillation) in warped disks

[11] In Fu and Lai (2009) an approximation has been introduced in treatment of vertical behavior of oscillations. Examinations how much this approximation is valid will be worthwhile.

have been examined numerically by Ferreria and Ogilvie (2008) and Oktariani et al. (2010) by different ways. Their results, however, do not always coincide. More studies will be required.

12.5 Possible Excitation of High-Frequency QPOs in X-Ray Binaries: II Two-Armed Deformed Disks

The origins of kHz quasi-periodic oscillations observed in neutron-star LMXBs and of HFQPOs observed in black-hole LMXBs are still in debate, and there are no widely accepted models. Here, we restrict our attention to disk oscillation models. Then, two-armed c-mode (and vertical p-mode) oscillations seem to be one of possible candidates well describing some characteristics of twin kHz QPOs of neutron-star X-ray sources (chapter 7, see also Kato 2012a) and HFQPOs in black-hole X-ray sources (Kato 2012b).

The next problem is then whether excitation of such oscillations are really possible in disks. We assume here that at very high state (steep power-law state), low frequency global disk deformations are present, triggered by formation of tori or magnetic activities. The disk deformations assumed here are two-armed ($m_{\rm D}=2$) and low-frequency ($\omega_{\rm D}\sim 0$) ones.[12] Concerning $n_{\rm D}$, we consider separately two cases of $n_{\rm D}=0$ and $n_{\rm D}=1$ in Sects. 12.5.1 and 12.5.2, where $n_{\rm D}=0$ is plane-symmetric deformation, while $n_{\rm D}=1$ is plane-asymmetric deformation.

12.5.1 *Excitation of Two-Armed, c-Mode Oscillations in Two-Armed Deformed Disks with $n_{\rm D}=0$*

First, we consider the case where disk deformation is plane-symmetric with $n_{\rm D}=0$, i.e.,

$$m_{\rm D}=-2,\ n_{\rm D}=0,\ \omega_{\rm D}\simeq 0. \tag{12.24}$$

As ω_1-oscillation, we consider two-armed c-mode oscillation with $n_1=1$ with $\omega_1>0$. As counter-oscillation of this oscillation we take axisymmetric g-mode oscillation. That is, we adopt

- $m_1=2, n_1=1, \omega_1>0$ (two-armed c-mode)
- $m_2=0, n_2=1, \omega_2<0$ (axisymmetric g-mode)

---

[12]Observations sometimes show that high-frequency QPOs and low-frequency QPOs are correlated in X-ray binaries. The disk deformation of $\omega_{\rm D}$ might correspond to the low-frequency QPOs.

The propagation region of the above c-mode oscillation is given by $(\omega_1 - 2\Omega)^2 - \Omega_\perp^2 > 0$. This propagation region is separated into two regions: $\omega_1 > 2\Omega + \Omega_\perp$ and $\omega_1 < 2\Omega - \Omega_\perp$. Oscillations in the former region are not trapped in disks, because the propagation region extends far outside. We are thus interested in oscillations propagating in the latter region. The trapping of an oscillation ($\omega_1 > 0$) in this region is shown schematically in Fig. 12.5. This oscillation has $E_1/\omega_1 < 0$, because $\omega_1 - 2\Omega < 0$.

Next, let us consider ω_2-oscillation. A necessary condition of resonant coupling, $m_1 + m_2 + m_D = 0$, is satisfied by $m_2 = 0$. The propagation region of this oscillation is $\omega_2^2 < \tilde{\kappa}^2$. We consider oscillations with negative frequency ($\omega_2 < 0$). The propagation region of the oscillation in the case of $b^2 = 0.3$ is like that schematically shown in Fig. 12.5 (quantitative examination of trapping condition is not made in drawing the wavy line). The value of E_2/ω_2 of this oscillation is negative and the resonant condition of $(E_1/\omega_1)(E_2/\omega_2) > 0$ is satisfied. As is also shown in Fig. 12.5, the radial propagation regions of the ω_1- and ω_2-oscillations are overlapped and thus their resonant excitation is expected, if the condition of $\omega_1 + \omega_2 \simeq 0$ is realized. The frequency ω_1 for which $\omega_1 + \omega_2 \sim 0$ is realized depends on values of parameters such as disk structures (spin, magnetic fields and so on) and oscillation modes (node numbers of oscillations in the radial direction).

It is noted that if there is no magnetic field (i.e., $b^2 = 0$), $-\tilde{\kappa}(r)$ is equal to $-\kappa(r)$ and has a minimum value. (The value of $\kappa_{\max}$ is around 600 Hz in the case shown in Fig. 12.5 ($M = 1.4 M_\odot$ and $a_* = 0.3$).) The frequency is too low compared with typical trapped vertical p-mode oscillations, and thus the resonant condition, $\omega_1 + \omega_2 \simeq 0$ will not be realized, unless high overtones (large value of n_r) of vertical p-mode oscillations are considered. If there are poloidal magnetic fields ($b^2 \neq 0$), however, $\omega_1 + \omega_2 \simeq 0$ will be realized without considering such higher overtones of vertical p-mode oscillations, as schematically shown in Fig. 12.5. Quantitative examination is worthwhile.

12.5.2 *Excitation of Two-Armed, c-Mode and Vertical p-Mode Oscillations in Two-Armed Deformed Disks with $n_D = 1$*

In this subsection, we consider plane-asymmetric ($n_D = 1$) two-armed disk deformation:

$$m_D = -2, \ n_D = 1, \ \omega_D \simeq 0. \tag{12.25}$$

First, as disk oscillations we consider the following set:

- $m_1 = 2, n_1 = 2$, and $\omega_1 > 0$ (two-armed, vertical p-mode oscillation),
- $m_2 = 0, n_2 = 1$, and $\omega_2 < 0$ (axisymmetric g-mode oscillation)

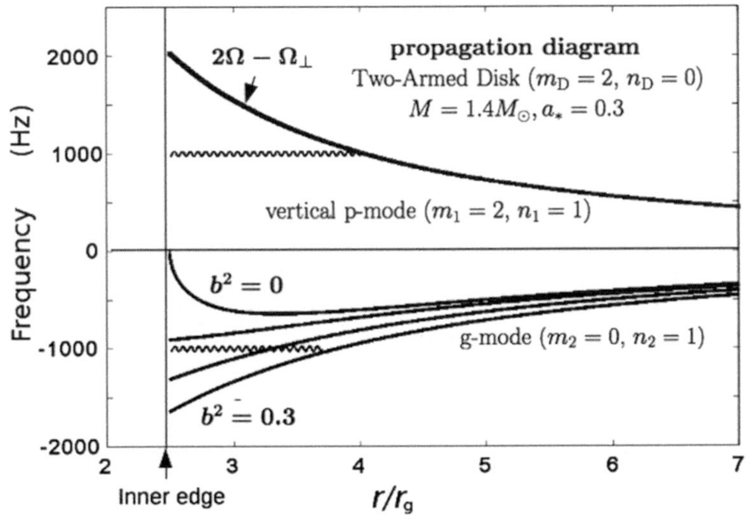

Fig. 12.5 Propagation diagram showing propagation regions of two-armed c-mode oscillation ($m_1 = 2$, $n_1 = 1$) and axisymmetric g-mode oscillation ($m_2 = 0$, $n_2 = 1$) in the case where the mass of the central source is $1.4M_\odot$ with spin parameter $a_* = 0.3$. The *upper curve* is the radial distribution of $2\Omega - \Omega_\perp$. The *lower three curves* show the radial distribution of $-\tilde{\kappa}$ in three cases of $b^2 = 0$, 0.1, and 0.2. The horizontal wavy lines show qualitatively positions of c-mode and g-mode oscillations in the disks with $b^2 = 0.3$. The positions of trapped oscillations are not results of quantitative examination of the trapping condition, but are shown only for demonstration of qualitative image.

The above set of oscillations satisfies one of resonant condition of $m_1 + m_2 + m_D = 0$. Furthermore, since the ω_1-oscillation is symmetric with respect to the equatorial plane (i.e., $n_1 = 2$) and the ω_2-oscillation is asymmetric ($n_2 = 1$), they can couple each other through the asymmetric disk deformation ($n_D = 1$).

The propagation region of the ω_1-oscillation is given by $(\omega_1 - 2\Omega)^2 > 2\Omega_\perp^2$, which consists of two regions of $\omega_1 > 2\Omega + \sqrt{2}\Omega_\perp$ and of $\omega_1 < 2\Omega - \sqrt{2}\Omega_\perp$. We are interest here in an oscillation in the latter region, which is shown schematically in Fig. 12.6. This oscillation has $E_1/\omega_1 < 0$, since $\omega_1 - 2\Omega < 0$. Next, let us consider ω_2-oscillation, which is axisymmetric g-mode oscillation and its propagation region is specified by $\omega_2^2 < \tilde{\kappa}^2$. Since we are interested in oscillations with negative frequency, their propagation region is like that shown in Fig. 12.6. Wavy lines are drawn only for demonstration.

Second, we mention that the following set of disk oscillations:

- $m_1 = 2$, $n_1 = 0$, and $\omega_1 > 0$ (two-armed p-mode oscillation),
- $m_2 = 0$, $n_2 = 1$, and $\omega_2 < 0$ (axisymmetric g-mode oscillation)

can be also excited under the disk deformation given by equation (12.25). The reason is similar to that in the case considered just above.

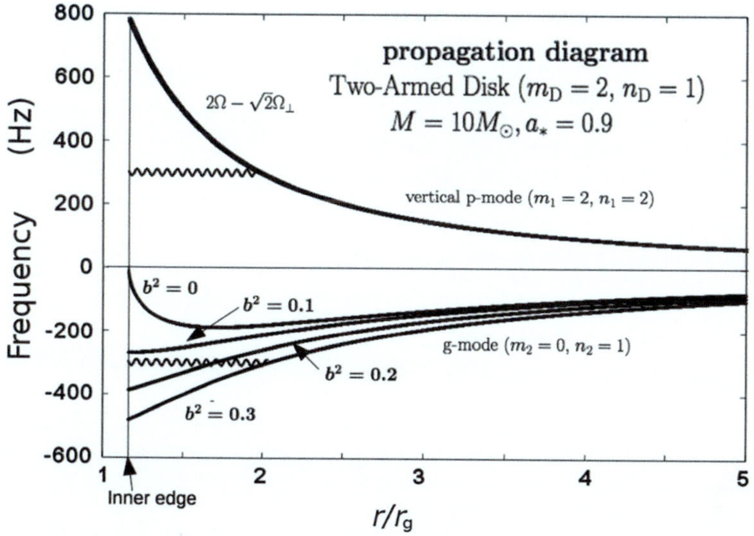

Fig. 12.6 Propagation diagram showing propagation regions of two-armed vertical p-mode oscillation ($m_1 = 2$, $n_1 = 2$) and axisymmetric g-mode oscillation ($m_2 = 0$, $n_2 = 1$) in the case where the mass of the central source is $10.0\,M_\odot$ with spin parameter $a_* = 0.9$. The *upper one curve* is the radial distribution of $\Omega - \sqrt{2}\Omega_\perp$. The *lower three curves* show the radial distribution of $-\tilde{\kappa}$ in three cases of $b^2 = 0, 0.1$, and 0.2. The horizontal wavy lines show qualitatively positions of the vertical p-mode and g-mode oscillations in the disks with $b^2 = 0.3$. The positions of trapped oscillations are not results derived from quantitative calculation of the trapping condition, but are shown only for demonstration of qualitative image.

References

Bardeen, J. M. & Petterson, J. A., 1975, Astrrophys. J., 195, L65
Ferreria, B. T., & Ogilvie, G. I. 2008, Mon. Not. R. Astron. Soc., 386, 2297
Ferreira, B. T. & Ogilvie, G. I. 2009, Mon. Not. R. Astron. Soc., 392, 428
Fu, W., & Lai, D. 2009, Astrophys. J., 690, 1386
Hirose, M., & Osaki, Y. 1990, Publ. Astron. Soc. Jpn., 42, 135
Kato, S. 2001, Publ. Astron. Soc. Jpn., 53, 1
Kato, S. 2012a, Publ. Astron. Soc. Jpn., 64, 62
Kato, S. 2012b, Publ. Astron. Soc. Jpn., 64, 139
Kato, S. 2013, Publ. Astron. Soc. Jpn., 65, 56
Kato, S. 2014, Publ. Astron. Soc. Jpn., 66, 21
Kato, S., Fukue, J., & Mineshige, S. 2008, *Black-Hole Accretion Disks — Towards a New paradigm —* (Kyoto University Press, Kyoto)
Kato, T., Imada, A., Umemura, M., & Nogami, D., et al. 2009, Publ. Astron. Soc. Jpn., 61, 395
Kato, T., & Osaki, Y. 2013, Publ. Astron. Soc. Jpn., 65, 97
Kato, T., & Osaki, Y. 2014, Publ. Astron. Soc. Jpn., 66,15
Lubow, S. H. 1991, Astrophys. J., 381, 259
Lubow, S. H. 1992, Astrophys. J., 398, 525
Machida, M. & Matsumoto, R. 2008, Publ. Astron. Soc. Jpn., 60, 613
Moritani, Y., Nogami, D., Okazaki, A. T., & Imada, A. 2013, Publ. Astron. Soc. Jpn., 65, 83

Ohshima, T., Kato, T., Pavlenko, E. P., & Ito, H., et al. 2012, Publ. Astron. Soc. Jpn., 64, 3
Okazaki, A. T., Kato, S., & Fukue, J. 1987, Publ. Astron. Soc. Jpn., 39, 457
Oktariani, F., Okazaki, A. T., & Kato, S. 2010, Publ. Astron. Soc. Jpn., 62, 709
Osaki, Y. 1985, Astron. Astrophys., 144, 369
Osaki, Y. 1996, Publ. Astron. Soc. Pac., 108, 390
Osaki, Y., & Kato. T. 2013a, Publ. Astron. Soc. Jpn., 65, 50
Osaki, Y., & Kato. T. 2013b, Publ. Astron. Soc. Jpn., 65, 95
Patterson, J., Jablonski, F., Koen, C., O'Donoghue, D., & Skillman, D. R. 1995, Publ. Astron. Soc. Pac., 107, 1183
Whitehurst, R. 1988, Mon. Not. R. Astron. Soc., 232, 35

Chapter 13
Sonic Point Instability and Stochastic Excitation of Oscillations by Turbulence

Abstract In this final chapter two possible excitation processes which are not mentioned yet are briefly introduced. One is a sonic point instability. In the case of relativistic disks, the accretion flows fall to central objects by passing a sonic point. In the case where the sonic point is a nodal-type critical point, accretion flows are locally unstable at the sonic point and waves propagating both inward and outward are generated there. This is an interesting excitation process of oscillations, but there is a slight doubt whether it is physical. The other is a possible excitation of oscillations by stochastic processes of turbulence. The stochastic processes are known to be the main processes of excitation of solar and stellar oscillations, but they are not examined yet in disks. This excitation process will be of importance, because many oscillations modes will be excited with small amplitudes. Examination of disk oscillations by stochastic processes will become in future one of important fields of diskoseismology. A brief outline of mathematical formulation of stochastic excitation of disk oscillations will be presented here.

Keywords Nodal-type critical point • Sonic point instability • Stochastic excitation • Turbulence

13.1 Type of Sonic Point and Instability

In Chap. 9, we have examined viscous overstability of disk oscillations. The results show that the innermost region of relativistic disks is a place favorable for viscous excitation of p-mode oscillations (see equation (9.48) and also see figure 12 of Kato 2001). This is because in the innermost region of relativistic disks, epicyclic frequency is much smaller than the frequency of the disk rotation by effects of the general relativity. The decrease of epicyclic frequency has an another effect on disk stability. The decrease of epicyclic frequency by the general relativity brings about the presence of inner edge of disks, as is well known in studies of relativistic disks. That is, inside a certain radius there is no circular orbit of free particles, and accretion flows become transonic.

The transonic radius (sonic point) is a critical point in the sense that in the differential equation describing radial flow u_r (i.e., $du_r/dr = N/D$), the denominator, D, vanishes at that point. Hence, for accretion flow to pass this sonic

S. Kato, *Oscillations of Disks*, Astrophysics and Space Science Library 437,
DOI 10.1007/978-4-431-56208-5_13

point the flow must have $N = 0$ as well as $D = 0$ at the sonic point (regularity condition). Since sonic point is a critical point, the topology around the point on the $u_r - r$ plane has a particular structure. In the case where the so-called Shakura-Sunyaev-type α-viscosity is adopted, there is a critical value of α, say α_c. When α is smaller than α_c, steady accretion flow passes a saddle-type critical point (radius), while in the case where α is larger than α_c, the flow passes a nodal-type critical point (Matsumoto et al. 1984). The critical value, α_c, is around 0.1 (Matsumoto et al. 1984; Muchotrzeb 1983; Muchotrzeb-Czerny 1986). Furthermore, these numerical simulations suggest that there is one-to-one correspondence between type of critical point and stability of transonic accretion flows. This is really shown by numerical simulations by Matsumoto et al. (1988, 1989) and analytically by Kato et al. (1988a,b). That is, for $\alpha > \alpha_c$ transonic flows pass a nodal-type critical point and are dynamically unstable, while flows with $\alpha < \alpha_c$ pass a saddle-type critical point and are stable.

Numerical simulations around the innermost region of relativistic disks with $\alpha > \alpha_c$ show that p-mode oscillations are generated at the sonic point and propagate away both inward and outward (Matsumoto et al. 1988, 1989). The generated waves have frequencies around $\kappa_{\rm max}$, where $\kappa_{\rm max}$ is the maximum value of epicyclic frequency. These results are confirmed by more careful simulations of non-isothermal cases by Honma et al. (1992), Chen and Taam (1995), and Milsom and Taam (1996, 1997). Recently, more extensive studies have been made by Miranda et al. (2015).

Typical results by Matsumoto et al. (1988) for isothermal perturbations in isothermal and vertically integrated disks are briefly presented here, because the essences of generation of oscillations are shown there. The type of viscosity adopted is the standard Shakura-Sunyaev-type α-viscosity, i.e., $T_{r\varphi} = -\alpha\Pi$, where $T_{r\varphi}$ and Π are vertically integrated $r\varphi$-component of stress tensor and vertically integrated pressure, respectively. Simulation results in the case of $\alpha = 0.3$ and $c_{\rm s}/c = 10^{-3}$ ($c_{\rm s}$ being acoustic speed) are presented in Fig. 13.1. Time variation of accretion rate slightly inside the sonic point is presented in Fig. 13.2 for a different parameter case.

Roughly speaking, axisymmetric p-mode oscillations excited have frequencies close to $\kappa_{\rm max}$ (see also figures 5 and 7 of Chen and Taam 1995) and are localized inside the radius of $\kappa_{\rm max}$. The reason why oscillations whose frequencies are close to $\kappa_{\rm max}$ are predominant seems not to be well understood yet. Frequency components higher than $\kappa_{\rm max}$ will be propagated away outside without being trapped. The oscillations whose frequencies are lower than $\kappa_{\rm max}$ will be trapped inside the disks.

Finally, some discussions are necessary in relation to the sonic point instability. As mentioned above, if the Shakura-Sunyaev-type α-viscosity is adopted, there is a critical value of α, above which the sonic point is nodal and oscillations are excited. If a standard viscosity-type expression for stress tensor is adopted, the nodal-type critical point does not appear for steady accretion flows (e.g., Abramowicz and Kato 1989), i.e., all transonic flows pass saddle-type critical point, i.e., no sonic type instability appears (Kato et al. 1993). This might suggest that the appearance of nodal-type critical point in the Shakura-Sunyaev-type α-viscosity is due to non-physical description of viscous flow near sonic point. It is noted, however, that adoption of standard viscosity-type expression for stress tensor is also not

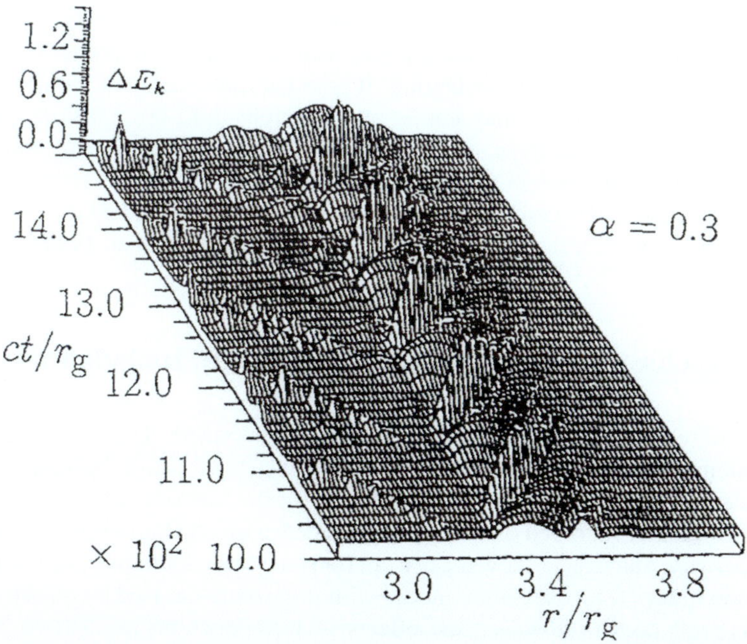

Fig. 13.1 Time variation of the kinetic energy $E_k(\equiv \Sigma v_r^2/2)$ of the radial motion (in units of $\Sigma_c c_s^2$, Σ_c being the surface density at the sonic point r_c). The parameters adopted are $\alpha = 0.3$, $c_s/c = 10^{-3}$, $\dot{M}/\dot{M}_{crit} = 1.6$, where $\dot{M}_{crit}$ is the criticl accretion rate defined by $\dot{M}_{crit} = L_E/c^2$ (Reprinted from Matsumoto et al. (1988) in *Physics of Neutron Stars and Black Holes*, ed. Y. Tanaka (University Academy Press, Tokyo), 155. Copyright ©).

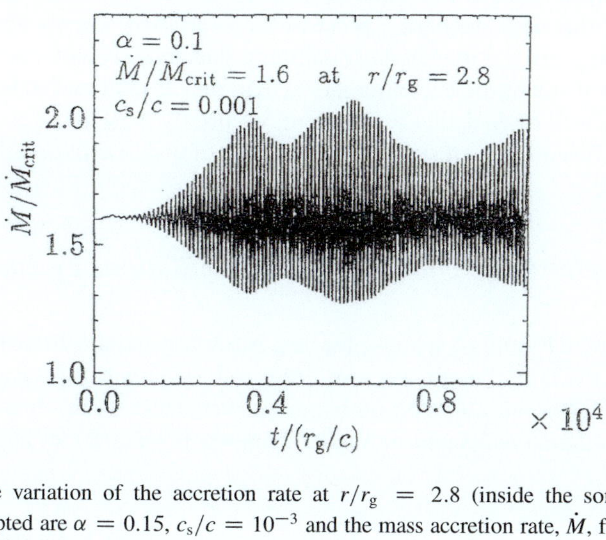

Fig. 13.2 Time variation of the accretion rate at $r/r_g = 2.8$ (inside the sonic point). The parameters adopted are $\alpha = 0.15$, $c_s/c = 10^{-3}$ and the mass accretion rate, $\dot{M}$, from the outside is $1.6\dot{M}_{crit}$ (Reprinted from Matsumoto et al. (1988) in *Physics of Neutron Stars and Black Holes*, ed. Y. Tanaka (University Academy Press, Tokyo), 155. Copyright ©).

rigorous when we consider transonic flows, since acoustic information is propagated from inside to outside through sonic point. More careful considerations might be necessary for the sonic point instability. It is noted, however, that the appearance of nodal-type critical point may not be always unphysical. For example, we hear that in the case of line-driven stellar winds in early type stars, linear momentum is transported by radiation pressure in non-diffusion forms. In this case, the sonic point of the wind can become nodal-type (e.g., Poe et al. 1990), and the nodal-type critical point is unstable against localized perturbations (Kato et al. 1993).

13.2 Stochastic Excitation of Oscillations by Turbulence

Effects of turbulence on global motions are complicated. The most simplified description of the effects is to introduce the concept of eddy viscosity. That is, turbulence is treated as viscosity when large scale motions are considered. In Chap. 9, we have examined the effects of turbulence on oscillations by introducing eddy viscosity. The results show that except for p-mode oscillations ($n = 0$) all other oscillation modes (i.e., oscillation modes with $n \neq 0$) are damped by turbulence.

As is well known, however, the effects of turbulence on oscillations are not simply evaluated by introducing the concept of eddy viscosity alone. Turbulence can excite oscillations outside the turbulent region (Lighthill 1952). A well-known astrophysical application is excitation of acoustic oscillations in the solar atmosphere and their propagation to the corona. Later, including the effects of viscous damping of oscillations due to turbulent viscosity, Goldreich and Keeley (1977) examined stochastic excitation of oscillations in turbulent media. They demonstrated that the stochastic processes of turbulent convection can excite the solar p-mode oscillations against damping due to turbulent viscosity. Recent developments of stochastic excitation of oscillations on solar and stellar convection zones are reviewed by Samadi (2011). Here, we briefly present the essential part of mathematical formulation of stochastic excitation of disk oscillations in disks.

13.2.1 *Stochastic Excitation of Disk Oscillations by Turbulence*

The mathematical formulation presented here is roughly parallel to that of Goldreich and Keeley (1977), although we start from a Lagrangian formulation of wave equation. We focus our attention on magnetic turbulence in the following formulations, but the final results will be similar even when we consider hydrodynamical turbulence.

The wave equation describing perturbations on MHD disks can be written as (see Chap. 3)

$$\frac{D_0^2\boldsymbol{\xi}}{Dt^2} = \delta\biggl(- \delta\psi - \frac{1}{\rho}\nabla p + \frac{1}{4\pi\rho}\mathrm{curl}\,\boldsymbol{B}\times\boldsymbol{B}\biggr), \tag{13.1}$$

where δ represents the Lagrangian variation and other notations have their usual meanings.

Here, for simplicity, we assume that magnetic fields are turbulent components alone, i.e., no global magnetic fields are assumed. Then, the basic equation describing displacement vector $\boldsymbol{\xi}$ associated with oscillations is (see Chap. 3)

$$\rho_0\frac{\partial^2\boldsymbol{\xi}}{\partial t^2} + 2\rho_0(\boldsymbol{u}_0\cdot\nabla)\frac{\partial\boldsymbol{\xi}}{\partial t} + \mathscr{L}(\boldsymbol{\xi}) = \boldsymbol{N}, \tag{13.2}$$

where $\mathscr{L}$ is a Hermite operator (see Chap. 3) and the left-hand side of this equation represents the linear wave equation. The right-hand side represents the source terms due to turbulence. The source term, $\boldsymbol{N}$, consists of hydrodynamical turbulence and magnetic turbulence. The hydrodynamical part can be found from equations (11.18) and (11.20), and the magnetic part is found easily from the last term on the right-hand side of equation (13.1). In the simplest case where the hydrodynamical turbulence occurs isothermally (i.e., $\Gamma_1 = 1$), the i-component of $\boldsymbol{N}$, i.e., N_i, is written as

$$N_i = \frac{\partial}{\partial r_j}\hat{N}_{ij}, \tag{13.3}$$

where

$$\hat{N}_{ij} = -p_0\frac{\partial\xi_{\mathrm{turb},k}}{\partial r_i}\frac{\partial\xi_{\mathrm{turb},j}}{\partial r_k} + \frac{1}{4\pi}\biggl(\frac{1}{2}b^2\delta_{ij} - b_ib_j\biggr), \tag{13.4}$$

where $\xi_{\mathrm{turb},i}$, for example, is the i-component of the displacement vector, $\boldsymbol{\xi}_{\mathrm{turb}}$, associated with turbulence, b_i is the i-component of the Eulerian variation of turbulent magnetic fields, $\boldsymbol{b}$, and b^2 is b_ib_i.

Let us write the solution of equation (13.2) by the sum of the normal modes of oscillations as

$$\boldsymbol{\xi} = \frac{1}{2}\sum_j A_j(t)\hat{\boldsymbol{\xi}}_j\exp(i\omega_jt) + \mathrm{c.c.}, \tag{13.5}$$

where c.c. denotes the complex conjugate, and $A_j(t)$ is the amplitude of the normal mode $\boldsymbol{\xi}_j$. The time change of amplitude $A_j(t)$ comes from the effects of turbulence.

Substitution of equation (13.5) into equation (13.2) leads to

$$\Re \sum_j 2\frac{dA_j}{dt}\rho_0[i\omega_j + (\boldsymbol{u}_0 \cdot \nabla)]\hat{\boldsymbol{\xi}}_j \exp(i\omega_j t) + \text{c.c.} = 2\boldsymbol{N}. \tag{13.6}$$

In deriving equation (13.6) the term of dA_j^2/dt^2 has been neglected, because the time change of amplitude A_j by turbulence is slow.

Now, equation (13.6) is multiplied by $\boldsymbol{\xi}_k^* \exp(-i\omega_k t)$ and the resulting equation is integrated over the whole volume. Then, we have

$$\begin{aligned}
&2i\frac{dA_k}{dt}\int \rho_0 \boldsymbol{\xi}_k^*[\omega_k - i(\boldsymbol{u}_0 \cdot \nabla)]\boldsymbol{\xi}_k d^3r \\
&+\sum_{j\neq k} 2i\frac{dA_j}{dt}(\omega_j - \omega_k)\exp[i(\omega_j - \omega_k)t]\int \rho_0 \boldsymbol{\xi}_k^* \cdot \boldsymbol{\xi}_j d^3r \\
&+\text{terms with } \exp[-i(\omega_j + \omega_k)t] \\
&= 2\int \exp(-i\omega_k t)\boldsymbol{\xi}_k^* \cdot \boldsymbol{N} d^3r.
\end{aligned} \tag{13.7}$$

In deriving the second term of equation (13.7) we have used the orthogonality relation (3.18). The frequencies of normal mode oscillations which we are interested in are usually on the order of the frequency of disk rotation, which is much higher than the timescale of the change of A_k. Since we are interested in a slow long-term variation of A_k, the terms with rapid time variations are also neglected. Then, equation (13.7) can be reduced to

$$\frac{dA_k}{dt} = -i\frac{\omega}{2E_k}\int \exp(-i\omega_k t)\boldsymbol{\xi}_k^* \cdot \boldsymbol{N} d^3r, \tag{13.8}$$

where E_k is the wave energy of normal mode of oscillation characterized by $\boldsymbol{\xi}_k$ (see equation (3.29)).

In deriving equation (13.8) the viscous damping of oscillations has been neglected. In order to take into account the effects of the viscous damping of oscillations we should add a term of $\eta_k A_k$ on the left-hand side of equation (13.8), where η_k is the viscous damping rate. Then, the equation is integrated with respect to time to obtain

$$A_k(t) = -i\frac{\omega_k}{2E_k}\int_{-\infty}^{t} \exp[\eta_k(t' - t) - i\omega_k t']dt' \int \hat{\boldsymbol{\xi}_k}^* \cdot \boldsymbol{N} d^3r. \tag{13.9}$$

Hereafter, the subscript k attached to specify oscillation mode is omitted, because in the followings we need a subscript to represent the component of vectors.

The expression for $A(t)$ cannot be directly evaluated, because $\boldsymbol{N}(\boldsymbol{r}, t)$ is a fluctuating vector field and only its statistical properties are known. The expectation value of $|A(t)|^2$, denoted by $\langle\langle |A(t)|^2\rangle\rangle$, is evaluated from

$$\langle\langle |A(t)|^2\rangle\rangle = \frac{\omega^2}{4E^2}\int_{-\infty}^{t}\int_{-\infty}^{t}\int\int dt'dt''d^3r'd^3r''\Big[\exp[\eta(t'+t''-2t)-i\omega(t'-t'')]\Big]$$
$$\times\hat{\boldsymbol{\xi}}(\boldsymbol{r}')\cdot\Big\langle\Big\langle \boldsymbol{N}(\boldsymbol{r}',t')\boldsymbol{N}(\boldsymbol{r}'',t'')\Big\rangle\Big\rangle\cdot\hat{\boldsymbol{\xi}}^*(\boldsymbol{r}'')\Big]. \tag{13.10}$$

It is convenient to perform the volume integrations in equation (13.10) by parts by using equation (13.3) to give

$$\langle\langle |A(t)|^2\rangle\rangle = \frac{\omega_k^2}{4E^2}\int_{-\infty}^{t}\int_{-\infty}^{t}\int\int dt'dt''d^3r'd^3r''\Big[\exp[\eta(t'+t''-2t)-i\omega(t'-t'')\Big]$$
$$\times\frac{\partial\hat{\xi}_i(\boldsymbol{r}')}{\partial r'_j}\Big\langle\Big\langle \hat{N}_{ij}(\boldsymbol{r}',t')\hat{N}_{k\ell}(\boldsymbol{r}'',t'')\Big\rangle\Big\rangle\frac{\partial\hat{\xi}_\ell(\boldsymbol{r}'')}{\partial r''_k}. \tag{13.11}$$

Let us now introduce variables $\boldsymbol{\lambda}$, $\boldsymbol{r}_0$, τ and t_0 defined by

$$\boldsymbol{\lambda} = \boldsymbol{r}''-\boldsymbol{r}', \quad \boldsymbol{r}_0 = \frac{1}{2}(\boldsymbol{r}'+\boldsymbol{r}''),$$
$$\tau = t''-t', \quad t_0 = \frac{1}{2}(t'+t''), \tag{13.12}$$

since $\langle\langle \hat{N}_{ij}(\boldsymbol{r}',t')\hat{N}_{k\ell}(\boldsymbol{r}'',t'')\rangle\rangle$ is a function of $\boldsymbol{\lambda}$ and τ in homogeneous steady turbulence. Then, $\langle\langle |A_k(t)|^2\rangle\rangle$ is written as

$$\langle\langle |A(t)|^2\rangle\rangle = \frac{\omega^2}{4E^2}\int_{-\infty}^{t}\exp[2\eta(t_0-t)]dt_0\int_{2(t_0-t)}^{(2(t-t_0)}d\tau\int\int d^3r_0d^3r$$
$$\times\exp(i\omega\tau)\frac{\partial\hat{\xi}_i(\boldsymbol{r}')}{\partial r'_j}\Big\langle\Big\langle \hat{N}_{ij}(\boldsymbol{r}',t')\hat{N}_{k\ell}(\boldsymbol{r}'',t'')\Big\rangle\Big\rangle\frac{\partial\hat{\xi}_k(\boldsymbol{r}'')}{\partial r''_\ell}. \tag{13.13}$$

Since the eddy lifetime is much shorter than the lifetime ($\sim 1/\eta$) of oscillations, we extend the integration over τ from $-\infty$ to ∞ (e.g., Goldreich and Keeley 1977; Samadi and Goupil 2001) and perform the integration to lead equation (13.13) to

$$\langle\langle |A(t)|^2\rangle\rangle E = \frac{\omega^2}{8\eta E}\int_{-\infty}^{\infty}d\tau\int\int d^3r_0d^3\lambda$$
$$\times\exp(i\omega\tau)\frac{\partial\hat{\xi}_i(\boldsymbol{r}')}{\partial r'_j}\Big\langle\Big\langle \hat{N}_{ij}(\boldsymbol{r}',t')\hat{N}_{k\ell}(\boldsymbol{r}'',t'')\Big\rangle\Big\rangle\frac{\partial\hat{\xi}_k(\boldsymbol{r}'')}{\partial r''_\ell}, \tag{13.14}$$

where $\boldsymbol{r}' = \boldsymbol{r}_0 - \boldsymbol{\lambda}/2$ and $\boldsymbol{r}'' = \boldsymbol{r}_0 + \boldsymbol{\lambda}/2$, and $t' = t_0 - \tau/2$ and $t'' = t_0 + \tau/2$. Equation (13.14) is a formal and basic expression for wave energy of excited oscillations.

13.2.2 Estimate of the Order of Amplitude of Excited Oscillations

To proceed from this point, we need information about $\langle\langle \hat{N}_{ij}(\boldsymbol{r}', t')\hat{N}_{k\ell}(\boldsymbol{r}'', t'')\rangle\rangle$. This is a difficult problem, since time-dependent structure of fourth-order turbulent correlation quantities is required. Detailed studies will be a subject in future. Here, we shall be satisfied by roughly estimating the order of $\langle\langle \hat{N}_{ij}(\boldsymbol{r}', t')\hat{N}_{k\ell}(\boldsymbol{r}'', t'')\rangle\rangle$ and thus the magnitude of $\langle\langle |A|^2\rangle\rangle E$.

Here, we simply assume that the correlation $\langle\langle N_{ij}(\boldsymbol{r}', t')N_{k\ell}(\boldsymbol{r}'', t'')\rangle\rangle$ consists of a τ-independent term and a term which decreases with increase of τ as

$$\langle\langle N_{ij}(\boldsymbol{r}', t')N_{k\ell}(\boldsymbol{r}'', t'')\rangle\rangle = \langle\langle N_{ij}(\boldsymbol{r}', t')N_{k\ell}(\boldsymbol{r}'', t'')\rangle\rangle_0 + \langle\langle N_{ij}(\boldsymbol{r}', t')N_{k\ell}(\boldsymbol{r}'', t'')\rangle\rangle_\tau \exp\biggl(-\frac{\tau^2}{\tau_0^2(\boldsymbol{\lambda})}\biggr), \quad (13.15)$$

where $\tau_0(\boldsymbol{\lambda})$ is a mean correlation time and depends in general on $\boldsymbol{\lambda}$. The subscripts 0 and τ attached to $\langle\langle\ \rangle\rangle$ denote, respectively, a τ-indenpendent and τ-dependent terms. Then, performing the integration with respect to τ in equation (13.14), we have

$$\langle\langle |A|^2\rangle\rangle E = \frac{\pi^{1/2}\omega^2}{8\eta E}\int\int d^3r_0 d^3\lambda \exp\biggl(-\frac{1}{4}\omega^2\tau_0^2(\boldsymbol{\lambda})\biggr)\tau_0(\boldsymbol{\lambda})T, \quad (13.16)$$

where T is given by

$$T = \frac{\partial\hat{\xi}_i(\boldsymbol{r}')}{\partial r_j'}\biggl\langle\biggl\langle \hat{N}_{ij}(\boldsymbol{r}', t')\hat{N}_{k\ell}(\boldsymbol{r}'', t'')\biggr\rangle\biggr\rangle_\tau \frac{\partial\hat{\xi}_k(\boldsymbol{r}'')}{\partial r_\ell''}. \quad (13.17)$$

It should be noted that the τ-independent term disappears in equation (13.16).

As shown in equation (13.4), $\hat{N}_{ij}$ consists of hydrodynamical and hydromagnetic terms. Here, we consider only the latter term, because MHD turbulence will be important in accretion disks. In both cases of hydrodynamical and magnetic turbulences, however, the amplitude of excited oscillations and its parameter dependences will be essentially similar, as fas as turbulent Alfvén speed, $(\langle\langle b^2\rangle\rangle/4\pi\rho_0)^{1/2}$, and turbulent fluid speed, $v_{\rm turb}$, are comparable.

In the case of the magnetic turbulence, we have (see equation (13.4))

$$\frac{1}{\rho_0(\boldsymbol{r}')\rho_0(\boldsymbol{r}'')}\left\langle\left\langle\hat{N}_{ij}(\boldsymbol{r}',t')\hat{N}_{k\ell}(\boldsymbol{r}'',t'')\right\rangle\right\rangle_\tau = \frac{1}{4}\left\langle\left\langle(\hat{b})_1^2(\hat{b})_2^2\right\rangle\right\rangle_\tau \delta_{ij}\delta_{k\ell}$$
$$-\frac{1}{2}\left\langle\left\langle(\hat{b})_1^2(\hat{b}_k\hat{b}_\ell)_2\right\rangle\right\rangle_\tau \delta_{ij} - \frac{1}{2}\left\langle\left\langle(\hat{b}_i\hat{b}_j)_1(\hat{b})_2^2\right\rangle\right\rangle_\tau \delta_{k\ell}$$
$$+\left\langle\left\langle(\hat{b}_i\hat{b}_j)_1(\hat{b}_k\hat{b}_\ell)_2\right\rangle\right\rangle_\tau, \tag{13.18}$$

where the subscripts 1 and 2 show the values at $(\boldsymbol{r}',t')$ and $(\boldsymbol{r}'',t'')$, respectively, and $\hat{b}_i$ is turbulent Alfven speed defined by $\hat{b}_i = b_i/(4\pi\rho_0)^{1/2}$.

As shown in equation (13.18) we need fourth-order correlations of turbulent quantities. A simple way for approximately evaluating fourth-order quantities is to introduce the quasi-normal distribution of turbulent quantities, which permits us to express fourth order moments in terms of second order ones. For example, we have

$$\langle\langle(\hat{b}_i\hat{b}_j)_1(\hat{b}_k\hat{b}_\ell)_2\rangle\rangle = \langle\langle(\hat{b}_i\hat{b}_j)_1\rangle\rangle\langle\langle(\hat{b}_k\hat{b}_\ell)_2\rangle\rangle$$
$$+\langle\langle(\hat{b}_i)_1(\hat{b}_k)_2\rangle\rangle\langle\langle(\hat{b}_j)_1(\hat{b}_\ell)_2\rangle\rangle + \langle\langle(\hat{b}_i)_1(\hat{b}_\ell)_2\rangle\rangle\langle\langle(\hat{b}_j)_1(\hat{b}_k)_2\rangle\rangle. \tag{13.19}$$

It is noted that in evaluation of $\langle\langle(\hat{b}_i\hat{b}_j)_1(\hat{b}_k\hat{b}_\ell)_2\rangle\rangle_\tau$, the first term on the right-hand side of equation (13.19) has no contribution, because the term is a τ-independent term.

To perform the volume integration in the λ-space in equation (13.16), we need to know the λ-dependences of $\langle\langle\hat{N}_{ij}(\boldsymbol{r}',t')\hat{N}_{k\ell}(\boldsymbol{r}'',t'')\rangle\rangle_\tau$ and of $\tau_0(\boldsymbol{\lambda})$. That is, we need spectra of turbulence and decay times of spectrum components. This is complicated problem, and beyond the scope of this chapter. We shall be satisfied simply by considering only energy-containing eddy. Its size and the correlation time are taken, respectively, to be $\lambda_{\rm eddy}$ and $\lambda_{\rm eddy}/\langle\langle\hat{b}^2\rangle\rangle^{1/2}$. We assume that $\lambda_{\rm eddy}\sim H$ (H being disk half thickness), when we estimate the order of amplitude of excited oscillations.

After the above preparations, we separately consider p-mode oscillations, and c and vertical p-mode oscillations.

13.2.2.1 p-Mode Oscillations

Let us denote the order of magnitude of $\hat{\xi}_r(\boldsymbol{r})$ by $\hat{\xi}$, and the order of the radial wavelength of oscillations by L. Then, the order of $\partial\xi_r/\partial r$ is ξ/L. The order of $\hat{\xi}_\varphi$ is the same as that of $\hat{\xi}_r$, but $\hat{\xi}_z$ is roughly zero in the case of p-mode oscillations. Considering them, we can approximately estimate the volume integration of T, using equations (13.17), (13.18), and (13.19), as

$$\int Td^3\lambda \sim \left(\frac{\xi}{L}\right)^2 \left(\rho_0\langle\langle\hat{b}^2\rangle\rangle\right)^2 H\lambda_{\rm eddy}^2 \sim \rho_0\langle\hat{b}^2\rangle \left(\frac{\xi}{L}\right)^2 \left(\frac{\lambda_{\rm eddy}}{D}\right)^2 E_{\rm turb}, \tag{13.20}$$

where $E_{\rm turb}$ is the turbulent energy in the wave propagation region and has been taken as $E_{\rm turb} \sim \rho_0\langle\hat{b}^2\rangle HD^2$, where D is the radial extend of the wave propagation region, which is comparable with L since we are interesting in trapped oscillations.

The order of the wave energy, $|E|$, is $\bar{\rho}_0\omega^2\xi^2HD^2$, where $\bar{\rho}_0$ represents an average value of ρ_0 in the wave propagation region. This wave energy is further estimated to be $\bar{\rho}_0c_{\rm s}^2\xi^2(D^2/H)$, when we consider oscillations whose frequency, ω, is on the oder of angular velocity of disk rotation, $\Omega_{\rm K}$, since $\Omega_{\rm K}H \sim c_{\rm s}$ ($c_{\rm s}$ being acoustic speed). Using equation (13.20) and the above estimate of magnitudes of various quantities, we have, from equations (13.16),

$$\frac{\langle\langle|A|^2\rangle\rangle|E|}{E_{\rm turb}} \sim \alpha\frac{\omega}{\eta}(\omega\tau_0)\exp\left(-\frac{1}{4}\omega^2\tau_0^2\right)\frac{\langle\langle\hat{b}^2\rangle\rangle}{c_{\rm s}^2}\left(\frac{\lambda_{\rm eddy}}{D}\right)^2\left(\frac{H}{L}\right)^2, \tag{13.21}$$

where α is a dimensionless constant which will be moderately smaller than unity.

Considering that $\tau_0 \sim \lambda_{\rm eddy}/\langle\langle\hat{b}^2\rangle\rangle^{1/2}$, we have $\omega\tau_0 \sim c_{\rm s}/\langle\langle\hat{b}^2\rangle\rangle^{1/2}(\lambda_{\rm eddy}/H)$, when the frequency, ω, of oscillations is close to $\Omega_{\rm K}$. That is, $\omega\tau_0$ can be taken to be around unity. If η is taken to be the inverse of the viscous time scale, we have $\eta \sim \langle\langle\hat{b}^2\rangle\rangle^{1/2}\lambda_{\rm eddy}/L^2$ since turbulent kinematic viscosity is $\sim \langle\langle\hat{b}^2\rangle\rangle^{1/2}\lambda_{\rm eddy}$. This gives $\omega/\eta \sim c_{\rm s}/\langle\langle\hat{b}^2\rangle\rangle^{1/2}(L^2/H\lambda_{\rm eddy})$. These consideration leads to

$$\frac{\langle\langle|A|^2\rangle\rangle|E|}{E_{\rm turb}} \sim \alpha\left(\frac{\lambda_{\rm eddy}}{D}\right)^2, \tag{13.22}$$

where α is an another dimensionless constant (different from α in equation (13.21)), which will be again moderately smaller than unity.

In the innermost region of relativistic disks, p-mode oscillations whose frequencies are on the order of Keplerian frequency in that region are trapped, if the inner edge of disks is a reflection boundary of the oscillations (see Chaps. 6 and 7). Equation (13.22) suggests that such oscillations can be sustained by turbulence, although their oscillation energy is much smaller than the turbulent energy by a factor of $\alpha(\lambda_{\rm eddy}/D)^2$.

13.2.2.2 c-Mode and Vertical p-Mode Oscillations

In oscillations other than p-modes, $\hat{\xi}_z$ is not negligible and comparable with or larger than $\hat{\xi}_r$ and $\hat{\xi}_\varphi$ in magnitude. Furthermore, $\hat{\xi}_r$, $\hat{\xi}_\varphi$ and $\hat{\xi}_z$ have z-dependences, i.e., $\hat{\xi}_r$ and $\hat{\xi}_\varphi$ are proportional to $\mathscr{H}_n(z/H)$ and $\hat{\xi}_z$ is proportional to $\mathscr{H}_{n-1}(z/H)$, where $\mathscr{H}_n(z/H)$ is the Hermite polynomials of order of n ($n \geq 1$) with argument z/H.

Hence, for example, $\partial\hat{\xi}_r/\partial z$ does not vanish and its magnitude is larger than that of $\partial\hat{\xi}_r/\partial r$ by the factor of L/H. Hence, we can expect that the volume integration of T is larger than that given by equation (13.20) by the factor of $(L/H)^2$.

Next, we should consider the term, η, representing turbulent viscous damping. In the case of oscillations with $n \neq 0$, the damping of oscillations due to viscosity is stronger than that in the case of p-mode oscillations, because the characteristic wavelength of oscillations is the vertical wavelength, which is on the order of H. Hence, η is larger than that adopted in the case of p-mode oscillations by a factor of $(L/H)^2$.

The two effects mentioned above act in the opposite directions when the wave amplitude of c-mode and vertical p-mode oscillations is considered. As the results, the amplitude of c-mode and vertical p-mode oscillations expected by stochastic processes might be comparable with that of p-mode oscillations.

The above considerations are based on knowledge of isotropic turbulence. In the case of magnetic turbulence in accretion disks, the origin of turbulence is magneto-rotational instability (MRI), and the turbulence may be anisotropic. Effects of anisotropic turbulence should be also considered in estimate of η. Thus, further studies will be necessary to derive more reliable amplitudes of oscillations.

13.3 Final Remarks

Before closing this final chapter, we shall briefly summarize author's personal views concerning what excitation processes can contribute to what kinds of oscillations, and what observed oscillatory phenomena are due to what kinds of disk oscillations.

In Table 13.1 we list excitation processes and oscillation modes which can be excited by them. For example, (i) viscous processes can excite p-mode oscillations (viscous overstability). The reason why p-mode oscillations are excited is that

Table 13.1 Excitation processes and oscillations modes excited.

Excitation process	Oscillation excited
Viscous overstability	p-modes; c-modes in tori (uncertain)
Corotation resonance	Non-axisymmetric ($m \neq 0$) p-modes
Wave-wave resonance (disk deformation)	
Tidal deformation	One-armed ($m = 1$) p-mode (precession mode)
	One-armed ($m = 1$) c-mode (tilt mode)
Warp ($m_D = 1, n_D = 1$)	A set of p-mode ($m = 1$) and g-mode ($m = 0$)
Two-armed deformation ($m_D = 2, n_D = 0$)	A set c-mode ($m = 2$) and g-mode ($m = 0$)
Two-armed deformation ($m_D = 2, n_D = 1$)	A set of p-mode ($m = 2$) and g-mode ($m = 0$)
	A set of vertical p-mode ($m = 2, n = 2$) and g-mode ($m = 0$)
Sonic point instability	p-modes
Stochastic processes (turbulence)	Many modes

the excitation process due to shear motions overcome the conventional viscous damping process. In other oscillation modes, however, situations are changed and they are damped. However, if turbulence is strongly anisotropic or disks are geometrically thick (i.e., tori), c-mode oscillations might be also excited in addition to p-mode oscillations. (ii) Corotation resonances are powerful excitation processes, but what they can excite are non-axisymmetric p-modes alone. Non-axisymmetric oscillations other than p-modes (i.e., g-modes, c-mode, and vertical p-modes) are damped by corotation resonance. (iii) For wave-wave resonant processes to work, disk deformation is necessary. What kinds of disk oscillations are excited depend on what kinds of disk deformations are expected. If various types of disk deformations are possible, many kinds of oscillations can be excited. (iv) Stochastic processes will excite many oscillation modes, but there is no detailed analytical examination up to the present.

Table 13.2 summarizes what excitation processes are possible candidates of what observed oscillatory phenomena.

Table 13.2 Periodic (quasi-periodic) phenomena in astrophysical objects, their interpretation by oscillation modes, and their possible excitation processes.

Oscillatory phenomena	Oscillation modes	Possible excitation processes
Be stars		
V/R variations	One-armed p-mode	Viscous overstability (Chap. 9)
Dwarf Novae		
Superhumps	One-armed p-mode	Tidal deformation (Chap. 12)
Negative superhumps	Tilt mode[a]	Tidal deformation (Chap. 12)
LMXBs		
kHz QPOs (NS)		
Correlation of twin QPOs	c-mode	Disk deformation (Chap. 12)
		Stochastic process (Chap. 13)
Order of frequencies	c-mode	Disk deformation (Chap. 12)
	p-modes	Viscous overstability (Chap. 9)
		Corotation resonance (Chap. 10)
		Disk deformation (Chap. 12)
HF QPOs (BH)		
3:2 pair	p-mode	Viscous overstability (Chap. 9)
		Corotation (Chap. 10)
		Disk deformation (Chap. 12)
	c-mode	Disk deformation (Chap. 12)
Frequencies	p-modes	Viscous overstability (chap. 9)
		Corotation (Chap. 10)
		Disk deformation (Chap. 12)
	c-mode	Disk deformation (Chap. 12)

Note: Stochastic processes are not mentioned except for a case in this table, but they will be one of possible candidates of excitation of many oscillatory phenomena
[a]Tilt mode is one-armed c-mode oscillation with $n = 1$

(i) V/R variations in Be stars

The V/R variations observed in Be stars will be due to excitation of one-armed p-mode oscillation (one-armed precession mode) in Be-star disks. The mode can be excited by viscous overstability (see Chap. 9). Observations show that in many objects the one-armed oscillations required to describe the observed V/R variation are prograde ones, although there seem to be exceptional cases (Chap. 1). A simplified analysis, however, shows that discrete one-armed oscillation modes are retrograde and not prograde (Chap. 8). For one-armed p-mode oscillations to have prograde precession, the oscillations need to be global in the sense that the effects of increase of disk thickness with radius are non-negligible in their propagation (Chap. 8).

(ii) Superhumps in dwarf novae

Positive superhumps observed in dwarf novae are recognized to be one-armed p-mode oscillations (one-armed precession mode (Osaki 1985)) excited by tidal instability (Whitehurst 1988a,b; Hirose and Osaki 1990; Lubow 1991) (Chap. 7). Lubow (1991) showed that the excitation is due to a mode-mode coupling process in tidally deformed disks. This excitation process can be interpreted as one of examples of more general wave-wave resonant instabilities in deformed disks (Chaps. 11 and 12).

Concerning negative superhumps, many possible origins have been proposed. One of possibilities is that they are tilt modes excited by the mode-mode coupling processes in tidally deformed disks (Lubow 1992). The wave-wave resonant processes (Chaps. 11 and 12) support this possibility and further suggest that tilt modes can be excited in various wave-wave resonances (not restricted to the so-called 3:1 resonance) (Chaps. 11 and 12). Distinct from the excitation of one-armed precession mode which acts at a particular stage of disk expansion, the excitation of tilt modes by resonance processes continues to work during a finite interval of disk expansion (Chap. 12).

(iii) KHz QPOs in neutron-star X-ray binaries

Origins of kHz QPOs in neutron-star X-ray binaries and HFQPOs in black-hole X-ray binaries are still in debate. Assuming that these QPOs are due to disk oscillations, we summarize our arguments about their possible origins.

First, we consider kHz QPOs. We pay our attention to two observational characteristics. One is correlated changes of twin QPOs, and the other is the order of frequencies. First, kHz QPOs often appear in pairs and their frequencies change with time. The change, however, occurs so that it is along a common curve on the frequency-frequency diagram (see Chap. 1 and Fig. 1.3). This correlated frequency change of twin QPOs can be better described by regarding them as two c-mode oscillations rather than by regarding them as two p-mode oscillations, as long as conventional parameters of neutron stars are adopted (see Chaps. 7 and 8). In other words, if the QPOs could come from the p-mode oscillations, their frequencies would be rather robust and might not change in a wide frequency range.

An issue to be worried about c-mode oscillations is their excitation. The excitation processes of c-mode oscillations are rather restricted compared with those of p-mode ones. We think that promising processes of their excitation will be wave-wave resonant instabilities in deformed disks (Chaps. 11 and 12), and stochastic processes of turbulence (Chap. 13).

The second point which we should take attention is their frequencies. As long as the frequencies are concerned, p-mode oscillations are good candidates of kHz QPOs as well as c-mode oscillations are. In the case of p-mode oscillations, there are many possible excitation processes as shown in Table 13.2.

(iv) HFQPOs in black-hole X-ray binaries

Any attempts interpreting HFQPOs as disk oscillations are not yet successful. Some BH sources have twin QPOs whose frequency ratio is close to 3:2, unchanged with time, although the number of sources with twin QPOs is restricted (see Chap. 1). By regarding this frequency ratio as essential, Kluźniak and Abramowicz (2001) and Abramowicz and Kluzńiak (2001) proposed a 3:2 resonance model. This model is quite attractive and had many contributions to subsequent developments in this field, but excitation of two oscillations with 3:2 frequency ratio seems to have no particular preference from theoretical points of view. We suppose that the appearance of the 3:2 frequency ratio is in chance. As shown in Fig. 7.2, the frequency ratio of the $m = 3$ mode to $m = 2$ one is close to 3:2 (see also Lai and Tsang 2009). The p-mode oscillations may be excited by many processes. Typical ones are viscous overstability (Chap. 9) and coronation resonance (Chap. 10).

One of important points where we should pay attention is that some black-hole sources with HFQPOs have extremely high spins. That is, comparisons between observations (K-line of Fe and continuous X-ray spectrum) and theoretical disk models show that some BH sources have extremely high spins close to $a_* = 1$ (McClintock et al. 2011). Any successful theoretical QPO models need to be able to describe such high spin cases. It is noted that a set of oscillations (c-mode and axisymmetric g-mode) which are resonantly excited in two-armed deformed disks seem to have possibility to describe a high spin of central sources (Sect. 12.5 and Kato 2012). A problem is whether such disk deformations required are really present in actual disks.

Finally, we should emphasize that in this monograph we have restricted our attention to geometrically thin disks, but this is only for simplicity. To qualitative comparison of frequencies of trapped oscillations with observational QPOs more realistic disk models will be required. Geometrically thick disks or tori where general relativistic effects are fully taken into account should be adopted to examine HFQPOs. Furthermore we like to say that among various possible excitation mechanisms, stochastic processes are attractive, although there are no detailed studies yet.

References

Abramowicz, M. A., & Kluzńiak, W. 2001, Astron. Astrophys., 374, L19
Abramowicz, M. A., & Kato, S. 1989, Astrophys. J., 336, 304
Chen, X., & Taam, R. E. 1995, Astrophys. J., 412, 254
Goldreich, P., & Keeley, D. A. 1977, Astrophys. J., 212, 243
Hirose, M., & Osaki, Y. 1990, Publ. Astron. Soc. Jpn., 42, 135
Honma, F., Matsumoto, R., & Kato, S. 1992, Publ. Astron. Soc. Jpn., 44, 529
Kato, S. 2001, Publ. Astron. Soc. Jpn., 53, 1
Kato, S. 2012, Publ. Astron. Soc. Jpn., 64, 139
Kato, S., Honma, F., & Matsumoto, R. 1988a, Mon. Not. R. Astron. Soc., 185, 629
Kato, S., Honma, F., & Matsumoto, R. 1988b, Publ. Astron. Soc. Jpn., 40, 709
Kato, S., Wu, X. B., Yang, L. T., & Yang, Z. L. 1993, Mon. Not. R. Astron. Soc., 260, 317
Kluźniak, W., & Abramowicz, M. A. 2001, Acta Phys., Pol. B32, 3605
Lighthill, M. 1952, Proc. R. Soc. A-Math. Phys. Eng. Sci., A211, 564
Lubow, S.H. 1992, Astrophys. J., 398, 525
McClintock, J. E., Narayan, R., Davis, S. W., Gou, L., Kulkarni, A., Orosz, J. A., Penna, R. F., Remillard, R. A., & Steiner, J. F. 2011, Class. Quantum Grav., Special Volume for GR19
Matsumoto, R., Kato, S., Fukue, J., & Okazaki, A. T. 1984, Publ. Astron. Soc. Jpn., 36, 71
Matsumoto, R., Kato, S., & Honma, F. 1988, in *Physics of Neutron Stars and Black Holes*, ed. Y. Tanaka (Universal Academy Press, Tokyo), p155
Matsumoto, R., Kato, S., & Honma, F. 1989, in *Theory of Accretion Disks*, ed. F. Meyer, W. J. Duschl, J. Frank, E. Meyer-Hofmeister (Kluwer Academic Publisher, Dordreicht), p167
Milsom, J. A., & Taam, R. E. 1996, Mon. Not. R. Astron. Soc., 283, 919
Milsom, J. A., & Taam, R. E. 1997, Mon. Not. R. Astron. Soc., 286, 358
Miranda, R., Horák, J., & Lai, D., 2015, Mon. Not. R. Astron. Soc., 446, 240
Muchotrzeb, B. 1983, Acta Astron., 33, 79
Muchotrzeb-Czerny, B. 1986, Acta Astron., 36,1
Osaki, Y. 1985, Astron. & Asrtrophys., 144, 369
Poe, C. H., Owicki, S. P., & Castor, J. I. 1990, Astrophys. J., 358, 199
Samadi, R. 2011, in *The Pulsations of the Sun and the Stars*, Lecture Notes in Physics, Volume 832, Springer-Verlag (Berlin Heidelberg) p. 305
Samadi, R., & Goupil, M.-J., 2001, Astron. Astrophys., 370, 136
Whitehurst, R. 1988a, Mon. Not. R. Astron. Soc., 232, 35
Whitehurst, R. 1988b, Mon. Not. R. Astron. Soc., 233, 529

Appendix A
Basic Hydromagnetic Equations Describing Perturbations

First, general forms of hydromagnetic (MHD) equations are summarized. Subsequently, equations describing small amplitude perturbations over steady states are presented, using cylindrical coordinates. The basic equations are presented in the Newtonian forms, although in some cases we are interested in general relativistic disks. In this book, effects of general relativity are taken into account, for simplicity, by introducing general relativistic expressions for epicyclic frequencies (see Appendix B), adopting in other parts the Newtonian expressions.

A.1 General Form

Basic equations under MHD approximations (frozen-in approximations) are summarized.

(a) **Equation of continuity**
The conservation of mass (equation of continuity) is

$$\frac{\partial \rho}{\partial t} + \mathrm{div}(\rho \boldsymbol{v}) = 0, \tag{A.1}$$

where ρ is the density and $\boldsymbol{v}$ the velocity vector. In terms of the Lagrange derivative:

$$\frac{d}{dt} = \frac{\partial}{\partial t} + (\boldsymbol{v} \cdot \nabla), \tag{A.2}$$

the continuity equation (A.1) is expressed as

$$\frac{d\rho}{dt} + \rho\, \mathrm{div}\boldsymbol{v} = 0. \tag{A.3}$$

S. Kato, *Oscillations of Disks*, Astrophysics and Space Science Library 437,
DOI 10.1007/978-4-431-56208-5

(b) **Equation of motion**

The equation of motion is described as

$$\rho\frac{d\boldsymbol{v}}{dt}=\rho\left[\frac{\partial\boldsymbol{v}}{\partial t}+(\boldsymbol{v}\cdot\nabla)\boldsymbol{v}\right]=-\rho\nabla\psi-\nabla p+\frac{1}{4\pi}\mathrm{curl}\,\boldsymbol{B}\times\boldsymbol{B}+\rho\boldsymbol{N}+\rho\boldsymbol{K}, \tag{A.4}$$

where ψ is the gravitational potential, p the pressure, $\boldsymbol{B}$ the magnetic flux density, $\boldsymbol{N}$ the viscous force per unit mass, and $\boldsymbol{K}$ the external force per unit mass.

The term of $(1/4\pi)\mathrm{curl}\,\boldsymbol{B}\times\boldsymbol{B}$ in equation (A.4) represents the electromagnetic force (the Lorentz force) and the time variation of $\boldsymbol{B}$ is governed by the induction equation, which is written below. The viscous force $\boldsymbol{N}$ is expressed as

$$\rho N_i=\frac{\partial t_{ik}}{\partial x_k}, \tag{A.5}$$

where t_{ik} is the viscous stress tensor:

$$t_{ik}\equiv\eta\left(\frac{\partial v_i}{\partial x_k}+\frac{\partial v_k}{\partial x_i}-\frac{2}{3}\delta_{ik}\frac{\partial v_j}{\partial x_j}\right)+\zeta\delta_{ik}\frac{\partial v_j}{\partial x_j}, \tag{A.6}$$

η being the dynamical viscosity (first viscosity) and ζ the bulk viscosity (second viscosity) (the latter is usually ignored).

The equation of motion can be expressed explicitly as

$$\rho\frac{dv_i}{dt}=\rho\left(\frac{\partial}{\partial t}+v_j\frac{\partial}{\partial x_j}\right)v_i=-\rho\frac{\partial\psi}{\partial x_i}-\frac{\partial p}{\partial x_i}+\rho K_i-\frac{1}{8\pi}\frac{\partial B^2}{\partial x_i}+\frac{1}{4\pi}B_j\frac{\partial B_i}{\partial x_j}$$
$$+\frac{\partial}{\partial x_k}\left[\eta\left(\frac{\partial v_i}{\partial x_k}+\frac{\partial v_k}{\partial x_i}-\frac{2}{3}\delta_{ik}\frac{\partial v_j}{\partial x_j}\right)+\zeta\delta_{ik}\frac{\partial v_j}{\partial x_j}\right]. \tag{A.7}$$

(c) **Induction equation**

The time variation of magnetic field $\boldsymbol{B}$ is governed by the induction equation:

$$\frac{\partial\boldsymbol{B}}{\partial t}=\mathrm{curl}(\boldsymbol{v}\times\boldsymbol{B})+\eta\nabla\boldsymbol{B}, \tag{A.8}$$

where $\eta[\equiv c^2/(4\pi\sigma_\mathrm{e}))]$ is the magnetic diffusivity.

(d) **Energy equation**

The conservation of thermal energy is described as

$$\rho T\frac{ds}{dt}=\rho T\left[\frac{\partial s}{\partial t}+(\boldsymbol{v}\cdot\nabla)s\right]=\Phi+\rho\epsilon-\mathrm{div}\boldsymbol{F}+\frac{J^2}{\sigma_\mathrm{e}}, \tag{A.9}$$

where T is the temperature, s the specific entropy, Φ the viscous dissipative function, ϵ the other heating rate (such as nuclear energy) per unit mass (i.e., $\rho\epsilon$ per unit volume), and $\boldsymbol{F}$ the energy flux (which includes, e.g., the radiative flux $\boldsymbol{F}_{\rm rad}$, the convective one $\boldsymbol{F}_{\rm conv}$, and the conductive one $\boldsymbol{F}_{\rm cond}$). The last term on the right-hand side of equation (A.9) is the Joule heating.

Using the first law of thermodynamics: $Tds = dU + pdV$, where U the internal energy per unit mass, we can rewrite the left-hand side of the energy equation (A.9) as $\rho Tds/dt = \rho dU/dt + p\,{\rm div}\,\boldsymbol{v}$. For an ideal gas, moreover, the internal energy U is expressed as $U = [1/(\gamma-1)](p/\rho)$, where $\gamma = C_p/C_v$ is the ratio of the specific heats. Hence, the left-hand side of the energy equation (A.9) becomes, with the help of the continuity equation (A.1),

$$\rho T\frac{ds}{dt} = \frac{1}{\gamma-1}\left(\frac{dp}{dt} - \gamma\frac{p}{\rho}\frac{d\rho}{dt}\right). \tag{A.10}$$

The viscous dissipative function Φ, which expresses the viscous heating rate per unit volume, is expressed as

$$\Phi = t_{ik}\frac{\partial v_i}{\partial x_k} = \eta\frac{1}{2}\left(\frac{\partial v_i}{\partial x_k} + \frac{\partial v_k}{\partial x_i}\right)^2 + \left(\zeta - \frac{2}{3}\eta\right)\left(\frac{\partial v_j}{\partial x_j}\right)^2. \tag{A.11}$$

The energy flux $\boldsymbol{F}$ is generally written in the form:

$$\boldsymbol{F} = -K\,{\rm grad}T, \tag{A.12}$$

where K is the 'conductivity'. For example, in optically thick regime, the radiative conductive flux $\boldsymbol{F}_{\rm rad}$ is

$$\boldsymbol{F}_{\rm rad} = -\frac{4acT^3}{3\bar{k}\rho}\,{\rm grad}T, \tag{A.13}$$

where a the radiation constant and $\bar{k}$ the (Rosseland-mean) total opacity.

Thus, the energy equation is finally expressed as

$$\frac{1}{\gamma-1}\left(\frac{dp}{dt} - \gamma\frac{p}{\rho}\frac{d\rho}{dt}\right) = \Phi + \rho\epsilon + {\rm div}\,(K\,{\rm grad}T) + \frac{J^2}{\sigma_{\rm e}}. \tag{A.14}$$

(e) **Equation of state**

The equation of state of optically thick systems is

$$p = \frac{\mathscr{R}}{\bar{\mu}}\rho T + \frac{1}{3}aT^4, \tag{A.15}$$

where $\mathscr{R}$ is the gas constant and $\bar{\mu}$ the mean molecular weight.

A.1.1 Expressions by Cylindrical Coordinates

We focus our attention only on geometrically thin accretion disks. Hence, for the convenience of the readers, we explicitly write down the basic equations in cylindrical coordinates (r, φ, z), where the z-axis is the axis of disk rotation.

(a) **Equation of continuity**

$$\frac{\partial \rho}{\partial t} + \frac{\partial}{r\partial r}(r\rho v_r) + \frac{\partial}{r\partial \varphi}(\rho v_\varphi) + \frac{\partial}{\partial z}(\rho v_z) = 0, \tag{A.16}$$

where (v_r, v_φ, v_z) are the components of velocity in the cylindrical coordinates.

(b) **Equation of motion**

$$\frac{\partial v_r}{\partial t} + v_r\frac{\partial v_r}{\partial r} + v_\varphi\frac{\partial v_r}{r\partial \varphi} + v_z\frac{\partial v_r}{\partial z} - \frac{v_\varphi^2}{r} = -\frac{\partial \psi}{\partial r} - \frac{1}{\rho}\frac{\partial p}{\partial r} + F_{\mathrm{B}r} + N_r + K_r, \tag{A.17}$$

$$\frac{\partial v_\varphi}{\partial t} + v_r\frac{\partial v_\varphi}{\partial r} + v_\varphi\frac{\partial v_\varphi}{r\partial \varphi} + v_z\frac{\partial v_\varphi}{\partial z} + \frac{v_r v_\varphi}{r} = -\frac{\partial \psi}{r\partial \varphi} - \frac{1}{\rho}\frac{\partial p}{r\partial \varphi} + F_{\mathrm{B}\varphi} + N_\varphi + K_\varphi, \tag{A.18}$$

$$\frac{\partial v_z}{\partial t} + v_r\frac{\partial v_z}{\partial r} + v_\varphi\frac{\partial v_z}{r\partial \varphi} + v_z\frac{\partial v_z}{\partial z} = -\frac{\partial \psi}{\partial z} - \frac{1}{\rho}\frac{\partial p}{\partial z} + F_{\mathrm{B}z} + N_z + K_z, \tag{A.19}$$

where

$$F_{\mathrm{B}r} = \frac{1}{4\pi}[\mathrm{rot}\,\boldsymbol{B}\times\boldsymbol{B}]_r = \frac{1}{4\pi}\left[\left(\frac{\partial B_r}{\partial z} - \frac{\partial B_z}{\partial r}\right)B_z - \left(\frac{\partial}{r\partial r}rB_\varphi - \frac{\partial B_r}{r\partial \varphi}\right)B_\varphi\right], \tag{A.20}$$

$$F_{\mathrm{B}\varphi} = \frac{1}{4\pi}[\mathrm{rot}\,\boldsymbol{B}\times\boldsymbol{B}]_\varphi = \frac{1}{4\pi}\left[\left(\frac{\partial}{r\partial r}rB_\varphi - \frac{\partial B_r}{r\partial \varphi}\right)B_r - \left(\frac{\partial B_z}{r\partial \varphi} - \frac{\partial B_\varphi}{\partial z}\right)B_z\right], \tag{A.21}$$

$$F_{\mathrm{B}z} = \frac{1}{4\pi}[\mathrm{rot}\,\boldsymbol{B}\times\boldsymbol{B}]_z = \frac{1}{4\pi}\left[\left(\frac{\partial B_z}{r\partial \varphi} - \frac{\partial B_\varphi}{\partial z}\right)B_\varphi - \left(\frac{\partial B_r}{\partial z} - \frac{\partial B_z}{\partial r}\right)B_r\right], \tag{A.22}$$

and

$$\rho N_r = \frac{1}{r}\frac{\partial}{\partial r}(rt_{rr}) + \frac{1}{r}\frac{\partial t_{r\varphi}}{\partial \varphi} - \frac{t_{\varphi\varphi}}{r} + \frac{\partial t_{rz}}{\partial z}, \tag{A.23}$$

$$\rho N_\varphi = \frac{1}{r^2}\frac{\partial}{\partial r}(r^2 t_{r\varphi}) + \frac{1}{r}\frac{\partial t_{\varphi\varphi}}{\partial \varphi} + \frac{\partial t_{z\varphi}}{\partial z}, \tag{A.24}$$

$$\rho N_z = \frac{1}{r}\frac{\partial}{\partial r}(rt_{rz}) + \frac{1}{r}\frac{\partial t_{\varphi z}}{\partial \varphi} + \frac{\partial t_{zz}}{\partial z}, \tag{A.25}$$

the viscous stress tensor t_{ik} being given by

$$t_{rr} = 2\eta\frac{\partial v_r}{\partial r} + \left(\zeta - \frac{2}{3}\eta\right)\mathrm{div}\boldsymbol{v}, \tag{A.26}$$

$$t_{r\varphi} = t_{\varphi r} = \eta\left[r\frac{\partial}{\partial r}\left(\frac{v_\varphi}{r}\right) + \frac{1}{r}\frac{\partial v_r}{\partial \varphi}\right], \tag{A.27}$$

$$t_{rz} = t_{zr} = \eta\left(\frac{\partial v_z}{\partial r} + \frac{\partial v_r}{\partial z}\right), \tag{A.28}$$

$$t_{\varphi\varphi} = 2\eta\left(\frac{1}{r}\frac{\partial v_\varphi}{\partial \varphi} + \frac{v_r}{r}\right) + \left(\zeta - \frac{2}{3}\eta\right)\mathrm{div}\boldsymbol{v}, \tag{A.29}$$

$$t_{\varphi z} = t_{z\varphi} = \eta\left(\frac{\partial v_\varphi}{\partial z} + \frac{1}{r}\frac{\partial v_z}{\partial \varphi}\right), \tag{A.30}$$

$$t_{zz} = 2\eta\frac{\partial v_z}{\partial z} + \left(\zeta - \frac{2}{3}\eta\right)\mathrm{div}\boldsymbol{v}, \tag{A.31}$$

$$\mathrm{div}\boldsymbol{v} = \frac{1}{r}\frac{\partial}{\partial r}(rv_r) + \frac{1}{r}\frac{\partial v_\varphi}{\partial \varphi} + \frac{\partial v_z}{\partial z}. \tag{A.32}$$

For the accretion disk, only the $r\varphi$-component of the viscous stress tensor, $t_{r\varphi}$, is dominant; therefore,

$$\rho N_\varphi = \frac{1}{r^2}\frac{\partial}{\partial r}\left(r^2 t_{r\varphi}\right) = \frac{1}{r^2}\frac{\partial}{\partial r}\left[\eta r^3\frac{\partial}{\partial r}\left(\frac{v_\varphi}{r}\right)\right], \tag{A.33}$$

where we introduce the angular speed $\Omega = v_\varphi/r$. It should be noted that the alpha prescription means that $t_{r\varphi} = \eta r d\Omega/dr = -\alpha p$.

(c) **Induction equation**

In the MHD approximation, we have

$$\frac{\partial B_r}{\partial t} = \frac{\partial}{r\partial\varphi}(v_r B_\varphi - v_\varphi B_r) - \frac{\partial}{\partial z}(v_z B_r - v_r B_z), \tag{A.34}$$

$$\frac{\partial B_\varphi}{\partial t} = \frac{\partial}{\partial z}(v_\varphi B_z - v_z B_\varphi) - \frac{\partial}{\partial r}(v_r B_\varphi - v_\varphi B_r), \tag{A.35}$$

$$\frac{\partial B_z}{\partial t} = \frac{\partial}{r\partial r}[r(v_z B_r - v_r B_z)] - \frac{\partial}{r\partial\varphi}.(v_\varphi B_z - v_z B_\varphi), \tag{A.36}$$

(d) **Energy equation**

$$\frac{1}{\gamma-1}\Bigg[\left(\frac{\partial}{\partial t}+v_r\frac{\partial}{\partial r}+v_\varphi\frac{\partial}{r\partial\varphi}+v_z\frac{\partial}{\partial z}\right)p$$
$$-\gamma\frac{p}{\rho}\left(\frac{\partial}{\partial t}+v_r\frac{\partial}{\partial r}+v_\varphi\frac{\partial}{r\partial\varphi}+v_z\frac{\partial}{\partial z}\right)\rho\Bigg]$$
$$=\Phi+\rho\varepsilon+\frac{\partial}{r\partial r}\left(rK\frac{\partial T}{\partial r}\right)$$
$$+\frac{\partial}{r\partial\varphi}\left(K\frac{\partial T}{r\partial\varphi}\right)+\frac{\partial}{\partial z}\left(K\frac{\partial T}{\partial z}\right)+\frac{j^2}{\sigma_c}. \tag{A.37}$$

In cylindrical coordinates the viscous dissipative function Φ is expressed as

$$\Phi=\eta\Bigg[2\left(\frac{\partial v_r}{\partial r}\right)^2+2\left(\frac{\partial v_\varphi}{r\partial\varphi}+\frac{\partial v_r}{\partial r}\right)^2+2\left(\frac{\partial v_z}{\partial z}\right)^2$$
$$+\left(\frac{\partial v_\varphi}{r\partial\varphi}+\frac{\partial v_\varphi}{\partial r}-\frac{v_\varphi}{r}\right)^2+\left(\frac{\partial v_\varphi}{\partial z}+\frac{\partial v_z}{r\partial\varphi}\right)^2+\left(\frac{\partial v_z}{\partial r}+\frac{\partial v_r}{\partial z}\right)^2\Bigg]$$
$$+\left(\zeta-\frac{2}{3}\eta\right)\left[\frac{\partial(rv_r)}{r\partial r}+\frac{\partial v_\varphi}{r\partial\varphi}+\frac{\partial v_z}{\partial z}\right]^2. \tag{A.38}$$

In the case of accretion disks where the shear due to disk rotation dominates over others, we have

$$\Phi=t_{r\varphi}\left(\frac{\partial v_\varphi}{\partial r}-\frac{v_\varphi}{r}\right)=\eta\left(\frac{\partial v_\varphi}{\partial r}-\frac{v_\varphi}{r}\right)^2=\eta r^2\left[\frac{\partial}{\partial r}\left(\frac{v_\varphi}{r}\right)\right]^2. \tag{A.39}$$

A.2 Equations Describing Small Amplitude Disk Oscillations

Small-amplitude perturbations are superposed on steady disks. The velocity perturbation over rotation is denoted by $(u_r,\ u_\varphi,\ u_z)$. If the unperturbed flow is rotation alone, we have

$$(v_r,v_\varphi,v_z)=(0,r\Omega,0)+(u_r,u_\varphi,u_z). \tag{A.40}$$

Furthermore, if we assume that in the unperturbed state the magnetic fields are purely toroidal, and the perturbed part is denoted by (b_r, b_φ, b_z), we have

$$(B_r, B_\varphi, B_z) = (0, B_0, 0) + (b_r, b_\varphi, b_z). \tag{A.41}$$

Then, basic equations describing adiabatic and inviscid motions are given as shown below.

(a) **Equation of continuity**

$$\left(\frac{\partial}{\partial t} + \Omega\frac{\partial}{\partial\varphi}\right)\rho_1 + \frac{\partial}{r\partial r}(r\rho_0 u_r) + \frac{\partial}{r\partial\varphi}(\rho_0 u_\varphi) + \frac{\partial}{\partial z}(\rho_0 u_z) = 0. \tag{A.42}$$

(b) **Equation of motion**
The r-, φ-, and z-components of the linearized equation of motions are written, respectively, as

$$\left(\frac{\partial}{\partial t} + \Omega\frac{\partial}{\partial\varphi}\right)u_r - 2\Omega u_\varphi$$
$$= -\frac{1}{\rho_0}\frac{\partial}{\partial r}\left(p_1 + \frac{B_0 b_\varphi}{4\pi}\right) + \frac{B_0}{4\pi\rho_0}\left(\frac{\partial b_r}{r\partial\varphi} - \frac{2b_\varphi}{r}\right) + \frac{\rho_1}{\rho_0^2}\left[\frac{\partial}{\partial r}\left(p_0 + \frac{B_0^2}{8\pi}\right) + \frac{B_0^2}{4\pi r}\right], \tag{A.43}$$

$$\left(\frac{\partial}{\partial t} + \Omega\frac{\partial}{\partial\varphi}\right)u_\varphi + \frac{\kappa^2}{2\Omega}u_r$$
$$= -\frac{1}{\rho_0}\frac{\partial}{r\partial\varphi}\left(p_1 + \frac{B_0 b_\varphi}{4\pi}\right) + \frac{B_0}{4\pi\rho_0}\left(\frac{\partial b_\varphi}{r\partial\varphi} + \frac{b_r}{r}\right) + \frac{1}{4\pi\rho_0}\left(b_r\frac{\partial}{\partial r} + b_z\frac{\partial}{\partial z}\right)B_0, \tag{A.44}$$

$$\left(\frac{\partial}{\partial t} + \Omega\frac{\partial}{\partial\varphi}\right)u_z$$
$$= -\frac{1}{\rho_0}\frac{\partial}{\partial z}\left(p_1 + \frac{B_0 b_\varphi}{4\pi}\right) + \frac{B_0}{4\pi\rho_0}\frac{\partial b_z}{r\partial\varphi} + \frac{\rho_1}{\rho_0^2}\frac{\partial}{\partial z}\left(p_0 + \frac{B_0^2}{8\pi}\right), \tag{A.45}$$

where p_1 and ρ_1 denote the perturbed parts of the pressure and density, respectively.

(c) **Induction equation**
The induction equation gives

$$\left(\frac{\partial}{\partial t}+\Omega\frac{\partial}{\partial\varphi}\right)b_r=B_0\frac{\partial u_r}{r\partial\varphi},\tag{A.46}$$

$$\left(\frac{\partial}{\partial t}+\Omega\frac{\partial}{\partial\varphi}\right)b_\varphi=r\frac{d\Omega}{dr}b_r-\frac{\partial}{\partial r}(B_0u_r)-\frac{\partial}{\partial z}(B_0u_z),\tag{A.47}$$

$$\left(\frac{\partial}{\partial t}+\Omega\frac{\partial}{\partial\varphi}\right)b_z=B_0\frac{\partial u_z}{r\partial\varphi}.\tag{A.48}$$

(d) **Adiabatic relation**
Another relation that we need here is a relation between p_1 and ρ_1. Considering isothermal perturbations, we adopt

$$p_1=\rho_1c_s^2.\tag{A.49}$$

Appendix B
Derivation of Relativistic Epicyclic Frequencies

General relativistic expressions for epicyclic frequencies were obtained by Aliev and Galtsov (1981). Later, from necessity of studying characteristic behaviors of disk oscillations in the innermost region of relativistic disks, especially from a viewpoint of wave trapping, Kato and Fukue (1980), Okazaki et al. (1987) and Kato (1990) derived horizontal and vertical epicyclic frequencies in the Kerr metric. Here, we derive the relativistic epicyclic frequencies in the Kerr metric, following Okazaki et al. (1987) and Kato (1990).

B.1 Basic Equations

We start from the Kerr metric expressed in terms of Boyer-Lindquist coordinates,

$$ds^2 = \frac{\Delta}{\rho^2}[cdt - a\sin^2\theta d\phi]^2 - \frac{\sin^2\theta}{\rho^2}[(r^2+a^2)d\phi - acdt]^2 - \frac{\rho^2}{\Delta}dr^2 - \rho^2 d\theta^2. \tag{B.1}$$

Here, functions Δ and ρ are defined by

$$\Delta = r^2 - rr_{\rm g} + a^2 \tag{B.2}$$

$$\rho^2 = r^2 + a^2\cos\theta, \tag{B.3}$$

where a is a parameter representing the angular momentum per unit mass of the Kerr black hole and in the range $0 \le a \le r_{\rm g}/2$.

The Euler equation of a free particle is given by

$$v^{\mu}_{;\alpha}v^{\alpha} = 0, \tag{B.4}$$

S. Kato, *Oscillations of Disks*, Astrophysics and Space Science Library 437,
DOI 10.1007/978-4-431-56208-5

where v^μ is the particle's 4-velocity dx^μ/ds, and $v^\mu_{;\alpha}$ denotes the covariant derivative of v^μ defined by

$$v^\mu_{;\alpha} = \frac{\partial v^\mu}{\partial x_\alpha} + \Gamma^\mu_{k\alpha} v^k, \tag{B.5}$$

$\Gamma^\mu_{k\alpha}$ being Christoffel symbols.

We first consider a circular motion in the equatorial plane ($\theta = \pi/2$), i.e., $v^\mu = (U^t, 0, U^\phi, 0)$. It then follows from equation (B.4) that the angular velocity, $\Omega_{\rm K}$, of a circular motion observed at infinity is

$$\Omega_{\rm K} = \frac{d\phi}{dt} = \frac{U^\phi}{U^t} = \pm\frac{(GM)^{1/2}}{r^{3/2} \pm a(r_{\rm g}/2)^{1/2}}. \tag{B.6}$$

Here and hereafter, the upper sign refers to the prograde orbit, while the lower sign to retrograde orbit. The redshift factor U^t is also found from equation (B.4) to be

$$cU^t = \frac{r^{3/2} \pm a(r_{\rm g}/2)^{1/2}}{r^{3/4}[r^{3/2} - 3r^{1/2}r_{\rm g}/2 \pm 2a(r_{\rm g}/2)^{1/2}]^{1/2}}. \tag{B.7}$$

We next consider a motion slightly perturbed from a circular orbit. The coordinate velocity is written as

$$\frac{dx^\mu}{dt} = (1, u^r, u^\theta, \Omega_{\rm K} + u^\phi), \tag{B.8}$$

where u^r, u^θ, u^ϕ are the velocity components associated with the infinitesimal perturbations. To derive linearized equations for u^r, u^θ, and u^ϕ, it is convenient to rewrite the Euler equation (B.4) in the form

$$\left[\left(\frac{dx^\mu}{dt}\right)_{;\nu} + \frac{\partial \ln u^t}{\partial x^\nu}\frac{dx^\mu}{dt}\right]\frac{dx^\nu}{dt} = 0. \tag{B.9}$$

Here,

$$u^t = U^t\left\{1 + U^t\left[\left(r^2 + a^2 + \frac{r_{\rm g}}{r}a^2\right)\Omega - \frac{r_{\rm g}}{r}ca\right]u^\phi\right\}, \tag{B.10}$$

which is obtained from a linearized form of $v_\mu v^\mu = 1$.

B.2 Horizontal Epicyclic Frequency

Let us consider a perturbed motion in the equatorial plane, i.e., a motion infinitesimally deviated from the circular one with $u^\theta = 0$. Then, substituting equations (B.8) and (B.10) into equation (B.9), we obtain linearized equations for u^r and u^ϕ:

$$\frac{\partial u^r}{\partial t} \mp (2\Delta/r^3)(rr_{\rm g}/2)^{1/2}u^\phi = 0, \tag{B.11}$$

$$\frac{\partial u^\phi}{\partial t} \pm \left[\frac{(rr_{\rm g}/2)^{1/2}(r^2 - 3rr_{\rm g} \pm 8a(rr_{\rm g}/2)^{1/2} - 3a^2)}{2\Delta(r^{3/2} \pm a(r_{\rm g}/2)^{1/2})^2}\right]u^r = 0. \tag{B.12}$$

In the above equations, the upper sign of $\mp$ or $\pm$ is for direct orbit, and the lower sign is for retrograde orbit. Eliminating u^r or u^ϕ we have

$$\left(\frac{\partial^2}{\partial t^2} + \kappa^2\right)\begin{pmatrix} u^r \\ u^\phi \end{pmatrix} = 0, \tag{B.13}$$

where

$$\kappa^2 = \frac{GM}{r^3}\frac{1 - 3r_{\rm g}/r \pm 8a_*(r_{\rm g}/2r)^{3/2} - 3a_*^2(r_{\rm g}/2r)^2}{[1 + a_*(r_{\rm g}/2r)^{3/2}]^2}. \tag{B.14}$$

Here, the dimensionless spin parameter, a_*, has been introduced, where a_* is defined by $a_* = a/(r_{\rm g}/2)$ and in the range of $0 \le a_* < 1$.

B.3 Vertical Epicyclic Frequency

Next, we consider a small amplitude perturbation of v^θ with $u^r = u^\phi = 0$. Then, from equation (B.9), after lengthy but straightforward calculations, we have

$$\left(\frac{\partial}{\partial t} + \Omega\frac{\partial}{\partial \phi}\right)u^\theta = -\Omega_\perp^2\delta\theta, \tag{B.15}$$

where

$$\Omega_\perp^2(r) = \Omega_{\rm K}^2\left[1 + \left(3 + 2\frac{r_{\rm g}}{r}\right)\frac{r_{\rm g}^2}{4r^2}a_*^2\right] \mp \Omega_{\rm K}\frac{c}{r}\left(\frac{r_{\rm g}}{r}\right)^2 a_*, \tag{B.16}$$

where the upper and lower signs of $\mp$ are direct and retrograde orbits, respectively, and $\delta\theta$ is the displacement of the particle in the θ-direction. Because the time derivative of $\delta\theta$ is u^θ, equation (B.16) shows that $\Omega_\perp$ is the vertical epicyclic frequency. Finally, using an expression for $\Omega_{\rm K}$ given by equation (B.6), we have an expression for the vertical epicyclic frequency:

$$\Omega_{\perp}^{2} = \Omega_{\mathrm{K}}^{2}\left[1 \mp 4\left(\frac{r_{\mathrm{g}}}{2r}\right)^{3/2} a_{*} + 3\left(\frac{r_{\mathrm{g}}}{2r}\right)^{2} a_{*}^{2}\right]. \tag{B.17}$$

References

Aliev, A. N., & Galtsov, D.V. 1981, Gen. Relativ. Gravit., 13, 899
Kato, S., & Fukue, J. 1980, Publ. Astron. Soc. Jpn., 32, 377
Kato, S. 1990, Publ. Astron. Soc. Jpn., 42, 99
Okazaki, A.T., Kato, S., & Fukue, J. 1987, Publ. Astron. Soc. Jpn., 39, 457

Appendix C
Wavetrain and Wave Action Conservation

It is widely known that wave action is conserved in wavetrain (Bretherton and Garret 1969). Here, we briefly summarize the concept of wavetrain and that of wave action conservation, following Bretherton and Garret (1969).

C.1 Wavetrain and Wave Action Conservation

A wavetrain is a system of almost sinusoidal propagating waves with a recognizable dominant local frequency ω, vector wavenumber $\boldsymbol{k}$, and amplitude a. These may vary with position $\boldsymbol{r}$ and time t, but only slowly, in the sense that appreciable changes are apparent only over many periods and wavelengths. The dominant frequency and wavenumber may be derived from a phase function $\theta(\boldsymbol{r}, t)$ by

$$\omega = \frac{\partial \theta}{\partial t}, \quad k_i = -\frac{\partial \theta}{\partial r_i} \tag{C.1}$$

and the wave crests are surfaces of constant θ. At each point, ω and $\boldsymbol{k}$ are connected by a dispersion relation

$$\omega = \Omega(\boldsymbol{k}, \lambda), \tag{C.2}$$

where the local properties of the medium are for convenience summarized in the parameter $\lambda(\boldsymbol{r}, t)$, and are also assumed to be slowly varying (Bretherton and Garret 1969). It is noted here that Ω in equation (C.2) is not confused with Ω of angular velocity of disk rotation .

We now define wave energy density e, which is usually written in the form of

$$e = a^2 F(\omega, \boldsymbol{k}, \lambda). \tag{C.3}$$

S. Kato, *Oscillations of Disks*, Astrophysics and Space Science Library 437,
DOI 10.1007/978-4-431-56208-5

Bretherton and Garret (1969) then showed that for a wide class of physical systems, we have a conservation relation:

$$\frac{\partial}{\partial t}\left(\frac{e}{\omega'}\right) + \mathrm{div}\left(\frac{e}{\omega'}\boldsymbol{v}_g\right) = 0, \tag{C.4}$$

where $\boldsymbol{v}_g$ is the group velocity defined by

$$(v_g)_i = \frac{\partial \Omega}{\partial k_i}, \tag{C.5}$$

and the intrinsic frequency ω' is the Doppler shifted one and related to ω by

$$\omega' = \omega - \boldsymbol{U} \cdot \boldsymbol{k}, \tag{C.6}$$

where $\boldsymbol{U}$ is the velocity relative to the observes. Equation (C.4) shows a conservation relation, and e/ω' and $e\omega'\boldsymbol{v}_g$ are called, respectively, wave action and wave action flux.

Our main concern in this book is differentially rotating disks with no radial flow. In this case, ω' is written as

$$\omega' = \omega - m\Omega, \tag{C.7}$$

and is the frequency observed in the corotating frame, $\tilde{\omega}$. It is noted that $\Omega(r)$ is the angular velocity of disk rotation and should not be confused with Ω in equation (C.5).

Reference

Bretherton, F. P. and Garret, C. J., 1969, Proc. R. Soc. A-Math. Phys. Eng. Sci., A302, 529

Appendix D
Modes of Tidal Waves

In a binary system, the disk around a primary star is deformed by tidal force of a secondary star. In such tidally deformed disks, a set of disk oscillations can be simultaneously excited by a wave-wave resonant process coupled with the disk deformation (see Chap. 11). A typical example of disk oscillations in tidally deformed disks is superhumps in dwarf novae (see Chap. 12). To examine what types of disk oscillations can be excited on tidally deformed disks, we must know what tidal waves are expected on tidally deformed disks. In this appendix, we examine tidal waves expected in tidally deformed disks (Kato 2014). Parameters specifying the tidal waves are (i) size of orbit of the secondary (binary separation, a), (ii) eccentricity of the orbit, e, and (iii) inclination of the orbital plane from the disk plane, δ.

D.1 Tidal Potential

We consider the tidal perturbations that are induced at a position P($\boldsymbol{r}$) on the disk of primary by scondary star of mass M_s. When the point P is at a distance $R[=(r^2+z^2)^{1/2}]$ from the center of the primary and the secondary star's zenith distance observed at the point P is ϑ (see Fig. D.1), the tidal gravitational potential $\psi_D(\boldsymbol{r},t)$ at the point P is given by (e.g., Lamb 1924)

$$\psi_D = -\frac{GM_s}{(D^2 - 2RD\cos\vartheta + R^2)^{1/2}} + \frac{GM_s}{D}^2 R\cos\vartheta, \tag{D.1}$$

where D is the distance between the primary and secondary stars at time t. The second term on the right-hand side represents the potential of a uniform field of force of the secondary acting on the primary.

S. Kato, *Oscillations of Disks*, Astrophysics and Space Science Library 437,
DOI 10.1007/978-4-431-56208-5

There are two ways to treat equation (D.1). One is to use the expansion of $(1 - 2\xi\cos\vartheta + \xi^2)^{-s}$ by a series of $\cos j\vartheta$ $(j = 0, 1, 2, \ldots)$ as

$$(1 - 2\xi\cos\vartheta + \xi^2)^{-s} = \sum_{j=0}^{\infty} b_s^{(j)}(\xi)\cos j\vartheta, \tag{D.2}$$

where $b_s^{(j)}$'s are called Laplace coefficients, and in the case of $s = 1/2$, we have (e.g., Araki 1980)

$$b_{1/2}^{(0)} = 1 + \frac{1}{4}\xi^2 + \frac{9}{64}\xi^4 + \ldots, \qquad b_{1/2}^{(1)} = \xi + \frac{3}{8}\xi^3 + \frac{15}{64}\xi^5 + \ldots,$$
$$b_{1/2}^{(2)} = \frac{3}{4}\xi^2 + \frac{5}{16}\xi^4 + \ldots, \qquad b_{1/2}^{(3)} = \frac{5}{8}\xi^3 + \frac{35}{128}\xi^5 + \ldots \tag{D.3}$$

The alternative way is to expand by a power series of ξ as

$$(1 - 2\xi\cos\vartheta + \xi^2)^{-s} = \sum_{j=0}^{\infty} C_j^s(\vartheta)\xi^j, \tag{D.4}$$

where $C_j^s(\vartheta)$ is the Gegenbauer polynomials, and in the case of $s = 1/2$, especially, they are Legendre polynomials, $P_j(\xi)$.

Here, we adopt the latter method, although the former expansion seems to be used in many cases. Then, we have

$$-\frac{\psi_{\rm D}}{GM_{\rm s}/D} = 1+\left(\frac{R}{D}\right)^2 P_2(\cos\vartheta)+\left(\frac{R}{D}\right)^3 P_3(\cos\vartheta) + \ldots, \tag{D.5}$$

where $P_2(\cos\vartheta)$ and $P_3(\cos\vartheta)$ are the Legendre polynomials P_ℓ of argument $\cos\vartheta$ with $\ell = 2$ and $\ell = 3$, respectively. In the case where the orbit of the secondary is elliptical, D is a function of time. Hence, it is convenient to normalize $\psi_{\rm D}$ by a time-independent quantity. Here, we normalize $\psi_{\rm D}$ by use of the gravitational potential $GM_{\rm s}/a$ at the the mean separation radius, a, as (the relation between D and a is shown later)

$$-\frac{\psi_{\rm D}}{GM_{\rm s}/a} = \frac{a}{D}\left[1+\left(\frac{R}{a}\right)^2\left(\frac{a}{D}\right)^2 P_2(\cos\vartheta)+\left(\frac{R}{a}\right)^3\left(\frac{a}{D}\right)^3 P_3(\cos\vartheta) + \ldots\right]. \tag{D.6}$$

D.1.1 $P_2(\cos\vartheta)$, $P_3(\cos\vartheta)$ Expressed in Terms of (φ, β) and (θ, γ)

The next problem is to represent $P_2(\cos\vartheta)$ (and $P_3(\cos\vartheta)$) in terms of the spherical coordinates (φ, β) of the point P and spherical coordinates (θ, γ) representing the position of the secondary star. As a preparation, we consider a unit sphere whose center is at the center of the primary, as shown in Fig. D.1. The poles of the sphere are taken in the direction perpendicular to the disk plane. The orbital plane of the secondary inclines to the disk plane by angle δ. Let us denote the spherical coordinates of the point P by φ and β, as shown in Fig. D.1. The angle φ is measured from the nodal point N. Then, using a formula of spherical trigonometry we have

$$\cos\vartheta = \sin\beta\,\sin\gamma + \cos\beta\,\cos\gamma\,\cos(\theta - \varphi). \tag{D.7}$$

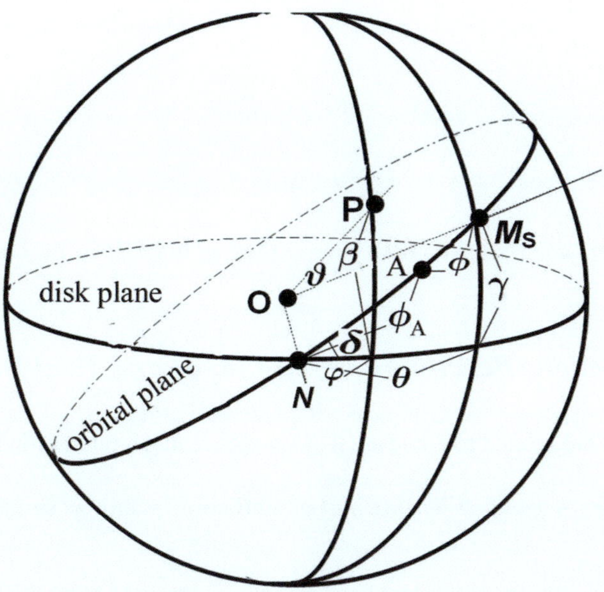

Fig. D.1 Relation between disk plane and orbital plane of secondary. Point O is the position of central star (primary star), and the disk plane and the orbital plane of secondary star is misaligned with inclination angle δ. Point N is the ascending nodal point of the secondary star, and Point A is the periastron of the secondary star. Point P is the position of a point on disk, whose spherical coordinates are (φ, β), and M_s is the position of secondary star and is represented by (θ, γ). (After S. Kato 2014, Publ. Astron. Soc. Jpn., 66, 21, PASJ ©).

Hence, simple algebraic calculations give

$$\begin{aligned} P_2(\cos\vartheta) &= \frac{1}{2}(3\cos^2\vartheta - 1) \\ &= \frac{1}{4}(3\sin^2\beta - 1)(3\sin^2\gamma - 1) + \frac{3}{4}\sin 2\beta \sin 2\gamma \cos(\theta - \varphi) \\ &\quad + \frac{3}{4}\cos^2\beta \cos^2\gamma \cos[2(\theta - \varphi)], \end{aligned} \tag{D.8}$$

and

$$\begin{aligned} P_3(\cos\vartheta) &= \frac{1}{2}(5\cos^3\vartheta - 3\cos\vartheta) \\ &= \frac{1}{4}\sin\beta \sin\gamma \Big[25\sin^2\beta \sin^2\gamma - 15(\sin^2\beta + \sin^2\gamma) + 9\Big] \\ &\quad + \frac{1}{8}\cos\beta \cos\gamma \Big[75\sin^2\beta \sin^2\gamma + 3 - 15(\sin^2\beta + \sin^2\gamma)\Big]\cos(\theta - \varphi) \\ &\quad + \frac{15}{4}\sin\beta \sin\gamma \cos^2\beta \cos^2\gamma \cos[2(\theta - \varphi)] \\ &\quad + \frac{5}{8}\cos^3\beta \cos^3\gamma \cos[3(\theta - \varphi)], \end{aligned} \tag{D.9}$$

where $P_2(\cos\vartheta)$ and $P_3(\cos\vartheta)$ have been expressed in terms of β, γ, and $\theta - \varphi$.

D.1.2 Relations Between (θ, γ) and (δ, ϕ) and Tidal Waves

Here, θ and γ are related to δ and $\phi_A + \phi$ by spherical trigonometric formulae:

$$\cos\gamma = \cos(\phi_A + \phi)\cos\theta + \sin(\phi_A + \phi)\sin\theta\cos\delta, \quad \sin\gamma = \sin(\phi_A + \phi)\sin\delta, \tag{D.10}$$

where ϕ_A is the angular direction of periastron, A, measured from the nodal point N along the orbit of the secondary, and ϕ is the position of the secondary on the orbit, measured from the periastron, A (see Fig. D.1). These relations (D.10) show that in the limit of $\delta = 0$, we have $\gamma = 0$ and $(\phi_A + \phi) = \theta$, as expected. Even when $\delta \neq 0$, the above relations (D.10) show that $\sin\gamma = \delta \sin(\phi_A + \phi)$ and $\phi_A + \phi = \theta$ until the order of δ^2. Hence, assuming that the misalignment between the disk and orbital planes is not large, we adopt

$$\theta = (\phi_A + \phi), \quad \sin\gamma = \delta\sin(\phi_A + \phi), \tag{D.11}$$

and the terms of the order of δ^2 are neglected in equations (D.8) and (D.9). Then, $P_2(\cos\vartheta)$ and $P_3(\cos\vartheta)$ are approximated as

$$P_2(\cos\vartheta) \sim \frac{3}{4}(1 - 3\sin^2\beta) + \frac{3}{2}\delta\sin 2\beta\left[\sin(2\phi + 2\phi_A - \varphi) + \sin\varphi\right] + \frac{3}{4}\cos^2\beta\cos\left[2(\phi + \phi_A - \varphi)\right], \tag{D.12}$$

and

$$P_3(\cos\vartheta) \sim -\frac{3}{4}\delta\sin\beta\,(5\sin^2\beta - 3)\sin(\phi + \phi_A) + \frac{1}{8}(3 - 15\sin^2\beta)\cos\beta\cos(\phi + \phi_A - \varphi) + \frac{15}{8}\delta\sin\beta\cos^2\beta\left[\sin(3\phi + 3\phi_A - 2\varphi) - \sin(\phi + \phi_A - 2\varphi)\right] + \frac{5}{8}\cos^3\beta\cos\left[3(\phi + \phi_A - \varphi)\right]. \tag{D.13}$$

In the limiting case where the secondary star's orbit is circular (i.e., $e = 0$), ϕ is obviously $\phi = \Omega^*_{\rm orb}t$. Hence, in this case, if the orbital plane coincides with the disk plane (i.e., $\delta = 0$), the tidal waves are two-armed ($m_D = 2$) with frequency $2\Omega^*_{\rm orb}$ (i.e., $\omega_D = 2\Omega^*_{\rm orb}$) (see equation (D.12) and Table 12.1), if the expansion in equation (D.5) is terminated by the second term on the right-hand side. If the expansion proceeds till the next term, one-armed ($m_D = 1$) tidal waves with frequency $\Omega^*_{\rm orb}$, and three-armed ($m_D = 3$) tidal waves with frequency $3\Omega^*_{\rm orb}$ appear (see equation (D.13) and Table 12.1), but the ratio of $\omega_{\rm orb}/m_D$ is still $\Omega^*_{\rm orb}$. In misaligned cases ($\delta \neq 0$), the situations are changed and even when $e = 0$, we have one-armed ($m_D = 1$) tidal waves with $\omega_{\rm orb} = 2\Omega^*_{\rm orb}$ (see equation (D.12) and Table 12.2), and two-armed ($m_D = 2$) waves with $\omega_D = \Omega^*_{\rm orb}$ and $\omega^*_D = 3\Omega^*_{\rm orb}$ (see equation (D.13) and Table 12.2). That is, in the case of $\delta \neq 0$, tidal waves with $\omega_{\rm orb}/m_D \neq \Omega^*_{\rm orb}$ appear.

The arguments in the above paragraph show that in the framework of circular orbits ($e = 0$) the tidal deformation does not bring about the tidal waves required for some resonant instability listed in Table 12.3. This means that for such resonant instability to be realized, the orbit of the secondary star is needed at least to be eccentric ($e \neq 0$). To know whether such tidal waves really appear if eccentric orbits are considered, we must examine (i) the deviation of $\phi(t)$ from $\phi = \Omega^*_{\rm orb}t$ and (ii) the time variation of a/D in the cases of eccentric orbits.

The functional form of $\phi(t)$ is well known in celestial mechanics. The main results concerning $\phi(t)$ are summarized as follows. Let the eccentricity and the mean radius of the orbit be e and a, respectively. Then, the distance of the secondary from the center of the primary, D, changes with a change of $\phi(t)$ as

$$D = \frac{a(1-e^2)}{1 + e\,\cos\phi}. \tag{D.14}$$

Now, we introduce an angle u (eccentric anomaly) defined by

$$D = a(1 - e\,\cos u). \tag{D.15}$$

Then, the equation of motion shows that the time variation of $u(t)$ is described by the Kepler equation:

$$u - e\,\sin u = \Omega^*_{\rm orb} t, \tag{D.16}$$

where $u = 0$ (and thus $\phi = 0$) is taken at $t = 0$. In the limit of the circular orbit ($e = 0$), we have $D = a$ and $u = \phi = \Omega^*_{\rm orb} t$.

Equation (D.16) is solved with respect to u by a power series of e, assuming that e is small. Then, we have (e.g., Araki 1980)

$$\begin{aligned} u = \; & \Omega^*_{\rm orb} t + e\,\sin(\Omega^*_{\rm orb} t) + \frac{1}{2}e^2 \sin(2\Omega^*_{\rm orb} t) \\ & + \frac{1}{8}e^3\Big[3\,\sin(3\Omega^*_{\rm orb} t) - \sin(\Omega^*_{\rm orb} t)\Big] + \ldots \end{aligned} \tag{D.17}$$

Combination of equations (D.14) and (D.15) gives $(1 + e\,\cos\phi)(1 - e\,\cos u) = (1 - e^2)$. From this equation and $\phi = \Omega^*_{\rm orb} t$ in the limit of $e = 0$, we obtain (e.g., Araki 1980)

$$\begin{aligned} \phi = \; & \Omega^*_{\rm orb} t + 2e\,\sin(\Omega^*_{\rm orb} t) + \frac{5}{4}e^2 \sin(2\Omega^*_{\rm orb} t) \\ & + \frac{1}{12}e^3\Big[13\sin(3\Omega^*_{\rm orb} t) - 3\sin(\Omega^*_{\rm orb} t)\Big] + \ldots \end{aligned} \tag{D.18}$$

Furthermore, $a/D = (1 - e\,\cos u)^{-1}$ is also expanded by a power series of e as

$$\frac{a}{D} = 1 + e\,\cos u + e^2\,\cos^2 u + e^3\,\cos^3 u + \ldots \tag{D.19}$$

Then, by using equation (D.17) we can write a/D explicitly as a function of $\Omega_{\rm orb} t$:

$$\begin{aligned} \frac{a}{D} = \; & 1 + e\,\cos(\Omega^*_{\rm orb} t) + e^2 \cos(2\Omega^*_{\rm orb} t) \\ & + \frac{1}{8}e^3\Big[9\,\cos(3\Omega^*_{\rm orb} t) - \cos(\Omega^*_{\rm orb} t)\Big] + \ldots \end{aligned} \tag{D.20}$$

If expressions for P_ℓ's (equations (D.12) and (D.13)), a/D (equation (D.20)), and ϕ (equation (D.18)) are substituted into equation of ψ_D (equation (D.6)), we have ψ_D directly expressed in terms of the coordinates of the observing point, (φ, β), the secondary's orbital parameters (Ω^*_{orb}, e, a) and the inclination, δ, between the disk and orbital planes. The tidal waves have generally forms of cos $[n\Omega^*_{orb} - m_D\varphi]$ or sin $[n\Omega^*_{orb} - m_D\varphi]$. The set of m_D and $n(\equiv \omega_D/\Omega^*_{orb})$ in various cases is shown in Tables 12.1 and 12.2, by taking $m_D > 0$.

References

Kato, S. 2014, Publ. Astron. Soc. Jpn, 66, 21
Lamb, H. 1924, *Hydrodynamics* (Cambridge University Press, Cambridge), p336
Araki, T. 1980, *Mécanique Céleste* (Kooseisha, Tokyo), pp. 85–86 (in Japanese)

MIX
Papier aus verantwortungsvollen Quellen
Paper from responsible sources
FSC® C105338

If you have any concerns about our products,
you can contact us on
ProductSafety@springernature.com

In case Publisher is established outside the EU,
the EU authorized representative is:
Springer Nature Customer Service Center GmbH
Europaplatz 3, 69115 Heidelberg, Germany

Printed by Libri Plureos GmbH
in Hamburg, Germany